Advanced Catalysis Processes in Petrochemicals and Petroleum Refining:

Emerging Research and Opportunities

Mohammed C. Al-Kinany
King Abdulaziz City for Science and Technology, Saudi Arabia

Saud A. Aldrees
King Abdulaziz City for Science and Technology, Saudi Arabia

A volume in the Advances in Chemical and Materials Engineering (ACME) Book Series

IGI Global
DISSEMINATOR OF KNOWLEDGE

Published in the United States of America by
IGI Global
Engineering Science Reference (an imprint of IGI Global)
701 E. Chocolate Avenue
Hershey PA, USA 17033
Tel: 717-533-8845
Fax: 717-533-8661
E-mail: cust@igi-global.com
Web site: http://www.igi-global.com

Library of Congress Cataloging-in-Publication Data

Names: Al-Kinany, Mohammed C., 1950- editor. | Aldrees, Saud A., 1978- editor.
Title: Advanced catalysis processes in petrochemicals and petroleum refining : emerging research and opportunities / edited by Mohammed C. Al-Kinany and Saud A. Aldrees.
Description: Hershey PA : Engineering Science Reference, [2019] | Includes bibliographical references.
Identifiers: LCCN 2018042437| ISBN 9781522580331 (hardcover) | ISBN 9781522580348 (ebook)
Subjects: LCSH: Petroleum--Refining. | Petroleum chemicals. | Catalysis.
Classification: LCC TP690 .A235 2019 | DDC 665.5/38--dc23 LC record available at https://lccn.loc.gov/2018042437

This book is published in the IGI Global book series Advances in Chemical and Materials Engineering (ACME) (ISSN: 2327-5448; eISSN: 2327-5456)

British Cataloguing in Publication Data
A Cataloguing in Publication record for this book is available from the British Library.

All work contributed to this book is new, previously-unpublished material.
The views expressed in this book are those of the authors, but not necessarily of the publisher.

For electronic access to this publication, please contact: eresources@igi-global.com.

Advances in Chemical and Materials Engineering (ACME) Book Series

ISSN:2327-5448
EISSN:2327-5456

Editor-in-Chief: J. Paulo Davim, University of Aveiro, Portugal

MISSION

The cross disciplinary approach of chemical and materials engineering is rapidly growing as it applies to the study of educational, scientific and industrial research activities by solving complex chemical problems using computational techniques and statistical methods.

The **Advances in Chemical and Materials Engineering (ACME) Book Series** provides research on the recent advances throughout computational and statistical methods of analysis and modeling. This series brings together collaboration between chemists, engineers, statisticians, and computer scientists and offers a wealth of knowledge and useful tools to academics, practitioners, and professionals through high quality publications.

COVERAGE

- Toxicity of Materials
- Artificial Intelligence Methods
- Chemical Engineering
- Ceramics and glasses
- Ductility and Crack-Resistance
- Thermo-Chemical Treatments
- Computational methods
- Sustainable Materials
- Materials to Renewable Energies
- Fatigue and Creep

The Advances in Chemical and Materials Engineering (ACME) Book Series (ISSN 2327-5448) is published by IGI Global, 701 E. Chocolate Avenue, Hershey, PA 17033-1240, USA, www.igi-global.com. This series is composed of titles available for purchase individually; each title is edited to be contextually exclusive from any other title within the series. For pricing and ordering information please visit http://www.igi-global.com/book-series/advances-chemical-materials-engineering/73687. Postmaster: Send all address changes to above address.

Titles in this Series

For a list of additional titles in this series, please visit:
https://www.igi-global.com/book-series/advances-chemical-materials-engineering/73687

Nanotechnology in Aerospace and Structural Mechanics
Noureddine Ramdani (Research and Development Institute of Industry and Defense Technologies, Algeria & Harbin Engineering University, China)
Engineering Science Reference • ©2019 • 356pp • H/C (ISBN: 9781522579212) • US $265.00

Polymer Nanocomposites for Advanced Engineering and Military Applications
Noureddine Ramdani (Research and Development Institute of Industry and Defense Technologies, Algeria & Harbin Engineering University, China)
Engineering Science Reference • ©2019 • 435pp • H/C (ISBN: 9781522578383) • US $225.00

Process Analysis, Design, and Intensification in Microfluidics and Chemical Engineering
Harrson Silva Santana (University of Campinas, Brazil) João Lameu da Silva Jr (Federal Institute of Education, Science, and Technology of South of Minas Gerais, Brazil) and Osvaldir Pereira Taranto (University of Campinas, Brazil)
Engineering Science Reference • ©2019 • 367pp • H/C (ISBN: 9781522571384) • US $215.00

Applications of Nanofluid Transportation and Heat Transfer Simulation
Mohsen Sheikholeslami (Babol Noshirvani University of Technology, Iran)
Engineering Science Reference • ©2019 • 692pp • H/C (ISBN: 9781522575955) • US $245.00

Strategic Applications of Measurement Technologies and Instrumentation
Soubantika Palchoudhury (University of Tennessee at Chattanooga, USA)
Engineering Science Reference • ©2019 • 223pp • H/C (ISBN: 9781522554066) • US $195.00

The Geometry of Higher-Dimensional Polytopes
Gennadiy Vladimirovich Zhizhin (Russian Academy of Sciences, Russia)
Engineering Science Reference • ©2019 • 286pp • H/C (ISBN: 9781522569688) • US $195.00

For an entire list of titles in this series, please visit:
https://www.igi-global.com/book-series/advances-chemical-materials-engineering/73687

IGI Global
DISSEMINATOR OF KNOWLEDGE

701 East Chocolate Avenue, Hershey, PA 17033, USA
Tel: 717-533-8845 x100 • Fax: 717-533-8661
E-Mail: cust@igi-global.com • www.igi-global.com

Table of Contents

Preface

Dear Readers,

We are delighted to present this IGI book on *Advanced Catalysis Processes in Petrochemicals and Petroleum Refining: Emerging Research and Opportunities*. An important aim of the book is to provide comprehensive overview the latest catalysis and catalytic processes in the field of Petrochemicals and Petroleum Refining. It is well known that catalysis and catalytic processes undoubtedly play a crucial and a very important role in modern petrochemical industries and petroleum refining. Crude oil is today the major energy source for transportation fuels worldwide and also a source for the production of a wide variety of chemical products. Many catalytic processes are necessary to convert crude oil and a heavy oil resides to final commercial products, such as, fuel additives, cleaner transportation fuels and basic petrochemical products.

This IGI book is organized in six chapters that cover important topics. the first one is catalysis in alkylation of benzene with ethene and propene to produce ethylbenzene and isopropylbenzene. It provides comprehensive overview about the historical of the alkylation, the development of homogeneous and heterogeneous Lewis and Brønsted acids and zeiolite catalysts; the liquid and gas phase alkylation processes; and the industrial technologies for ethylbenzene and cumene production.

It continues with the other five chapters dealing with the conversion of CO_2 to high value products, review on Fisher-Tropsch synthesis method in liquid fuel production, advanced catalysis and processes to convert heavy residues into fuels and high value chemicals, catalyst in biorefineries solution to promote environment sustainability in India, new catalytic approaches for producing alternative to MTBE Additives for reformulation of gasoline.

It is our hope and expectation that this book will provide a comprehensive knowledge to the students and researchers at universities and research centers. In addition, to the engineers in the petrochemical industries and petroleum refining sectors for a quick view on the latest catalysis and processes in this area.

Acknowledgment

We would like to take opportunity to express our special thanks of gratitude to the contributors for their hard working. Also, we would like to express our sincere thanks to our reviewers for their prompt action and constructive evaluations. Last but not the least, we would like to thank IGI global for giving us the chance to edit the book.

We hope this book's readers would enjoy the reading to have more knowledge in the latest advanced catalysis and processes in the field of petrochemical industries and petroleum refining.

Finally, while great care has been taken to eliminate errors, still some might have been crept in and may be not spotted. The editors will be glad if such errors are brought to their notice.

Yours sincerely

Editors

Mohammed Al-Kinany
King Abdul Aziz City for Science and Technology, Saudi Arabia

Saud Aldrees
King Abdul Aziz City for Science and Technology, Saudi Arabia

Chapter 1
Catalysis in Alkylation of Benzene With Ethene and Propene to Produce Ethylbenzene and Isopropylbenzene

Mohammed C. Al-Kinany
National Center for Petrochemical Technology (KACST), Saudi Arabia

Saeed M. Alshihri
National Center for Petrochemical Technology (KACST), Saudi Arabia

Saud A. Aldrees
National Center for Petrochemical Technology (KACST), Saudi Arabia

Eyad A. Alghilan
National Center for Petrochemical Technology (KACST), Saudi Arabia

Sami D. Aldrees
National Center for Petrochemical Technology (KACST), Saudi Arabia

Khawla M. Almalahi
National Center for Petrochemical Technology (KACST), Saudi Arabia

Norah H. Almousa
National Center for Petrochemical Technology (KACST), Saudi Arabia

Faisal M. Alotaibi
National Center for Petrochemical Technology (KACST), Saudi Arabia

Yousef I. Al-Rashed
National Center for Petrochemical Technology (KACST), Saudi Arabia

Feras A. A. Alshehri
National Center for Petrochemical Technology (KACST), Saudi Arabia

DOI: 10.4018/978-1-5225-8033-1.ch001

ABSTRACT

The alkylation of benzene with ethylene or propylene to form ethylbenzene (EB) or cumene is an industrially significant transformation. EB is used as an intermediate in the manufacture of styrene, which in turn is an important in the manufacture of many kinds of polymers. The primary use of cumene is in the co-production of phenol and acetone, which in turn are important in the manufacture of many kinds of chemicals and polymers. In industry, EB and cumene are mainly manufactured by the alkylation of benzene with ethene or propene via two methods, the gas and the liquid phase in the presence of Lewis and Brønsted acids. The development of efficient solid catalysts has gained much attention over the last decades. The objective of this chapter is to provide an overview of the history of the alkylation of benzene with ethene and propene, the development of homogeneous and heterogeneous Lewis and Brønsted acids and zeiolite catalysts, the liquid and gas phase alkylation processes, and the industrial technologies for EB and cumene production.

INTRODUCTION

Alkylation of aromatics: Alkylation is the transfer of an alkyl substituent from one molecule to another via an alkyl carbocation or carbonuim ion, a carbanion, a free radical, or a carbene, which are generated from electrophilic alkylating agents such as olefin or alkyl halide in the presence of catalyst, such as Brønsted acid, Lewis acid or zeolite.

The ethylene (ethene) feed used for ethylation of benzene is produced from the cracking of fractions obtained from distillation of natural gas and oil. The processes are:

(a) the steam cracking of ethane and propane (from natural gas and from crude oil),
(b) the steam cracking of naphtha from crude oil,
(c) the steam cracking of gas oil from crude oil.

The choice of feedstock depends on availability, price and what other products from cracking are needed for other chemical processes.

The alkylation of benzene with ethylene or propylene to form ethylbenzene or isopropylbenzene (cumene) over acid catalystisan industrially significant and the main commercial route for the production of important petrochemical intermediates, such as ethylbenzene(EB) as the key building block for manufacture polystyrene and isopropylbenzene (IPB - cumene), a precursor to solvents and chemical intermediates such as acetone and phenol; alkyl benzene sulfonates, used extensively in detergent building; and alkylnaphthalenes, precursors to advanced high strength polymers (Yurchak et al., 1979)

Ethylbenzene and isopropylbenzene are commercially the two largest volume derivatives of benzene. Combined, these two basic intermadiat chemicals account for nearly 75% of the world's consumption of petrochemical grade benzene.

The worldwide capacity amounting to around 30 million metric tons (MMT) per year, and in Chinese the value being about 5 million tons per year (Zhu et al., 2011). The global demand for ethylbenzene increased from 20.5m tons in 2000 to 25.1m tons in 2011 at annual growth rate of 1.9%. The increase in global demand for ethylbenzene has been due largely to ethylbenzene usage growth in developing countries, led by China, which has more than compensated for the declining demand in the US and large markets in Europe.

It was reported that in 2012 the world annual demand of EB was estimated to be 37 million metric tons per year, and expected to be about 38 million metric tons in 2016, with an average annual growth rate of approximately 3.5% from 2012 to 2016. In the other forecast period from 2011 to 2020, demand is expected to grow at a rate of 3.7%, with the increasing market dominance of Asia-Pacific region. Moreover furthermore, the overall EB demand in the world will expected to increase at an average annual rate of 3.9% from 2012 to 2022, and over 90% of the production of ethylbenzene is used in the manufacture of styrene. Figure (1) shows the world

Figure 1. World and Saudi Arabia Demand of Ethylbenzene

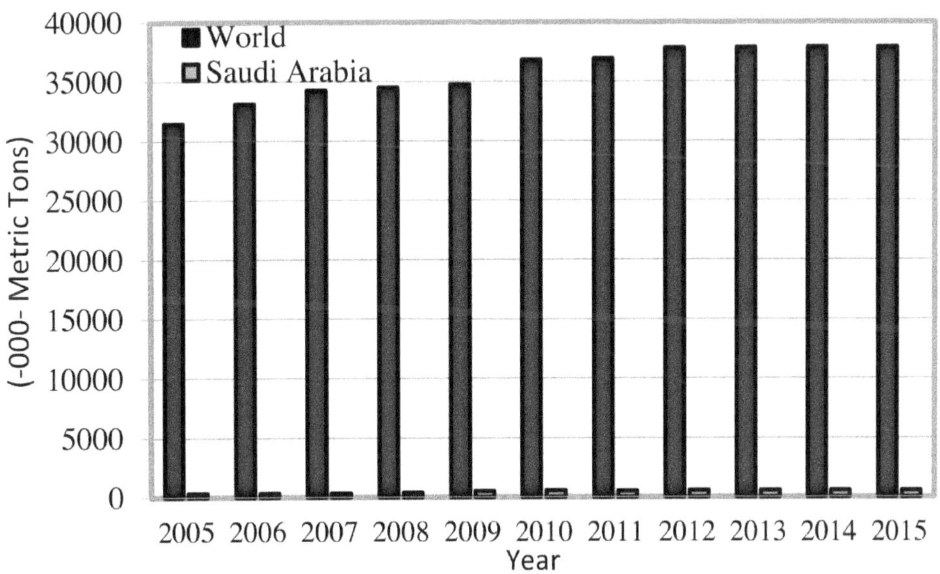

and Saudi Arabia demand of ethylbenzene (http://www.cmai global.com). (http://www.marketresearchreports.biz/analysis/153794) (https://www.ihs.com/products/ethylbenzene-chemical-economicshandbook.html).

The Importance of Ethylbenzene

Ethylbenzene (EB) is an organic compound with the formula $C_6H_5CH_2CH_3$.It is one of the most important chemical intermediate products, is the primary feedstock for the synthesis of styrene monomer which is polymerized into polystyrene, one of the most important polymers in the chemical industry (Hancock, 1975).

Global ethylbenzene market is expected to expand significantly over the forecast period owing to increasing applications in styrene production which is applied in expanding synthetic rubber and other end-use industries. Growing chemical industries in emerging economies are expected to foster ethyl benzene demand in diverse applications. Over 90% of the world's production of ethylbenzene is used in the manufacture of styrene which in turn is used in production of:

- polystyrene foam: used in food service products, food packaging material along with electrical and electronic devices. building insulation and refurbishment of old buildings.
- Composite products: used in tub and shower enclosures, automobile body panels, wind turbine parts, boats and many other applications.
- Styrene-Butadiene Latex (SBL): used in many paper coatings and in more than 90% of the broadloom carpeting made in the United States to attach carpet fibers to a backing material.
- Styrene-Butadiene Rubber (SBR): which reduces dependence on natural rubbers and provides improved performance in applications such as vehicle tires, leading to improved fuel efficiency.
- Styrene-Acrylonitrile(SAN) resins: used for food containers and optical fibers.
- Styrene-Polyester Acrylonitrile butadiene styrene (ABS): used in refrigerator liners, medical devices, small household appliances and luggage.
- Styrene plastic protective coatings and other products. (Faghihian et al., 2014; Gao et al., 2009)

Ethylbenzene also serves as an intermediate in the production of numerous chemicals. Coating industry extensively uses ethylbenzene for the production of paints, wood stains, varnishes and lacquers. It is also used as a solvent in the manufacturing rubber, machinery, paper coating and electroplating for aluminium bromide in anhydrous electro deposition of aluminium. (Vincent et al., 2005; Yurchak et al.,1979; Mitchell et al., 2012). It is also used in agricultural industry as an active ingredient in pesticides production. Increasing population and corresponding food demand is expected to boost agricultural and pesticides industries.

Moreover furthermore, EB is also used in the chemical industry in the manufacture of acetophenone, cellulose acetate, diethylbenzene (DEB), anthraquinone, ethylbenzene sulfonic acids, and alfa-methylbenzyl alcohol. A mixture of EB and xylene is used as a solvent in the paint industry and for producing pesticides in agriculture (ATSDR, 1999). In addition, EB is utilized as an additive in gasoline, which is witnessing growing popularity as an anti-knocking agent. Increasing ethylbenzene applications in natural gas extraction owing to increasing energy needs is also anticipated to drive global market over the forecast period.

The by-product of diethylbenzene from the ethylation of benzene is an important raw material for the conversion of divinylbenzene monomer. It is also prepared by vapor phase alkylation of ethylbenzene with ethylene using zeolitic catalysts. In another process, the efficiency of dodecatungsotphosphoric acid supported on acid treated clay has been evaluated in the alkylation of ethylbenzene with 80% ethanol (Bokade and Yadav, 2008).

Alkylation of Benzene With Ethylene (Ethene)

Technology of ethylbenzene production by ethylation of benzene with ethylene includes mainly two stages of alkylation and transalkylation. The control of technological parameters such as benzene-to-ethylene ratio in the feed, pressure, space velocity, contact time, type of reactor and temperature lets to conduct the process under the optimum conditions and monitor the output of ethylbenzene (Meyers, 2005; Ebrahimi et al, 2011).

Generally, benzene alkylation process, for the production of ethylbenzene, consists of the following four steps:

1. Alkylation step, in which benzene reacts with ethylene under optimum conditions.
2. separation step, in which the polyethylbenzenes are separated and recycled back to the alkylation or transalkylation reactor.
3. Transalkylation step, in which polyethylbenzenes such as diethylbenzenes (*o*-, *m*- and *p*-diethylbenzene isomers) and triethylbenzenes (1,3,5-, 1,2,3- and 1,2,4- isomers) in presence of benzene are converted to ethylbenzene on a reverse alkylation reaction.
4. Separation step, in which unreacted benzene, polyethylbenzenes and other components are separated from each other and high purity ethylbenzene is produced (Hamid et al, 2004)

Ethylation Reactions of Benzene

Ethylbenzene is produced by alkylation of benzene with ethylene. The activity and selectivity to the monoethylated product are very important in ethylbenzene synthesis. The EB produced may undergo polyethylation to di- and other polyethylbenzenes. Under the reaction conditions, the produced ethylbenzene further reacts with ethylene to form polyethylbenzenes, such as diethylbenzenes (*o*-, *m*- and *p*- isomers) and triethylbenzenes(1,3,5-, 1,2,3- and 1,2,4- isomers), tetraethylbenzene (1,2,3,5-, 1,2,3,4-and 1,2,4,5-isomers, pentaethylbenzene isomers and hexaethylbenzene. The formation of polyalkylated benzenes is depends on the technology of the process.

The industrial plant operates continuously and uses two different reactors for alkylation and transalkyation, as shown in figure 2. In ethylation reactor ethylene in its liquid phase goes through almost a complete reaction accompanied by the excess of benzene, which is necessary to form ethylbenzene.

Ethylation of Benzene

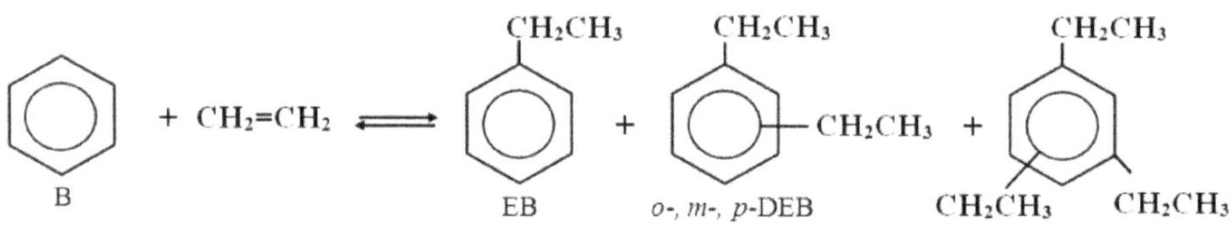

Then EB converts to produce small amounts of polyethylbenzenes. These reactions are the main side reactions. polyethylbenzenes are separated from ethylbenzene and then converted to ethylbenzene in the transalkyation reactor as shown in figure 2. (Fetisova et al,2010; Odedairo et al, 2011; Booth et al,1987).

Aluminium chloride catalyzed benzene alkylation was the first process to be developed for the production of ethylbenzene, and then solid-state acids such as solid phosphoric acid and zeolites were used as alkylation catalysts. Four main types of

Figure 2. Alkylator and transalkylator (single liquid phase)

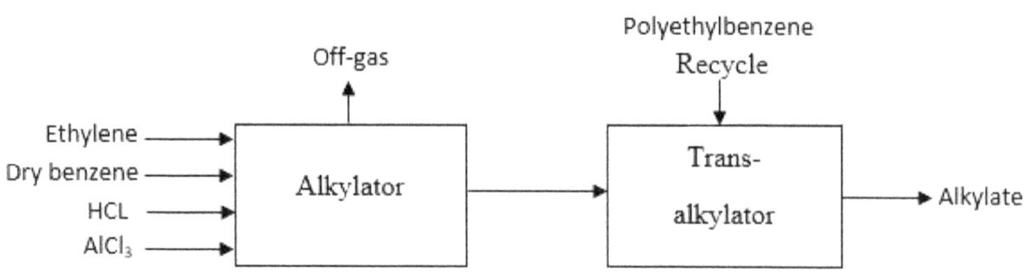

reactions include: transalkylation, isomerization, and disproportianation reactions of ethylated benzenes occurring during the alkylation of benzene with ethylene. Transalkylation of ethylbenzene isomers with benzene is performed to increase the yield of ethylbenzene. The addition of a transalkylation reactor in ethylbenzene production increases the yield of ethylbenzene by about 10%.

Transalkylation reaction of diethylbenzene with benzene (Al-Kinany et al, 1998, 1999, 2002, 2005, 2011, 2012; Al-Zahrani et al, 2002;2003).

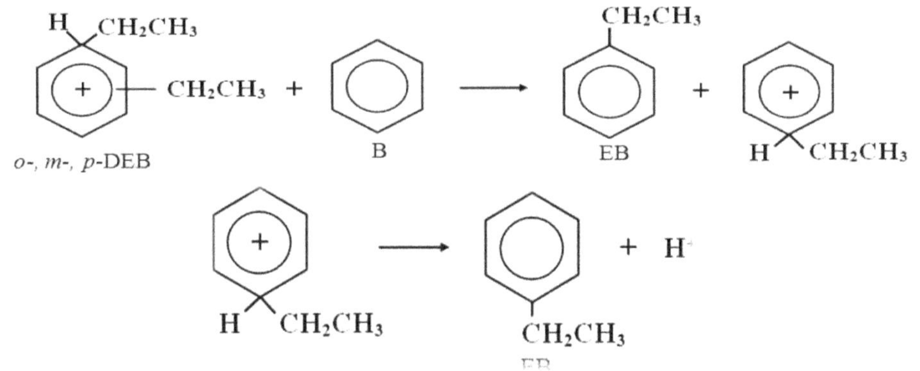

Isomerization of diethylbenzene (Al-Kinany et al, 1998, 1999, 2002, 2005, 2011, 2012; Al-Zahrani et al, 2003),

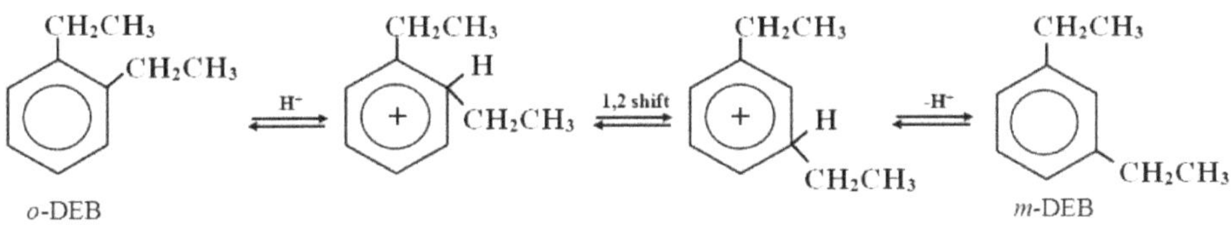

Disproportionation of diethylbenzene (Al-Kinany et al, 1998, 1999, 2002,2005, 2011,2012; Al-Zahrani et al 2003),

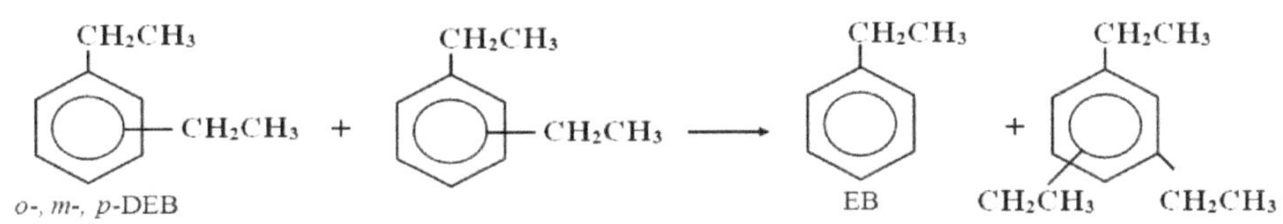

History of Development of Alkylation of Benzene With Ethylene

The benzene alkylation is referred to as key point of the ethylbenzene manufacturing and is essential for a wide range of industrial technologies (Yang et al., 2016).

It is well known that Lewis and Brønsted acids catalyze the alkylation reaction of benzene with olefins to give alkylated benzene products. In 1887 Charles Friedel and James Mason Crafts isolated amylbenzene after the treatment of amyl chloride with $AlCl_3$ in benzene (Friedel & Crafts, 1877).

This was not only one of the first descriptions of a Lewis acid used in organic synthesis but also the first example of what was later to be called Friedel–Crafts alkylation after its inventors. Today Friedel–Crafts alkylations remain the method of choice for the alkylation of benzene and other aromatic compounds.

In the 1930s, EB was first produced on the commercial scale in Germany and the United States by the alkylation of benzene with ethylene in the presence of a Friedel–Crafts catalyst (i.e. $AlCl_3$–HCl) at mild conditions. Its production was increased during World War II due to the high demand for synthetic polystyrene, i.e. butadiene rubber. In the 1950s, nearly 40% of the world's ethylbenzene capacity still utilizes Friedel- Crafts liquid-phase alkylation process with aluminium chloride-based $AlCl_3$ catalysts (Degnan et al, 2001).

In 1958, UOP developed a new process for alkylation of benzene with ethylene in presence of aluminium chloride catalyst. In 1960s the $AlCl_3$based catalysts were proposed to be replaced bysupported catalysts such as BF_3/Al_2O_3 and supported phosphoric acid (SPA) (Perego & Ingallina, 2002).Most production processes have approximately 95% conversion. These encompass the gas-phase reaction, liquid-phase process using an aluminium chloride Lewis acid catalyst, and a gas phase reaction in fixed bed reactors using a zeolite catalyst.

Despite the fact that zeolite-using alkylation technologies are more promising, many of refineries still use conventional liquid acidic catalysts. The possibility of transition to heterogeneous zeolite catalysts in a gas-phase process or in a liquid-phase process is not always economically practical due to mainly high operating pressure and temperature; and short life time (Srirattnai et al, 2002; Kharlampidi et al, 2000)

Over the intervening years many other Lewis acids including BF_3, $BeCl_2$, $TiCl_4$, $SbCl_5$ or $SnCl_4$ have been described as catalysts for the Friedel–Crafts alkylation. Furthermore, strong Brønsted-acids including sulfuric acid, hydrofluoric acid or super acids such as HF/SbF_5, HSO_3F/SbF_5, CF_3SO_3H, CF_3SO_3H/SbF_5(Al-Kinany et al,1998; Booth et al, 1987) and Nafion have also been shown to accelerate the alkylation reaction.

Disadvantages Liquid Phase

In many industrial processes alkylations are still performed with catalysts showing drawbacks. Often such catalysts are strong mineral acids or Lewis acids (e.g. HF, H_2SO_4, SPA and $AlCl_3$ based Friedel-Craft catalysts), which are highly toxic and non-environmental friendly catalysts; short catalyst life, the processes suffer from serious environmental, complex formation with polyethylated products and corrosion problems. They are dangerous to handle and to transport as they are capable of corroding storage and disposal containers. The catalyst is rejected after use and could not be recycled. Often, the products need to be separated from the acid with a difficult and energy consuming process (Hoeldrich, 2000). High lighted that the disposal of unavoidable side products formed in stoichiometric reactions (Perego & Angelina, 2002).

Moreover, the catalyst-recycling step in solid phosphoric acid catalysed processes raises difficulties in the process arrangement. The necessity for complete removal of traces of catalyst from the product after catalyst deactivation, to avoid the deterioration of the product, is a further challenge faced in the processes (Mitsutani, 2002). Additionally, it occurs frequently that these acids are neutralized at the end of the reaction and, therefore, the correspondent salts have to be disposed.

So there is ongoing research to develop an environmental friendly solid acid catalyst for alkylation processes to avoid the above mentioned problems. Zeolite is one of the candidate which is a crystalline structure used in the petroleum industry as a heterogeneous catalyst for hydrocarbon conversion reactions. Solid catalyst like zeolite utilizes the alkylation reaction processes for the preparation of alkylbenzenes related to the alkylation, isomerization, disproportionation, and transalkylation (Bellussi, 2004).

Catalysts and Processes Developments

Catalytic systems of the reaction of benzene with ethylene could be grouped into four processes:

- **Liquid-phase ethylation process using ($AlCl_3$-HCl) catalyst.** This process represents about 25% of the global market. The $AlCl_3$ ethylation process was licensed in 1986. Industrial-grade benzene, range of 98–99 wt% purity, has been used in several of these alkylation processes.
- **Vapor-phase ethylation process using zeolite catalyst such as ZSM-5.** This process represents about 50% of global EB market. The most recent vapor-phase plant was licensed in 1995. This process is also known to produce less than 1.0% of xylene in the product, which is economically undesirable.
- **Mixed-phase alkylation ethylation process using zeolite catalyst.** This method represents about 2% of the global market share. It can use dilute ethylene feed in range of polymer-grade ethylene
- **Liquid-phase ethylation process such as zeolite beta catalyst.** This method represents about 23% of global market share it is the prevailing alkylation process since 1995

Catalysts Employed in the Ethylation of Benzene

Catalysis played a central role during the second half of the 20th century in the development of the refining and petrochemical industries (Dahl & Kolboe, 1994). Today over 90% of all industrial chemicals are produced with the aid of catalysts (Michael et al, 1994; Wang et al, 2014). The traditional technologies for the alkylation/ transalkylation using strong mineral acids or Lewis acids (e.g. HF, H_2SO_4, $AlCl_3$) as catalysts has been used.

Metal Halide Catalysts (Lewis Acids)

Anhydrous aluminium chloride is the most widely used catalyst in the liquid phase alkylation of benzene with ethylene. There have been many attempts to establish an order of activity for the wide variety if Lewis acids which catalyse the ethylation of benzene Ipatieff et al (Ipatieff et al, 1937) reported the following order, based on the temperature at which ethylation of benzene occurred.

$AlCl_3 > ZrCl_4 > TaCl_5 > BF_3 > SbCl_3 > TiCl_4$

A more generalized table of order of "activity" for most of the Lewis acid type catalysts in this store of reaction has been compiled from widely scattered literature data (Olah, 1964, 1973).

$AlBr_3 > AlCl_3 > FeCl_3 > ZrCl_4 > TaCl_5 > BF_3 > UCl_4 > TiCl_3 > WCl_6 > ZnCl_2 > SnCl_4 > TiCl_4 > BeCl_2 > SbCl_5 > HgCl_2 > CuCl_2 > BiCl_3 > AsF_3$

Co-catalysts which acts as promotors, such as hydrogen chloride or a small amount of water with aluminium chloride, or hydrogen bromide with aluminium bromide etc. are usually needed to obtain efficient alkylation.

The poor solubility of aluminium chloride in benzene and the formation of additional compounds, (lower layer sludge) and olefinic hydrocarbons (present as reactants during the reaction), usually result in a decrease in the activity and life of the catalyst. Many of these difficulties can be avoided by the use of an inert solvent(Schmerling, 1946).

The advantage of adding in inert solvent is (a) to achieve a homogeneous medium, and (b) to produce metal halide solutions of high catalytic activity. Thus, a solvent can serve as a means of dispersing aluminium chloride in benzene which leads to minimisation of the formation of higher alkylated products, and less sludge formation.

The common inert solvents used are nitro paraffins (mainly, nitro-methane, ethane and propane), nitrobenzene, and chloro paraffins (methylchloride, ethylchloride and tetrachloroethane). As a result of the complex formation between the nitro alkane and aluminium chloride, $AlCl_3$, RNO_2, the formation of by-products is considerably depressed. However, one major disadvantage of using a solvent in the industrial processes is the high cost of recyclisation.

The possibility of intercalating metal halides such as $AlCl_3$, $AlBr_3$, SbF_5 and $FeCl_3$ in graphite and the use of these intercalation compound such as Friedel-Craft catalysts has been explored by Lalancette and Lafontaine (1973); Rodewald (1975). The application of these solid-supported catalysts for the heterogeneous vapour phase alkylation of benzene started to gain importance only in seventies.

The catalytic activity of intercalated aluminium chloride for aromatic alkylations of liquid hydrocarbons with alkylhalides or alkenes as alkylating agents has been studied by Lalancette et al. (1974).

He concluded that the intercalated aluminium chloride is a milder catalyst than neat $AlCl_3$ and the rate of reaction is lower, but the tendency to form polyalkylbenzene is reduced. One of the major problems encountered is that the aluminium chloride slowly leaches out.

Following on from these original studies by Lalancetteet al. (1974); Olahet al.(1977) has examined the use of graphite intercalation catalysts based on $AlCl_3$, $AlBl_3$, SbF_5 and $FeCl_3$ for the alkylation of benzene with ethylene and also benzene with diethylbenzenes. The results indicate the following order of activity, based on the percentage conversion of ethylene: $AlCl_3$, $AlBl_3$ and then SbF_5.

The major problems encountered with these catalysts are their relatively short life times, and losses of 30% of the overall aluminium chloride content by desorption of $AlCl_3$ from the catalytically active exposed area, and hydrolysis of $AlCl_3$ by small amounts of moisture in the feed causing complete deactivation of the catalyst after 6 h of operation. There is also obtained in the transalkylation reaction of benzene with diethylbenzenes.

Boron Trifluoride

Complex acid mixtures of boron trifluoride and promotors such as hydrogen fluoride, sulphuric acid, phosphoric acid, water, etc. have been studied and found to be very active Friedel-Crafts catalysts. Thus, the various catalyst systems BF_3/H_3PO_4 (Axe, 1946). BF_3/difluorophosphoricacid (Slaughter & McCauley, 1959). BF_3/FSO_3H (Ipatieff & Linn,1948). BF_3/H_2O (Ipatieff & Grosse, 1936; Axe, 1946), and BF_3 supported on inorganic metal oxide (i.e. Al_2O_3) have been investigated for the ethylation of benzene (Hervert & Linn, 1960). The most widely used systems studied are BF_3/H_2O (Uitti, 1977) and BF_3/Al_2O_3 (Universal Oil Products Co., 1971; Ward, 1971).

Boron trifluoride is considered to be one of the most powerful acceptor molecules. Solutions of boron triflouride in strong protonic acids are very active Friedel-Crafts catalysts and the catalytic activity of these complexes are attributed to their ability to donate protons to olefins. The relative reactivity of the boron trifluoride complexes is not strictly determined by the acid strength of the co-catalyst, although the activity of the co-catalysts appears to decrease in the order: HF> H_2SO_4> H_3PO_4 > H_2O

Protonic Acid Catalysts (Brønsted Acids)

The alkylation of benzene takes place in a strongly acidic medium. The type of acid used, its strength and the molar ratio of acid to the reactant exert a strong influence on the ethylation process. Several inorganic protonic acid catalysts such as sulphuric, phosphoric, hydrofluoric acid and mixture of sulphuric and phosphoric acids are generally of lower activity in Friedal-Craft reactions than aluminium chloride and other very active metal halides.

Sulphuric Acid

It has been shown by Ipatieff et al, (1936) and Wunderly et al,(1936) that concentrated sulphuric acid does not catalyse the reaction of benzene with ethylene at either atmospheric or super atmospheric pressure and moderate temperature. The use of higher temperature and higher acid concentration (e.g. 98%-100%) is impracticable (Komarewsky & Ruther, 1950) because sulphuric acid acts as a sulphonating agent and both benzene and the ethylated benzene react under these conditions to form sulphonic acids.

Hydrofluoric Acid

Hydrofluoric acid is the most selective acid catalyst for ethylation and it is more active than the other Brønsted acid (i.e. H_3PO_4, 98% H_2SO_4. the yield of ethylbenzene is low at a temperature of 5°C and at 116 psig (Bloch, 1972).

Hydrofluoric acid has the particular advantages that considerably higher yield of ethylated products is obtained and tar formation is considerably reduced, and this volatile catalyst (b.p.19.5°C) can be easily recovered. Extensive work has been carried out on ethylation of benzene using a combination of hydrofluoric acid with metal halide (i.e. HF/SbF_5) or protonic acid (i.e. HF/FSO_3H, HF/CF_3SO_3H and HF/SF_5) (Booth et al,1987).

Phosphoric Acid

In 1936 liquid phosphoric acid was first shown by Ipatieff et al (1936) to be a very effective catalyst for the ethylation of benzene. The benzene conversion of 70% to mono-, di-, and triethylbenzenes was obtained under specific conditions. One year later Ipatieff, (1937, 1946) developed the use of supported solid phosphoric acid catalyston. Kieselguhr and investigated the ethylation of benzene at a temperature of 280°C and pressure of 900 p.s.i.g. with a 10:1 molar ratio of benzene to ethylene and space velocity of 2.0. This gave an ethylene conversion of 90% and a product mixture consisting of momo-, di-, and polyethylbenzenes. In the early commercial operations (Mattox, 1945) using phosphoric acid on Kieselguhr, only a 50% ethylene conversion was obtained at 276°C, a pressure of 900 psig. and space velocity $1.52 cm^3.h^{-1}$ with a 16:1 molar ratio of benzene: ethylene and the products contained 78% ethylbenzene and 22% of polyethylbenzenes. Since 1958 different type of solid phase phosphoric acid catalyst has been employed in the commercial process (Maier & Ward, 1969).

Super Acid Catalysts

Trifluoromethanesulfonic acid (CF_3SO_3H) is a super acid and is also known as triflic acid. It is about a thousand times stronger than sulphuric acid. The strongest superacids are prepared by the combination of two components, a strong Brønsted acid Lewis acid and a strong, (CF_3SO_3H / SbF_5). Triflic acid and triflic acid with antimony pentafluoride systems were tested in both alkylation of benzene with ethylene to produce ethylbenzene and with propylene to produce isopropylbenzene. It was found very reactive catalyst for both reactions. The conversion of benzene in both reactions exceeds 75% under mild conditions (Booth et al, 1987; Al-Kinany et al,1997,1998,1999, 2002, 2005, 2011).

Solid Superacid Catalysts

Efforts have been directed to the possibility of applying solid super acid catalysts to the heterogeneous alkylation of benzene. These investigations have been reviewed by Olah et al,(1979).It has been suggested by Bloch, (1972) that the combination of SbF_5/HF or SbF_5/ FSO_3H on a porous inert solid support, such as activated charcoal (Rodewald, 1975) is an effective solid catalyst for catalysing the alkylation of benzene with ethylene. The ratio of the antimony pentafluoride to hydrogen fluoride or fluorosulphonic acid used is about 0.5:1. The ethylation reaction was carried out in a continuous flow fixed-bed pilot plant of conventional design, with a 10:1 molar ratio of benzene: ethylene at a pressure 300 psig. and a preferred temperature of about 0-75°C but detailed results have not been published.

Solid perfluorinated resin sulphonic acid catalysts, such as those based on the acid form of Dupont's Nafion ion membrane resin CF_3 $(CF_2)_X$ –O-CF_2-CF_2SO_3H, and higher perfluoroalkanesulphonic acids, such as perfluorodecanesulphonic acid or perfluorinated aliphatic sulphonic acid polymers, have been found to be effective catalysts for the Friedel-Crafts alkylation reaction. For example, it has been found by McClure (1977) that an unsupported perfluorinated polymer containing pendant sulphonic acid groups, is an effective catalyst for alkylation of benzene with ethylene in the liquid phase at the temperature range of 125-225°C. It has also been demonstrated that the process may be carried out either as a batch or continuous type of operation. In continuous operation the ethylation reaction is carried out in a tubular reactor with inert supports either above or below the catalyst, or mixed with the catalyst at a temperature 175°C.

Also, another work (Olah, Kaspi, & Bukala, 1977) on the alkylation of benzene with ethylene employing Nafion-H shows that the catalyst has good thermal stability compared with supported Lewis acids and good catalytic activity. It has been found that the conversion of ethylene on this catalyst increased from 10-44% when the temperature increased from 100-190°C with a 4:1 molar ratio of benzene: ethylene.

Solid Acid: Zeolitic Catalysts

Acid catalysts are widely used in the petrochemical industry, including oil refining, the production of petrochemicals and fine chemicals (Chang & Grover, 1977). Traditionally, ethylbenzene is produced by homogeneous Friedel–Crafts-type mineral acidic or Lewis acidic catalysts. But disadvantages of this process include: (Al-khattaf et al,2010; 2014; Perego & Ingallina, 2002):

- High energy consumption since the transalkylation is carried out at temperature higher than 180°C.
- Complex formation with polyethylated products.
- Short catalyst life.
- The catalyst is toxic and corrosive.
- Needs large quantity of catalyst (continuous addition and removal)
- The catalyst is rejected after use and could not be recycled.
- Pollution problem in disposing the waste catalyst complex

In order to overcome on these disadvantages, recently different kinds of catalysts, mainly zeolite, have been proposed as alternative to the $AlCl_3$. They are heterogeneous catalysts which present clear advantages including ease of handling, simplicity and versatility of process engineering and catalyst regeneration, exhibit robust shape and size selectivity through regular micropores, decreasing reactor and plant corrosion problems, and environmentally-safe disposal (Stocker, 1999),high surface area as well as uniform microporous structure coupled to shape selectivity, and thus they are better acidic catalysts when applied in refinery and petrochemical industry (Perego & Ingallina, 2002; Yang et al, 2016;Kaeding, et al, 1984; Chen & Garwood, 1986).

Perhaps the main characteristic of solid acids compared to liquid acids, is that solid acids possess specific acid sites differing markedly in their nature and strength, thus allowing a high degree of control of catalytic reactions (Socha et al, 1987; Hellring, 1987). Therefore, zeolite catalysts are advantageous in almost every way over the older Lewis and Brønsted acid catalysts, even though they operate through similar chemical mechanisms. In the production of ethylbenzene, by alkylation of benzene with ethylene, the substitution of the traditional $AlCl_3$ catalyst was already industrially performed with success, using a zeolite based catalyst. Most manufacturers moved toward zeolite catalyzed processes. Zeolite catalyzed processes are licensed by Mobil–Badger, Lummus–UOP, CDTech, and Dow Chemical.

It is well known that metal oxide catalysts are of importance in heterogeneous catalysis (Satterfield, 1991) especially chemical processes in industry (Bartholomew & Farrauto, 2011; Twigg, 1996). Solid acid catalysts are generally categorized by a combination of their Brønsted and/or Lewis acidity, the strength and number of their catalytic sites, and the morphology of the supporting material (e.g. surface area, pore size, chemical structure). The synthesis of pure Brønsted and pure Lewis acid catalysts continues to attract a great deal of academic and industrial interest.

Brønsted acid site is able to transfer a proton from the catalytic solid to the adsorbed molecule, whilst the Lewis acid site is able to accept an electron pair from the adsorbed molecule and a coordinative bond with the surface is thereby formed (Vora et al., 1997). These active sites in solid acid catalysts are usually formed either by ammonium or polyvalent cation exchange followed by calcination as shown in figure (3) (Kolboe et al,1993). Among the wide range of solid acid catalysts, zeolites have found many large-scale applications (Svelle et al, 2006). Their success is based on the ability to routinely and consistently function as high-volume, high-performance catalysts with strong Brønsted acidity that can be carefully controlled, and which can robustly withstand high reaction temperatures and catalyst regeneration conditions (Stocker, 1999).

A zeolite is best defined as a hydrated aluminosilicate that possesses a three-dimensional framework structure, formed by $[SiO_4]^{4-}$ and $[AlO_4]^{5-}$tetrahedra that are connected by sharing an oxygen atom. When an AlO_4 tetrahedron is substituted for a SiO_4 tetrahedron, a negative charge appears over the framework structure which is neutralized by the incorporation of extra framework cationic species (Na^+, NH_4^+, K^+, Ca^{2+}, Mg^{2+} *etc.*) (Choudhary et al, 2005; Chang et al,1979).

In general, zeolites have the following properties that make them particularly useful for a wide range of applications (Chang et al, 1979):

Figure 3. schematic representation of the generation of acid sites in zeolites (Zaidi and Pant, 2008)

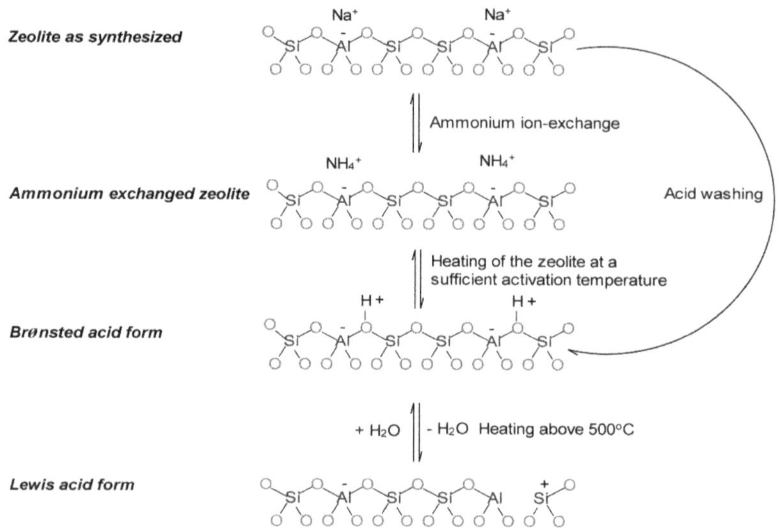

- Low cost, High surface area.
- High thermal and chemical stability.
- High selectivity of sorption.
- Tremendous diversity: >190 frameworks type are known and each can be further modified by introducing other elements, metal cations and molecular species.
- Uniform size and high density of pores.
- Possibility of modifying the electronic properties of the active sites.
- Non-toxic and environmentally friendly; recyclable.

Zeolite Structures

From both industrial and academic perspectives, the three most widely studied and used zeolites are zeolite Y, ZSM-5 and Beta. This is due to their unique crystal structures, their high thermal stability, their acidity and shape-selective properties. All of these properties match the requirements for many applications in oil refining, the petrochemical industry and environmental catalysis. Their properties can be tuned to the specific processes of diverse industrial applications(Teketel et al., 2012; Zhang et al., 2010). Briefly summarize the major features of the three zeolites investigated.

- Zeolite Y has the faujasite **(FAU)**framework; built by linking sodalite cages through double six-membered rings. This creates a large cavity called the super cage which is accessible by a three-dimensional 12-ring pore system (Kolboe et al, 1993).
- ZSM-5 zeolite **(MFI)** framework; built from five-membered rings. The pore structure consist of a two-dimensional 10 ring pore system. It contains cavities and comprises of two different pore systems, one consisting of zig-zag channels of near-circular cross-section and another of straight channels of elliptical shape (Kolboe et al, 1993; Teketel et al, 2009)
- Zeolite Beta**(BEA)** framework; a large pore, high-silica zeolite. The pore structure consists of channels of three-dimensional 12 rings pore systems interconnected by cages and constituted by the intersections of the channels (Vora et al, 1997; Freeman, 2002).

Zeolites are widely used throughout the oil refinery petrochemical and industry in a range of acid catalyzed processes. The analysis of the catalyst market and the position of zeolites therein is a striking indication of their continued strategic importance (Stocker, 1999).

The focus on cleaner fuels, lower cost petrochemicals and the issue of global sustainability has pushed the refining and petrochemical industries towards improvements in the performance of conventional catalysts. Zeolites offer the basis of many of these technological solutions building on efficient and balanced cooperation between industry and academia (Stocker, 1999; Yurchak et al, 1979).

Zeolitic catalysts later used industrially to produce ethylbenzene by ethylation of benzene with ethylene (Al-Kinany et al, 2016; Yang et al, 2016; Galadima & Muraza, 2015). However, since the 70s of the last century it became possible to produce EB with the use of heterogeneous zeolite catalyst on a commercial scale. Heterogeneous catalytic process gradually started to replace the old homogeneous catalytic technology. Nowadays most of the existing ethylbenzene production plants operate using the zeolite catalyst (Degnan et al, 2001; Perego & Ingallina, 2002).

A Wide variety of heterogeneous catalysts has been investigated by Moreno, (1988) so far, such as rare earth X and Y zeolites. The investigation of the ethylation of benzene with ethylene over the faujasite zeolite has been described by Namuangruk et al,(2004). More types of zeolitic catalysts also used in the alkylation of benzene with olefins such as ZSM-4, ZSM-5, ZSM-20, ZSM-38, ZSM-12, mordenite and offertite (Young & Skillman, 1979, 1981; Da, Han et al,2001; Ming-Yuan et al, 1988) silicates with functionalized surfaces (Hernández et al, 2010; Aguilar et al., 2008) and silicotungstic acid supported on alumina and on silica-alumina (Sebuluky & Henke, 1971).

In 1942 started the use of an amorphous aluminosilicate in alkylation of benzene with ethylene and propylene (Franck, 1988; O'Kelly, 1947). Later it was shown, that zeolites of type X and Y were effective for alkylation of aromatics with olefins (Ventuto et al, 1966). Cheng et al, (1998) reported that, beta, MCM-22, USY-1, and USY-2 zeolites have been shown to efficiently and USY-2 zeolites have been shown to efficiently activate the ethylation reaction of benzene with ethylene activate the ethylation reaction of benzene with ethylene for the ethylation of benzene with ethylene in the liquid phase method, under pressure and at low temperatures.

Al-Kinany et al, (2017) developed a method of preparing modified zeolite and a method of one cycle alkylating benzene with ethylene in presence of modified zeolite catalyst. The modified zeolite catalyst includes zeolite with ratio of silica to alumina ranging between 5% to 95%, kaolinite and transition metal oxide. The catalyst used in a gas phase process for alkylation of benzene with ethylene under variable temperatures and different ratios of benzene to ethylene. The results showed that the catalyst is very selective ethylbenzene (96%).

The zeolite MCM-49 showed a high selectivity to ethylbenzene with no formation of polyethylated products, such as diethylbenzene isomers and triethylbenzene isomers; and a correlation has been found between the activity and the content of the Brønsted acidic sites on the zeolite surface and in its lattice (Shi et al, 2015). In the production of ethylbenzene also catalysts of type MSA, MCM and BEA were used (Bellussi, 1991; Perego, 1999). In another study, Graf and Lefferts, (2009) were reported that, the oxidative ethylation of benzene with ethylene is effective over the ZSM-5 zeolite. Shi et al,(2015) showed that, the ethylbenzene selectivity in the liquid phase ethylation process of benzene with ethylene in the presence of H-TMP-16 zeolite reaches 99% (Xing et al, 2016).

An interesting study was also carried out by Du, et al, (2002) on the ethylation of benzene with ethylene in the presence of MCM-49 zeolites. Recently the main technical innovation in this field has been made by the introduction of new porous solid catalysts with the purpose to improve the reaction selectivity and to comply with the stringent environmental legislation (Kolesnikov, 2013).The mesoporous zeolites are more active in the ethylation of benzene with ethylene in the liquid phase process and more selective towards ethylbenzene (90–97%), compared to conventional zeolites (Groenet al., 2007; Lu et al, 2006).The first industrial application based on MCM-22 catalyst for ethylation of benzene in the liquid phase process to produce ethylbenzene by alkylation of benzene with ethylene was achieved in 1997(Perego & Ingallina, 2002).

Utilizing zeolites as catalyst in ethylation of benzene processes issues several problems such as high operating temperature and pressure (Christensen et al., 2003). The high reaction temperature usually leads to high tendency for coke formation, short life time, poor transalkylation capabilities and also there are high operating cost since the plant consists of two or more reactors and high degree of environmental pollution, because of considerable energy consumption in the mentioned processes.

Industrial Alkylation of Benzene with Ethylene Processes

The industrial alkylation process of benzene with ethylene, outlined in 1877 by the studies of Friedel, Crafts, and Gustavson, is conducted on polyalkylbenzene in the presence of an aluminium chloride complex. The alkylation of aromatic hydrocarbons with olefins is applied on a large scale in the chemical industry (Franck & Stadelhofer, 1988). Ethylbenzene is typically formed by one or more aromatic conversion processes involving the alkylation of benzene with ethylene. In industry, EB is mainly manufactured by the alkylation of benzene with ethylene via two methods, i.e., the gas-phase method and the liquid-phase method as shown in figure (4) (Kartal & Onal, 2008; Al-Kinany et al., 1998, 2012; Cejka et al, 1991; Hansen et al, 2008).

The alkylations of aromatic hydrocarbons with different low molecular weight olefins are applied on alarge scale in the chemical industry. Zeolite catalyzed processes arelicensed by UOP/CEPSA process, Universal Oil Products' (UOP) EBOne process, Mobil–Badger ethylbenzene (1976), Mobil–Badger EBMax1995, Lummus–UOP (1989), Mobil–Badger dilute ethylene 1987 CDTech 1995, SINOPEC 1993 and Dow Chemical (1998), Mobil–Raytheon's EBMax process (Vogt et al, 2015; Al-Khattaf et al, 2010; Čejka & Wichterlová, 2002; Busca, 2007).

Figure 4. Liquid-phase process for ethylbenzene production

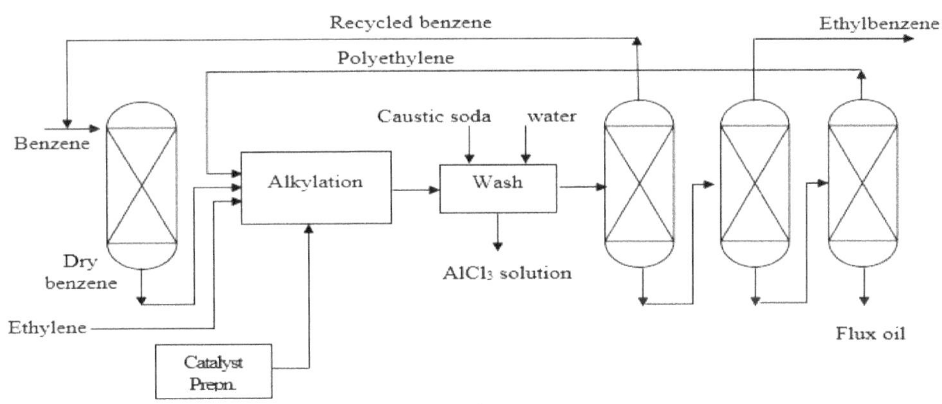

Mobil–Badger Ethylbenzene Process

Despite the huge amount of research activity, the first industrial application of a zeolite catalyst occurred in 1976 by Mobil–Badger. In the production of ethylbenzene, by alkylation of benzene with ethylene, the substitution of the traditional aluminuim chloride catalyst was industrially performed, using a zeolite based catalyst. Mobil-Badger first commercialized a process in 1980, using a solid acid catalyst containing a ZSM-5 type zeolite and operating in gas-phase (Chen & Garwood, 1986; Wang, 1993).

The gas-phase reaction does have some positive aspects. The zeolite catalyst is safer environmentally and disposal costs are not as high. Also, the catalyst is non-corrosive like miniral acid and Lewis acids. However, the catalyst life problem is a major issue with this process. The catalyst requires regeneration is necessary every four to six weeks of operation. Regeneration of catalyst takes about 36. To maintain steady state operation, it is typical to run two reactors in parallel.

Analogously to the "$AlCl_3$–HCl process", the polyethylbenzenes are recycled back to the reactor to undergo transalkylation. With the third generation process, commercialized in 1990s, a second reactor was added, primarily for transalkylation, significantly improving the performance as shown in figure (5).First introduced in 1980, the Mobil–Badger vapor phase process is still the most widely used zeolite catalyzed ethylbenzene manufacturing process. Since its initial commercial application

Figure 5. Alkylation of benzene with ethylene

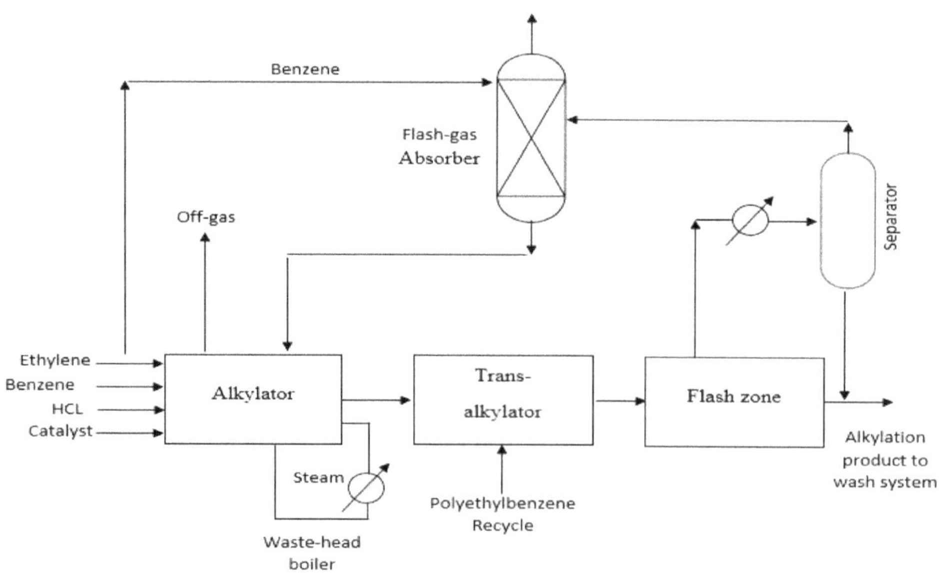

in 1981, Mobil–Badger offers both liquid phase and vapor phase transalkylation of diethylbenzene with benzene processes.

Currently, companies such as UOP, Mobil badger, etc. are using zeolites for production of ethylbenzene (Perego & Ingallina, 2002). This new process was commercialized in 1980. ZSM-5 has a low coke-forming tendency and, therefore, life cycles between regenerations up to 40–60 days are possible.

Lummus–UOP- Unocal Processes

In 1989 company Lumus, UOP, Unocal opened plant for the production of ethylbenzene in liquid phase by zeolite of type Y. There are several patents for this type of reaction, however, the Lummus/UOPEBOne is the recent and provides the most benefits over the gas phase process. A further improvement on Mobil–Badger ZSM-5 type zeolite was obtained by Unocal through the application of a Y type zeolite in the liquid-phase process (Wang, 1993; Wight, 1979).

The EBMax technology is presently available for license from Badger Licensing LLC. The EBMax technology was commercialized in 1995. There are a number of units in operation with nameplate capacities ranging from 290-1,400 thousand metric tons per year. By the end of 2008, Badger had licensed 31 plants representing a total of 16.4 million metric tons of EBMax capacity.

In 1989, Unocal–ABB Lummus Crest introduced a liquid-phase process based on a modified zeolite Y catalyst. The process design is very similar to the Mobil–Badger EBMax process, and uses two reactors, one for benzene alkylation and the other for diethylbenzene transalkylation. It is believed that the Lummus–UOP liquid phase ethylbenzene technology has recently shifted to the use of catalysts based on zeolite beta.

CDTech Process

The CDTech process uses the company's proprietary catalytic distillation (CD) technology. CDTech's CD column consists of two sections. The upper section is packed with the required catalysts and the lower section is equipped with distillation trays. The CDTech process uses zeolite Y as a catalyst (Horsley, 1997) principal attributes of the CDTech process reportedly are its long catalyst life and high product selectivity.

Albene Process

In 1992, the Indian Petrochemicals Corporation (IPCL) introduced a process that uses dilute ethanol rather than ethylene as an alkylating agent. Ethanol has no distinct advantage over ethylene except that it can be derived from waste streams or from biomass. The process reportedly uses [Fe]ZSM-5 as the catalyst. There are no reported commercial units that use this process.

Dow Chemical's Dilute Ethylene Process

Ethylbenzene was first produced on a commercial scale in the 1930s by Dow Chemical and by BASF in Germany. The ethylbenzene/ styrene industry remained relatively insignificant until World War II. Dow Chemical has introduced a dual stage ethane dehydrogenation to benzene alkylationethylbenzene process. This process mimics the dilute ethylene process referred to earlier. Ethane is dehydrogenated over a mordenite catalyst that contains zinc or gallium or one of the platinum group metals. In one Dow Chemical patent example (Progue, 1996) gallium-exchanged mordenite produced 86% ethyleneselectivity at 50% conversion per pass. The same patent cites a number of zeolites as possible alkylationcatalysts including beta, Y and ZSM-5.

Other Processes

In 1996, DSM introduced a process for converting butadiene into ethylbenzene. Chiyoda acquired rights to the process for license to third parties. The two-step process converts butadiene into vinyl cyclohexene and then into ethylbenzene.

In the Mobil–Badger process, benzene is alkylated in a fixed-bed reactor with ethylene in the gas phase using a ZSM-5 based catalyst (Dwyer, 1981; Chen et al, 1996). Together with the Badger Technology Center of Raytheon Engineers & Constructors, Mobil introduced second and third generation processes in 1986 and 1991. These next generation processes provided improved cycle lengths and higher yields (Degnan et al, 2001).

The Mobil–Badger process has also been modified to use dilute ethylene from FCC off-gas streams as a feedstock. A further improvement was obtained by Unocal through the application of a Y type zeolite in liquid-phase (Wang, 1993; Wight, 1979). The liquid-phase operation has the advantage of a better thermal control and longer catalysts life.

Isopropylation of Benzene to Produce Cumene

Cumene is an organic compound known as isopropylbenzene. The other names of cumene are 1-methylethylbenzene, 2-phenylpropane. Cumene C_9H_{12}, is an aromatic compound. It is a clear liquid at ambient conditions. It is used primarily for the manufacturing of phenol and its co-product acetone, which are two important petrochemicals with many uses in the chemical and polymer industries.

The alkylation of aromatic hydrocarbons with olefins has been applied on a large scale in chemical industry. Isopropylation of benzene with propylene to produce isopropylbenzene (cumene) is a well-known reaction. Commercial industrial-scale production began in 1942 by Friedel–Crafts alkylation of benzene with propylene to supply high-performance jet fuel for military aircraft. A vapor-phase process with a solid catalyst of phosphoric acid on kieselguhr and a liquid-phase alkylation in the presence of sulfuric acid was used.

The Importance of Cumene

The cumene-to-phenol route contributes more than 90% of phenol capacity in the world (Bao et al, 2015; Jiang et al, 2014). In recent years, the rapidly growing market of phenol increases the demand for cumene. The manufacture of cumene as a blending agent for high-octane gasoline and jet fuel was carried during the Second World War on a very large scale. Nowadays isopropylbenzene is the primary intermediate for the co-production of phenol and acetone (Perego & Ingallina, 2002). Cumene is firstly oxidized to cumene hydroperoxide, and then phenol and acetone are produced by the acid arrangement of cumene hydroperoxide (Huang et al, 2010; Yadav & Asthana, 2003).

Phenolic resins and bisphenol A are the main end uses for phenol. Bisphenol A, which is produced from phenol and acetone, has been the main driving force behind increased phenol demand. Its end use applications are in polycarbonate and epoxy resins.

Cumene is used extensively for the production of phenols, aceton (as a byproduct), (Weber et al, 2004) and its derivatives as shown in figure (6). Some of the phenol derivatives phenol resins, caprolactam, bisphenol A, and alkyl phenols. Bisphenol A is the largest derivative of phenol which is further utilized for the production of polycarbonate and epoxy resins.

The polycarbonate resins find its application in security (sheets & glazing), and architectural outlets. The polycarbonate is also utilized for the production of versatile and compact discs. Epoxy resins are utilized in high performance, adhesives, electrical laminates, coatings, paving, and flooring applications.

The phenol resins find its usage in home construction sector in the form of adhesives for structured wood panels, and in the production of mineral wool insulations as binders. Caprolactam another derivative of phenol, which is primarily utilized for the manufacturing of nylon 6.

Acetone is another important chemical produced by means of cumene. Acetone is used as a solvent and to produce many kinds of derivatives include methyl methacrylate, solvent, bisphenol A, and aldol chemicals. The methyl methacrylate

Figure 6. The major industrial applications of cumene

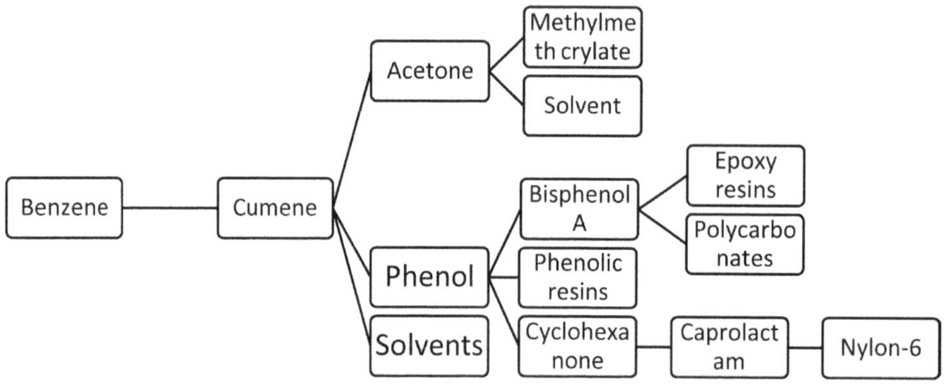

is mainly employed for the production of copolymers and homopolymers. These polymers are used in the manufacturing of liquid crystal displays and flat screen television sets.

World Annual Capacities of Cumene

Overall isopropylbenzene demand in the world is increased at an average annual rate of 4.5% from 2013 to 2017. Resulting in global isopropylbenzene demand of nearly (8,648,000) metric tonsin2013, and expected to be about (9,560,000) metric tons in 2017.

Figure 7. World and Saudi Arabia Average Annual Capacities of Cumene

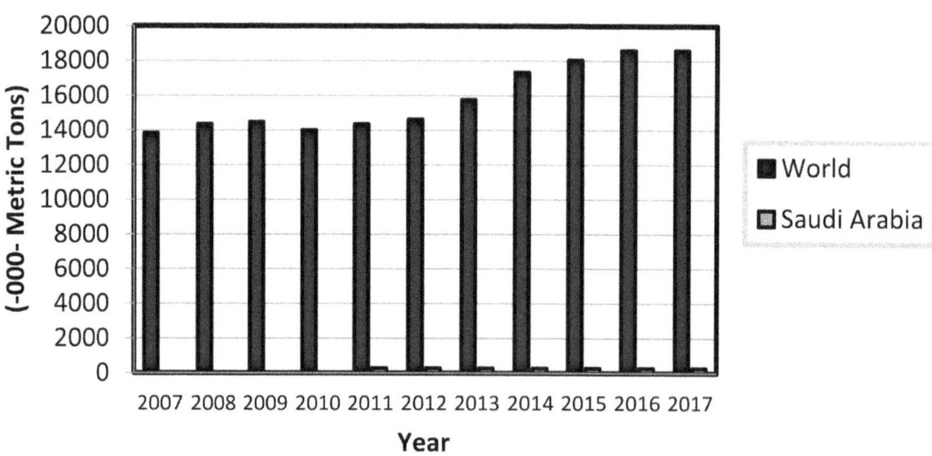

The largest world consumption of cumene market is Northeast Asia (43%), followed by North America and Western Europe(22% each). The Middle East will be one of the fastest-growing markets, with consumption increasing at an average annual rate of almost 15%, driven by the Petro-Rabigh plant addition in 2017. Capacity is expected to grow even faster, at an average annual growth rate of approximately 18.5% to 2020. However, the cumene market is driven by phenol and acetone market dynamics and the new capacity additions for cumene/phenol/acetone planned for the next five years will exceed the expected demand growth and cause operating rates to drop from 98% to 83%. https://www.ihs.com/products/cumene-chemical-economics-handbook.html

Isopropylation Reactions of Benzene

The reactions for cumene production from benzene and propylene are as follows:

Generally, benzene alkylation in a liquid phase or a gas phase process, for the production of isopropylbenzene, consists of the following three steps:

1. Isopropylation step, in which benzene reacts with propylene and produce isopropylbenzene and polyisopropylbenzenes.
2. Transalkylation step, in which polyisopropylbenzenes (generally *o*-, *p*- and *m*-diisopropylbenzene isomers;1,2,4-, 1,2,3- and 1,3,5-triisopropylbenzene isomers) at presence of benzene are converted to isopropylbenzene on a reverse alkylation process.
3. Separation step, in which unreacted benzene, polyisopropylbenzenes and other components are separated from each other and high purity isopropylbenzeneis produced.

Isopropylation of Benzene Reactions

The alkylation of benzene with propylene is an electrophilic aromatic substitution reaction in which the reaction involves the protonation of the catalyst acidic sites leading to isopropylbenzene, and further *o*-, *p*- and *m*-diisopropylbenzene isomers and triisopropylbenzene isomers. By the isomerization reactions some n- propylbenzene forms, which is undesirable as byproduct. The presence of stronger acid sites favors the formation of propylene dimers and trimers. Therefore, high selectivity of the catalyst to cumene is very important. It is remarkable that the polyisopropylated benzene by products such as diisopropylbenzene isomers and triisopropylbenzene isomers can be reconverted to cumene by the transalkylation reaction with benzene. Below, the following chemical reactions carried out which depend on the reaction conditions, type of catalyst and processes:

1. Isopropylation of benzene and further alkylation of cumene, diisopropylbenzene isomers and triisopropylbenzene isomers
2. Isomerization of diisopropylbenzenes and triisopropylbenzene.
3. Transalkylation of diisopropylbenzenes and triisopropylbenzenes with benzene.
4. Disproportionation of cumene with triisopropylbenzenes.
5. Side Reactions such as dimerization, trimerization and oligomerization of propylene; and alkylation of dimers and trimers with benzene.

Isopropylation Reactions (Al-Kinany, et al., 1997; 2000; 2001; 2013; 2015)

1. **Alkylation:**

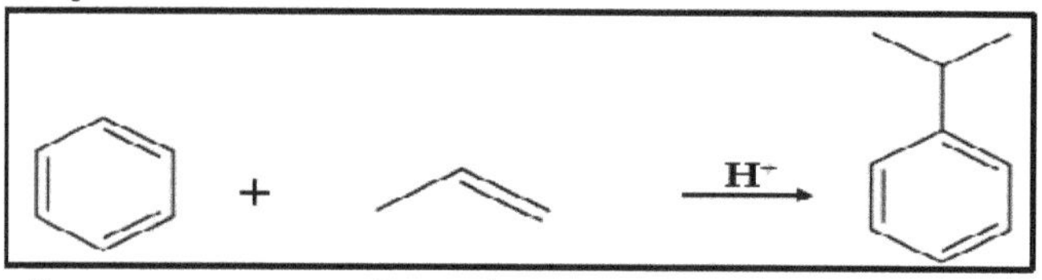

Benzene Propylene Cumene

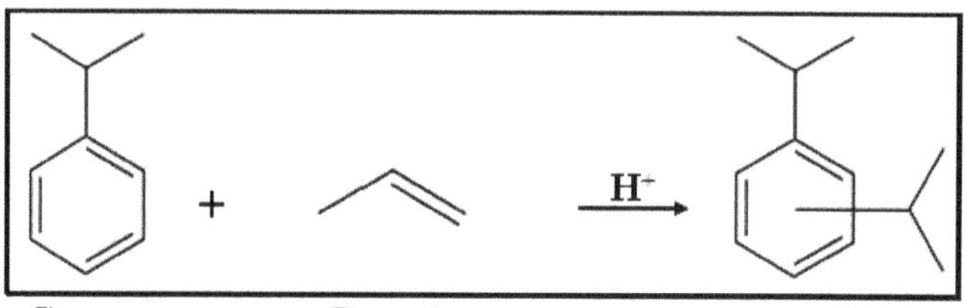

Cumene Propylene *o,m,p*-DIPB

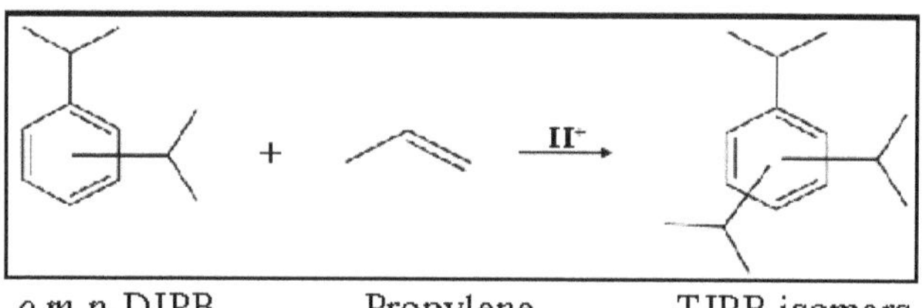

o,m,p-DIPB Propylene TIPB isomers

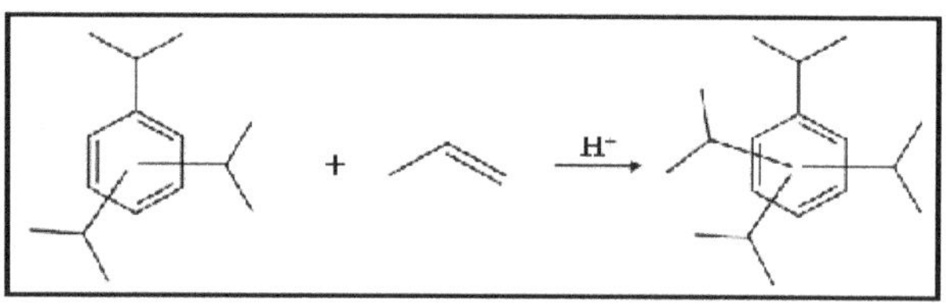

TIPB isomers Propylene TetIPB isomers

2. **Isomerization:**

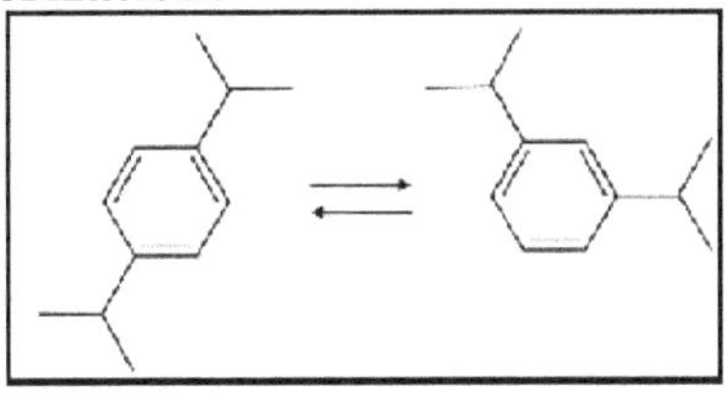

p-DIPB m-DIPB

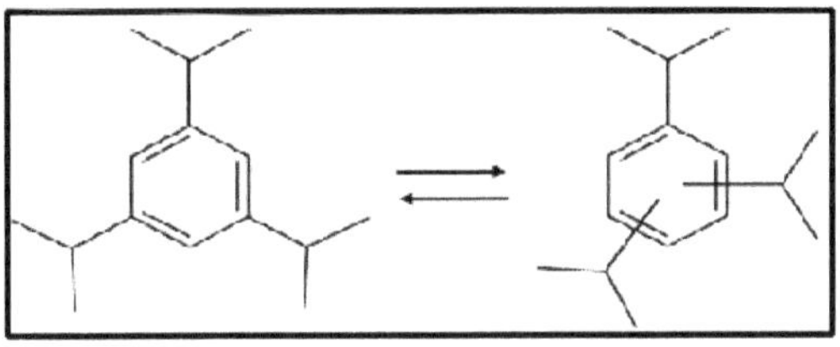

1,3,5-TIPB TIPB isomers

3. **Transalkylation:**

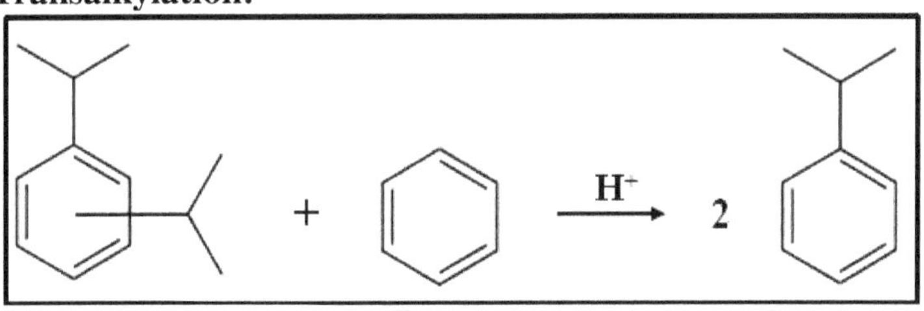

o,m,p-DIPB Benzene Cumene

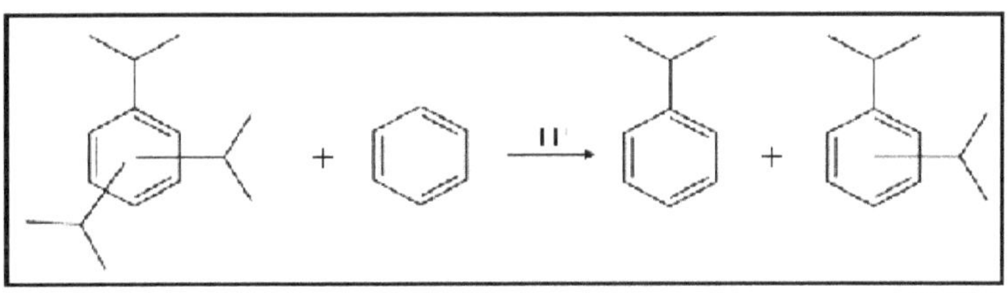

TIPB isomers Benzene Cumene o,m,p-DIPB

4. **Disproportionation:**

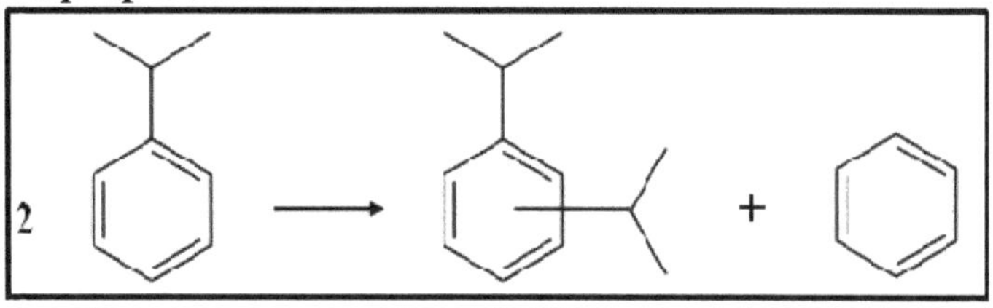

Cumene *o,m,p*-DIPB Benzene

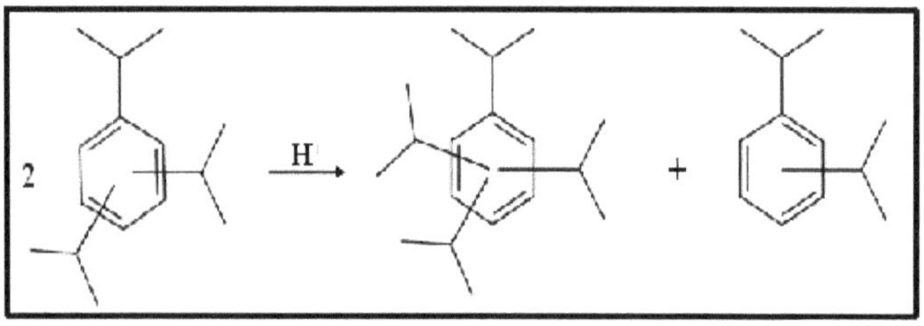

TIPB isomers TetIPB isomers *o,m,p*-DIPB

5. **Side Reactions:**

Dimerization, trimerization, oligomerization, alkylation....

$$CH_2{=}CH{-}CH_3 + CH_2{=}CH{-}CH_3 \xrightarrow{H^+} H_3C{-}CH{=}CH{-}(CH_2)_2{-}CH_3$$

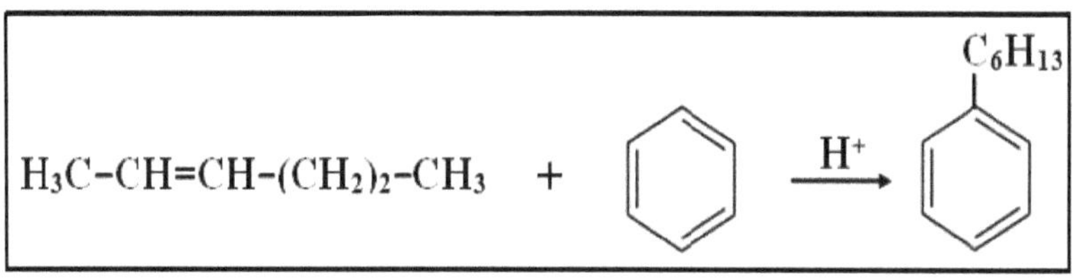

$$C_6H_{12} + C_6H_{13} \longrightarrow H_3C{-}CH{=}CH{-}(CH_2)_5{-}CH_3$$

Catalysts and Processes Developments for Cumene Synthesis

The typical reaction to manufacture cumene was isopropylation of benzene with propylene using sulphuric acid as catalyst. However, this processlead to corrosion problems. However, in recent years the production processes of cumene have been changed by discovering Zeolites catalyst. Process technology for cumene production by alkylation of benzene with propylene has been moving away from conventional aluminium chloride and phosphoric acid catalyzed Friedel–Crafts alkylation of benzene, toward zeolite-based processes.

Prior to 1992, virtually all cumene was produced by isopropylation of benzene with propylene using either solid phosphoric acid or aluminium chloride as alkylation catalysts. The SPA process, licensed by UOP, was developed in the 1940's primarily to produce cumene for aviation fuels (Ipatieff, 1945; Pujado, 1976).

The solid phosphoric acid catalyst consists of a complex mixture of orthosiliconphosphate, pyrosiliconphosphate, and polyphosphoric acid supported on kieselguhr. To maintain the desired level of activity and selectivity, small amounts of water are continuously fed into the reactor. The water continually liberates H_3PO_4 causing some downstream corrosion.

The efforts to find catalysts with high activity and selectivity, low deactivation, thermal stability and long life time have motivated the study of new materials. Various types of microporous zeolites were used by De Almeida et al (1994); Boveri et al (2006); Craciun et al (2007); Han et al (2003). Acidic clays (De Stefanis & Tomlinson, 2006), heteropoly acids (Hu et al., 1999; Okuhara et al, 1996), and zirconia supported 2-tungstophosphoric acid (Devassy et al, 2005) have already been used for benzene alkylation.

The traditional technologies suffer from a variety of drawbacks such as: low yields, corrosive catalysts and formation of oligomers and other isopropylated benzene complexes; and environmental concerns associated with the disposal of substantial amounts of phosphoric acid supported on kieselguhr or aluminuim chloride are eliminated by the use of zeolites because they are regenerable and can be safely disposed.

Various zeolite catalyst has been proposed as a possible solution to these problems since 1965, The zeolite-based processes produce higher cumene yields than the conventional solid phosphoric acid process because most byproducts of the diisopropylbenzene isomers and triisopropylbenzenes are converted to isopropylbenzene in separate transalkylation reactors or processes. In seventies, zeolite catalysts have begun to displace the conventional $AlCl_3$andsolid phosphoric acid Friedel–Crafts catalysts used in cumene production. More than 50% of the worldwide cumene capacity has converted to zeolite catalysts in the last15 years (Degnan et al, 2001).

In the early 1980s, Monsanto introduced an aluminuim chloride process based on the same chemistry used in the ethylbenzene process. Although many patents have issued over the last forty years, only in the last two decades the commercial application of several new alkylation processes, operating in liquid-phase and based on zeolite catalysts, such as mordenite, MCM-22, USY have been announced. (Girotti & Cappellazzo, 1997). The liquid-phase operation has the advantage of a better thermal control and longer catalysts life.

Problems related with reactants and products diffusion not allowed the use of ZSM-5 in the liquid-phase isopropylation of benzene with propylene (Kaeding & Holland, 1988). On the contrary good activity, selectivity and thermal stability of a Y type and beta type zeolite catalysts for the same isopropylation reaction in liquid-phase was reported by Harper et al (1977).

Industrial Processes of Cumene Production

The production of isopropylbenzene in industry is usually carried out byisopropylation of benzene with propylene as an alkylating agent in fixed bed reactors. Since the solid phosphoric acid and $AlCl_3$ catalysts cause corrosion, not recyclable and environmental problems, several zeolites based catalysts have been developed for producing cumene by isopropylation of benzene with propylene (Han et al, 2003; Tecza et al, 2003; Lei et al 2009).

Over the last 20 years, cumene producers have begun to convert to the more environmentally friendly and more efficient zeolite-based processes. Commercial cumene processes have been developed respectively by Dow (1992), CDTech (1995), Mobil–Badger (1993), Goelzer (1993), Enichem (1995), UOP (1995), and Bentham,(1996).Some more cumene production processes have been commercialized using catalyst based on zeolite by UOP/Q-Max process, Dow/Kellogg process, EniChem process, and Mobil/Badger process (Tecza & Brzozowski, 2003).

Mobil–Badger Cumene Process

First commercialized at Georgia Gulf, Pasadena, TX plant in 1994, the Mobil–Badger Cumene process consists of a fixed-bed alkylation reactor, a fixed-bed transalkylation reactor and a separation section (Goelzer et al, (1993). The process used MCM-22 and zeolite beta zeolites.

The Mobil–Badger process produces very pure cumene, 99.97 wt.% at 99.7 wt.% yield. The high cumene purity is primarily attributable to the high cumene selectivity of the MCM-22 catalyst with the life more than 5 years. The selectivity of the MCM-22 catalyst to isopropylbenzene reduces the size requirements for the fractionation section, and also reduces the coke forming reactions, which tend to shorten the cycle length.

Enichem Process

The Enichem cumene process was announced in 1995 and then commercialized at Enichem's Porto Torres, Sardinia plant in March 1996. Cumene yields and selectivities are both greater than 99%. Enichem's process uses a modified beta catalyst (Horsley, 1997; Meimam et al, 1996).

UOP Q-Max Process

UOP's Q-Max process was first commercialized at BTL Specialty Resin, Blue Island, IL plant in 1996 (Bentham et al, 1997). The process has also been licensed to Chevron where it has been in use at Chevron, Port Arthur, TX plant since 1997. UOP has not revealed the composition of the catalyst used in its process. UOP is reportedly working on a MgAPSO-31 catalyst that is very active and selective for cumene (Horsley, 1997).

The UOP/Q-Max process used the metal ion-containing beta-zeolite as the alkylation catalyst (Gajda, 1995). UOP has been able to optimize the acidity and reportedly eliminate the formation of *n*-propylbenzene.

Dow Chemical 3-DDM Process

Dow's 3-DDM cumene process uses a highly dealuminated mordenite catalyst in a two-stage fixed-bed alkylation and transalkylation process. The mordenite-based transalkylation stage of this process has operated commercially since 1992 at Dow's Terneuzen, The Netherlands plant (Meima, 1998). Dow's mordenite component reportedly has a silicon-to-aluminium ratio of 70 corresponding to azeolite with a low acid site density, but high acid sitestrength.

CDTech CD Cumene Process

CDTech's CD Cumene process involves the application of CD in a process that uses virtually the same design as the aforementioned CDTech ethylbenzene process (Smith et al, 1993). Cumene yields in excess of 99% are claimed for this dual stage process. Catalyst lifetimes are greater than two years in Mobil's and Dow's processes, Diisopropylbenzene produced in the isopropylation stage are converted to cumene in a second stage transalkylation reactor. The nature of CDTech's alklyation and transalkylation catalysts has not been disclosed, although CDTech patents refer to zeolites Y, omega and beta.

Other Processes

Lummus–Unocal reportedly have been developing a cumene process that is similar to their commercial aforementioned ethylbenzene process. The process, which has been run on a pilot-plant scale, uses a modified trilobe Y zeolitecatalyst that produces improved product selectivities (Horsley, 1997).Du Pont's nafion–silica composites have shown very good activity for liquid phase alkylation under very mild conditions of pressure and temperatures. The catalyst consists of nanometer-sized nafion particles dispersed in a highly porous silica matrix (Harmer, 1996).

CONCLUSION

Alkylation of benzene with ethene to produce ethylbenzene and with propene to produce isopropylbenzene (cumene) are widely used in the petrochemical industry. Both ethylbenzene and cumene are an important raw material in the petrochemical industry. In industry, ethylbenzene and cumene are mainly manufactured by the alkylation of benzene with ethene and propene via two processes, i.e. the liquid-phase process and gas-phase process.

The conventional catalyst for this alkylation process is Lwis acid ($AlCl_3$), which accounted for 40% of the worldwide ethylbenzene production in the past ten years. As utilization of this catalyst involves problems with separation, handling, safety and corrosion, immense efforts have been devoted to the development of new solid catalysts for ethylbenzene and isopropylbenzene production.

Compared with $AlCl_3$, zeolites possess advantages in easy separation, low corrosion, high surface area as well as uniform microporous structure coupled to shape selectivity, and thus they are better acidic catalysts when applied in petrochemical industry.

REFERENCES

Agency for Toxic Substances & Disease Registry (ATSDR). (1999). Toxicological Profile for Ethylbenzene (CAS# 100-41-4). Atlanta, GA: U. S. Department of health and human services, public health service.

Aguilar, J., Pergher, S. B. C., Detoni, C., Corna, A., Melo, F. V., & Sastre, E. (2008). Alkylation of biphenyl with propene using MCM-22 and ITQ-2 zeolites. *Catalysis Today*, 33–135, 667–672.

Al-Khattaf, S., Ali, M., & Čejka, J. (2010). Recent Development in Transformations of Aromatic Hydrocarbons over Zeolites. Zeolites and Catalysis, Synthesis. *Reactions and Applications*, *2*, 623–648.

Al-Khattaf, S., Ali, S., Aitani, A., Zilkova, N., Kubicka, D., & Čejka, J. (2014). Recent advances in reactions of alkylbenzenes over novel zeolites: The effects of zeolite structure and morphology. *Catalysis Reviews. Science and Engineering*, *56*(4), 333–402. doi:10.1080/01614940.2014.946846

Al-Kinany, M. C., & Al-Khowaiter, S. H. (1997). Ethylation and Isopropylation of Benzene using superacid catalyst, presented at the 15th World Petroleum Congress, Beijing, China, October 12-16.

Al-Kinany, M. C., & Al-Khowaiter, S. H. (1997) Ethylation and Isopropylation of Benzene using superacid catalyst. Presented at the *15th World Petroleum Congress*, Beijing, China, October 12-16, 1997.

Al-Kinany, M. C., & Al-Khowaiter, S. H. (1998). Isomerization and disproportionation of diethylbenzenesin the presence of trifluoromethane-sulphonic acid as catalyst.P resented at the Middle East PETROTECH'98, Manama, Bahrain.

Al-Kinany, M. C., & Al-Khowaiter, S. H. (1999). Synthesis of Ethylbenzene from Diethylbenzenes in the presence of Benzene using Trifilic Acid as Catalyst. *Studies in Surface Science and Catalysis*, *122*, 375–384. doi:10.1016/S0167-2991(99)80169-4

Al-Kinany, M. C., Al-Khowaiter, S. H., & Al-Malaki, F. (2000). Synthesis of Isopropylbenzene (Cumene) from Diisopropylbenzenes using Trifluoro-methanesulphonic acid as catalyst, Presented at *16th World Petroleum Congress*, Calgary, Canada, June 11-15.

Al-Kinany, M. C., Al-Khowaiter, S. H., & Al-Malaki, F. (2000). Synthesis of Isopropylbenzene (Cumene) from Diisopropylbenzenes using Trifluoro-methanesulphonic acid as catalyst. Presented at *16th World Petroleum Congress*, Calgary, Canada, June 11-15, 2000.

Al-Kinany, M. C., Al-Khowaiter, S. H., & Al-Malki, F. H. (2001). Synthesis of cumene (isopropylbenzene) from Diisopropylbenzenes in the presence of benzene using triflic acid as catalyst at room temperature. *Studies in Surface Science and Catalysis*, *133*, 459–464. doi:10.1016/S0167-2991(01)81995-9

Al-Kinany, M. C., Al-Khowaiter, S. H., Al-Rasheed, R. H., Al-Drees, S. A., & Al-Kharji, A. A. (2002). Low temperature synthesis of Ethylbenzene by transalkylation of o-Diethylbenzene with benzene using triflic acid as catalyst. The 5th International Conference and Exhibition on Chemistry and Industry, Manama, Bahrain.

Al-Kinany, M. C., Al-Megren, H. A., Al-Ghilan, E. A., Edwards, P. P., Xiao, T., Al-Shammari, A. S., & Al-Drees, S. A. (2012). Selective Zeolite Catalyst for Alkylation of Benzene with Ethylene to Produce Ethylbenzene. *Applied Petrochemical Research*, *2*(3-4), 73–83. doi:10.100713203-012-0022-6

Al-Kinany, M. C., Al-Megren, H. A., Aldrees, S., Al-Ghilan, E., Al-Dress, S., Al-Ghamdi, A., … Xiao, T. (2017). Method of preparing a modified zeolite catalyst and preparing ethylbenzene using one cycle process. US 0233308 A.

Al-Kinany, M. C., Al-Megren, H. A., Alghilan, E., Edwards, P., Xiao, T., & Al-Drees, S. (2016). *Al-Ghamdi, K.* Milan, Italy: Solid Acid Zeolite Catalysts for Benzene/Ethylene Alkylation Reactions. IRPC.

Al-Kinany, M. C., Al-Megren, H. A., Edwards, P., Xiao, T., Al-Drees, S., & Alghilan, E. … Al-Ghamdi, K. (2013). Isopropylbenzene (Cumene) Production by Alkylation of Benzene with Propylene Using Super acid Catalysts. 3rd KACST-Oxford Petrochemicals Forum "Disruptive Innovations for Modern Petroleum Refining and Petrochemical Industry", Oxford, UK.

Al-Kinany, M. C., Al-Megren, H. A., Edwards, P., Xiao, T., Al-Shammari, A. S., Al-Drees, S. A., & Alghilan, E. (2012). Synthesis of Ethylbenzene by Alkylation of Benzene with Ethylene over Modified Zeolite Catalyst in a Gas Phase Process. 2ndKACST-Oxford Petrochemical Forum "Sustainability Through Innovation in the Moderen Petrochemical Industries", Riyadh, Saudi Arabia.

Al-Kinany, M. C., Al-Megren, H. A., Edwards, P. P., Xiao, T., Al-Drees, S. A., & Alghilan, E. A. & Liu, B. (2015). Isopropylation of Benzene over Modified Zeolite with Metal Oxide to Produce Isopropylbenzene (Cumene). Pacifichem, Honolulu, HI. USA 15th-20th December.

Al-Kinany, M. C., Al-Megren, H. A., Edwards, P. P., Xiao, T., Al-Drees, S. A., Alghilan, E. A., . . . Alghamdi, K. (2013). Isopropylbenzene (Cumene) Production by Alkylation of Benzene with Propylene Using Super acid Catalysts. 3rd KACST-Oxford Petrochemicals Forum 2013. Disruptive Innovations for Modern Petroleum Refining and Petrochemical Industry, Oxford, UK, August 12-15, 2013.

Al-Kinany, M. C., Al-Megren, H. A., Edwards, P. P., Xiao, T., Al-Shammari, A. S., Al-Drees, S. A., & Al-Ghilan, E. A. (2011). Selective Catalysis of Zeolite in Alkylation of Benzene with Ethylene to Produce Ethylbenzene. The 1stKACST-Oxford Petrochemical Forum "Petroleum and Petrochemical Industries: Building the Future in a Sustainable Way", Oxford, UK.

Al-Kinany, M. C., Jibril, B. Y., Al-Khowaiter, S. H., Al-Dosari, M. A., Al-Megren, H. A., Al-Zahrani, S. M., & Al-Humaizi, K. I. (2005). Low temperature transalkylation of o-diethylbenzene with benzene to ethylbenzene using triflic acid as a catalyst. *Chemical Engineering and Processing*, *44*(8), 841–846. doi:10.1016/j.cep.2004.08.011

Al-Zahrani, S. M., Al-Kinany, M. C., Al-Humaizi, K. I., & Al-Khowaiter, S. H. (2002). Transalkylation and isomerization of ortho-diethylbenzene with benzene using trifluoromethanesulphonic acid catalyst: Kinetic analysis. *Chemical Engineering and Processing*, *41*(4), 321–327. doi:10.1016/S0255-2701(01)00148-9

Al-Zahrani, S. M., Al-Kinany, M. C., Al-Humaizi, K. I., & Al-Khowaiter, S. H. (2003). Kinetics and Mechanisms of Transalkylation and disproportionation of meta-Diethylbenzene by Triflic Acid Catalyst. *International Journal of Chemical Kinetics*, *35*(11), 555–563. doi:10.1002/kin.10152

Axe, W. N. (1946a). Alkenylation Process. Patent 2,403,963.

Axe, W. N. (1946b). Process for Reacton of Aromatic Hydorocarbons with normally Gaseous Unsaturated Hydrocarbons. Patent 2,412,595.

Bao, Y. H., Jiang, H., Xing, W. H., Chen, R. Z., & Fan, Y. Q. (2015). Liquid phase hydroxylation of benzene to phenol over vanadylacetylacetonate supported on amine functionalized SBA-15. *Reaction Kinetics, Mechanisms and Catalysis*, *116*(2), 535–547. doi:10.100711144-015-0898-0

Bartholomew, C. H., & Farrauto, R. J. (2011). *Fundamentals of Industrial Catalytic Processes*. New York: John Wiley & Sons.

Bellussi, G. (2004). Zeolite catalysts for the production of chemical commodities: BTX derivatives. Studies in Surface Science and Catalysis, 154(Part A), 53-65.

Bellussi, G., Clerici, M. G., Carati, A., & Cavani, F. (1991). US Patent 5 049 536. US Patent Office.

Bentham, M. F., Gadja, G. J., Jensen, R. H., & Zinner, H. A. (1997). Development and commercialization of solid acid catalysts. *ErdoelErdgasKohle*, *113*(2), 84–88.

Bentham, M. F., Gajda, G. J., Jensen, R. H., & Zinnen, H. A. (1996). *Development and commercialization of solid acid catalysts. DGMKConference*. Berlin, Germany: Catalysis on Solid Acids and Bases.

Bloch, S. H., & Pat, U. S. (1972). hydrocarbon conversion catalyst and process. *Chemical Abstracts*, 77.

Bokade, V. V., & Yadav, G. D. (2008). Heteropolyacid supported on acidic clay: A novel efficient catalyst for alkylation of ethylbenzene with dilute ethanol to diethylbenzene in presence of C8 aromatics. *Journal of Molecular Catalysis A Chemical, 285*(1-2), 155–161. doi:10.1016/j.molcata.2007.11.018

Booth, B. L., Al-Kinany, M.C., & Khosrow, L. (1987). Trifluormethanesulphonic Acid as a catalyst for Ethylation of Benzene. Chemical Society, Perkin Transactions, 2049-2055.

Boveri, M., Marquez-Alvarez, C., Laborde, M. A., & Sastre, E. (2006). Steam and acid dealumination of mordenite: Characterization and influence on the catalytic performance in linear alkylbenzene synthesis. *Catalysis Today, 114*(2-3), 217–225. doi:10.1016/j.cattod.2006.01.012

Busca, G. (2007). Acid catalysts in industrial hydrocarbon chemistry. [PubMed]. *Chemical Reviews, 107*(11), 5366–5410. doi:10.1021/cr068042e

Čejka, J., Corma, A., & Zones, S. I. (2010). *Zeolites and Catalysis: Synthesis, Reactions and Applications*. Wiley-VCH; doi:10.1002/9783527630295

Čejka, J., & Wichterlova, B. (2002). Acid-catalyzed synthesis of mono- and dialkyl benzenes over zeolites: Active sites, zeolite topology, and reaction mechanisms. *Catalysis Reviews. Science and Engineering, 44*(3), 375–421. doi:10.1081/CR-120005741

Cejka, J., Wichterlova, B., & Bednarova, S. (1991). Alkylation of toluene with ethene over H-ZSM-5 zeolites. *Applied Catalysis A, General, 79*(2), 215–226. doi:10.1016/0926-860X(91)80007-L

Chang, C. D., Lang, W. H., & Smith, R. L. (1979). The Conversion of Methanol and Other O-Compounds to Hydrocarbons over Zeolite Catalysts. *Journal of Catalysis, 56*(2), 169–173. doi:10.1016/0021-9517(79)90103-9

Chang, & Grover, J. J., (1977). US4058576A. US Patent Office.

Chen, N. Y., Garwood, W. A., & Dwyer, F. G. (1996). *Shape Selective Catalysis in Industrial Applications*. New York: Marcel Dekker.

Chen, N. Y., & Garwood, W. E. (1986). Industrial application of shape-selective catalysis. *Catalysis Reviews. Science and Engineering, 28*(2-3), 185–264. doi:10.1080/01614948608082251

Cheng, J. C., Degnan, T. F., Beck, J. S., Huang, Y. Y., Kalyanaraman, M. J., Kowalski, A., ... Mazzone, D. N. (1998). A comparison of zeolites MCM-22, beta, and USY for liquid phase alkylation of benzene with ethylene. *Studies in Surface Science and Catalysis*, *121*, 53–60. doi:10.1016/S0167-2991(99)80044-5

Choudhary, V. R., Mondal, K. C., & Mulla, S. A. R. (2005). Simultaneous Conversion of Methane and Methanol into Gasoline over Bifunctional Ga-, Zn-, In-, and/or Mo-Modified ZSM-5 Zeolites. [PubMed]. *Angewandte Chemie International Edition*, *44*(28), 4381–4385. doi:10.1002/anie.200500694

Christensen, C.H., Johannsen, K., &Schmidt, I. (2003). Catalytic benzene alkylation over mesoporous zeolite single crystals: improving activity and selectivity with a new family of porous materials. American Chemical Society, 125(44), 13370-13371.

Craciun, I., Reyniers, M.-F., Guy, B., & Marin, G. B. (2007). Effects of acid properties of Y zeolites on the liquid-phase alkylation of benzene with 1-octene: A reaction path analysis. *Journal of Molecular Catalysis A Chemical*, *277*(1-2), 1–14. doi:10.1016/j.molcata.2007.06.027

Da, Z., Han, Z., Magnoux, P., & Guisnet, M. (2001). Liquid-phase alkylation of toluene with long-chain alkenes over HFAU and HBEA zeolites. *Applied Catalysis A, General*, *219*(1-2), 45–52. doi:10.1016/S0926-860X(01)00662-7

Dahl, I. M., & Kolboe, S. (1994). On the Reaction Mechanism for Hydrocarbon Formation from Methanol over SAPO-34: I. Isotopic Labeling Studies of the Co-Reaction of Ethene and Methanol. *Journal of Catalysis*, *149*(2), 458–464. doi:10.1006/jcat.1994.1312

De Almeida, L. G., Dufaux, M., Taarit, Y. B., & Naccache, C. (1994). Liner alkylbenzene. J.American Oil Chemists'. *Society*, *71*(7), 675–694.

De Stefanis, A., & Tomlinson, A. A. G. (2006). Towards designing pillared clays for catalysis. *Catalysis Today*, *114*(2-3), 126–141. doi:10.1016/j.cattod.2006.01.019

Degnan, T. F. Jr, Smith, C. M., & Venkat, C. R. (2001). Alkylation of aromatics with ethylene and propylene: Recent developments in commercial processes. *Applied Catalysis A, General*, *221*(1-2), 283–294. doi:10.1016/S0926-860X(01)00807-9

Devassy, B. M., Lefebvre, F., & Halligudi, S. B. (2005). Zirconia Supported 12-Tungstophosphoric acid as Solid Catalyst for the Synthesis of Linear Alkyl Benzenes. *Journal of Catalysis*, *231*(1), 1–10. doi:10.1016/j.jcat.2004.09.024

Du, Y., Wang, H., & Chen, S. (2002). Study on alkylation of benzene with ethylene over β-zeolite catalyst to ethylbenzene by in situ IR.Molecular Catalysis A. *Chemical, 179*(1-2), 253–261.

Dwyer, F. G., & Walker, O. (1981). Bargaining in an asymmetrical power structure Catalysis in Organic Reactions. Chemical Industries, Marcel Dekker, New York, 37, 39–47.

Ebrahimi, A. N., & Ashkan, S. (2011). Modification and optimization of benzene alkylation process for production of ethylbenzene. *Chemical Engineering and Processing: Process Intensification, 50*(1), 31–36. doi:10.1016/j.cep.2010.10.011

Faghihian, H., & Mohammadi, M. H. (2014). Acid activation effect on the catalytic performance of Al-pillared bentonite in alkylation of benzene with olefins. *Applied Clay Science, 93-94*, 1–7. doi:10.1016/j.clay.2014.02.026

Fetisova, V. A., Ivashkina, E. D., Ivanchina, E. N., & Kravtsov, A. V. (2010). Mathematical model for the process of benzene alkylation by higher olefines. *Catalysis in Industry, 2*(1), 55–61. doi:10.1134/S2070050410010095

Franck, H. G., & Stadelhofer, J. W. (1988). *Industrial Aromatic Chemistry, Raw Materials-Processes- Products.* Heidelberg, Germany: Springer-Verlag; doi:10.1007/978-3-642-73432-8

Freeman, D., Wells, R. P. K., & Hutchings, G. J. (2002, January). Ga2O3/H-ZSM-5 and Ga2O3/WO3 Catalysts. *Journal of Catalysis, 205*(2), 358–365. doi:10.1006/jcat.2001.3446

Friedel, C., & Crafts, J. M. (1877). A New General Synthetical Method of Production Hydrocarbons. *Journal of the Chemical Society, 32*, 725–791.

Gajda, P. (1995). *Wilson* (p. 5434326). US: Discrete Molecular Sieve and Use in Aromatic- Olefin Alkylation. UOP Inc.

Galadima, A., & Muraza, O. (2015). Role of zeolite catalysts for benzene removal from gasoline via alkylation: A review. *Microporous and Mesoporous Materials, 213*, 169–180. doi:10.1016/j.micromeso.2015.02.029

Gao, J., Zhang, L., Hu, J., Li, W., & Wang, J. (2009). Effect of zinc salt on the synthesis of ZSM- 5 for alkylation of benzene with ethanol. *Catalysis Communications, 10*(12), 1615–1619. doi:10.1016/j.catcom.2009.04.029

Girotti, G., & Cappellazzo, O. (1997). Catalyst for the alkylation or transalkylation of aromatic compounds. *Eur. Pat. Appl., 687*, 500.

Goelzer, A. R., Ram, S., Hernandez, A., Chin, A. A., Harandi, M. N., & Morris, C. M. (1993). Mobil-Badger Technologies for Benzene Reduction in Gasoline. Paper presented at the 1993 NPRA Annual Meeting, San Antonio, TX.

Graf, P. O., & Lefferts, L. (2009). Reactive separation of ethylene from the effluent gas of methane oxidative coupling via alkylation of benzene to ethylbenzene on ZSM-5. *Chemical Engineering Science*, *64*(12), 2773–2780. doi:10.1016/j.ces.2009.02.031

Groen, J. C., Sano, T., Moulijn, J. A., & Pérez-Ramírez, J. (2007). Alkaline-mediated mesoporousmordenite zeolites for acid-catalyzed conversions. *Journal of Catalysis*, *251*(1), 21–27. doi:10.1016/j.jcat.2007.07.020

Hamid, G., Ahari, S., Farshi, J. A., & Majid, K. (2004). Modelling and Simulation Of Benzene Alkylation Process Reactors For Production Of Ethylbenzene. *Petroleum and Coal*, *46*(1), 55–63.

Han, M. H., Lin, S. X., & Roduner, E. (2003). Study on the alkylation of benzene with propylene over Hβ zeolite. *Applied Catalysis A, General*, *243*(1), 175–184. doi:10.1016/S0926-860X(02)00565-3

Han, M. Z., Cui, Z., Xu, C., Chen, W., & Jin, Y. (2003). Synthesis of linear alkylbenzene catalyzed by Hβ-zeolite. *Applied Catalysis A, General*, *238*(1), 99–107. doi:10.1016/S0926-860X(02)00334-4

Hancock, E. G. (Ed.). (1975). *Benzene and its industrial derivatives*. Ernest Benn Limited.

Hansen, N., Bruggemannt, T., Bell, A., & Keil, F. (2008). Theoretical investigation of benzene alkylation with ethene over H-ZSM5. *The Journal of Physical Chemistry C*, *112*(39), 15402–15411. doi:10.1021/jp8036022

Harmer, M. A., Farneth, W. E., & Sun, Q. (1996). High Surface Area Nafion Resin/Silica Nanocomposites: A New Class of Solid Acid Catalyst. *Journal of the American Chemical Society*, *118*(33), 7708–7715. doi:10.1021/ja9541950

Harper, E. F., Ko, D. Y., Lese, H. K., Sabourin, E. T., & Williamson, R. C. (1977). *Industrial and Laboratory Alkylations*. American Chemical Society.

Hellring, S. D., Schmitt, K. D., & Chang, C. D. (1987). Synthesis and decomposition of trimethyloxonium ZSM-5, a purported intermediate in methanol conversion into gasoline. *Journal of the Chemical Society. Chemical Communications*, (17): 1320–1322. doi:10.1039/c39870001320

Hernández-Cortez, J. G., Martinez, L., Soto, L., López, A., Navarrete, J., Manríquez, M., ... López-Salinas, E. (2010). Liquid phase alkylation of benzene with dec-1-ene catalyzed on supported 1.2-tungstophosphoric acid. *Catalysis Today*, *150*(3-4), 346–352. doi:10.1016/j.cattod.2009.12.003

Hervert, G. L., Grove, D., & Linn, C. B. (1960). Alkylation OF Aromatic Hydrocarrons. US Patent 2,939,89.

Hoeldrich, W. F. (2000). Environmentally benign manufacturing of fine and intermediate chemicals. *Catalysis Today*, *62*(1), 115–130. doi:10.1016/S0920-5861(00)00413-2

Horsley, J. A. (1997). Producing bulk and fine chemicals using solid acids. *Chemtech*, *27*(10), 45–49.

Hu, C., Zhang, Y., Xu, L., & Peng, G. (1999). Continuous syntheses of octyl phenol or nonane phenol on supported heteropoly acid catalysts. *Applied Catalysis A, General*, *177*(2), 237–244. doi:10.1016/S0926-860X(98)00269-5

Huang, C. C., Peng, J. J., Wu, S. H., Hou, H. Y., You, M. L., & Shu, C. M. (2010). Effects of cumenehydroperoxideon phenol and acetone manufacturing by DSC and VSP2. *Journal of Thermal Analysis and Calorimetry*, *102*(2), 579–585. doi:10.100710973-010-0953-z

Ipatieff, V. N. (1937). Treatment of Hydrorocarbons. US Patent 2,067,764.

Ipatieff, V. N. (1945). US Patent 2,382,318. US Patent Office.

Ipatieff, V. N., Corson, B. B., & Pines, H. J. (1936). Influence of Sulfuric Acid Concentration upon Reaction between Olefins and Benzene. *Journal of the American Chemical Society*, *58*(6), 919–922. doi:10.1021/ja01297a021

Ipatieff, V. N., & Grosse, A. V. (1936). Alkylation of aromatics with olefins in the presence of boron fluoride. *Journal of the American Chemical Society*, *58*(11), 2339. doi:10.1021/ja01302a505

Ipatieff, V. N., & Linn, C. B. (1947). Alkylation of Aromatics. US Patent 2,428,279.

Ipatieff, V. N., Pines, H., & Komarewsky, V. I. (1936). Phosphoric acid as the catalyst for alkylation of aromatic hydrocarbons. *Industrial & Engineering Chemistry*, *28*(2), 222–223. doi:10.1021/ie50314a019

Ipatieff, V. N., & Schmerling, L. (1946). Ethylation of Benzene in Presence of Solid Phosphoric Acid. *Industrial & Engineering Chemistry*, *38*(4), 400–402. doi:10.1021/ie50436a017

Jiang, H., She, F., Du, Y., Chen, R. Z., & Xing, W. H. (2014). One-step continuous phenol synthesis technology via selective hydroxylation of benzene over ultrafine TS-1 in a submerged ceramic membrane reactor. *Chinese Journal of Chemical Engineering*, *22*(11–12), 1199–1207. doi:10.1016/j.cjche.2014.09.016

Kaeding, W., Barile, G., & Wu, M. (1984). Mobil Zeolite Catalysts for Monomers. *Catalysis Reviews. Science and Engineering*, *26*(3-4), 597–612. doi:10.1080/01614948408064727

Kaeding, W. W., & Holland, R. E. (1988). Shape-selective reactions with zeolite catalysts: VI. Alkylation of benzene with propylene to produce cumene. *Journal of Catalysis*, *109*(1), 212–216. doi:10.1016/0021-9517(88)90200-X

Kaeding, W. W., & Young, L. B. (1978). U.S. Patent 4,086,287. US Patent Office.

Kartal, O. E., & Onal, I. (2008). Synthesis of ZSM-5 from modified clinoptilolite and its catalytic activity in alkylation of benzene to ethylbenzene. *Chemical Engineering Communications*, *195*(8), 1043–1057. doi:10.1080/00986440801907078

Kharlampidi, K. E., Izmailov, I. R., & Batyrshin, N. N. (2000). Alkylation of benzene with 1-butene in the presence of AlCl3. *Russian Journal of General Chemistry*, *70*(2), 278–280.

Kolboe, S. (1993). Hydrocarbon Formation from Methanol/Dimethyl Ether Over Protonated Zeolites and Molecular Sieves. New Insights from Recent Experiments. *Studies in Surface Science and Catalysis*, *75*, 449–461. doi:10.1016/S0167-2991(08)64030-6

Kolboe, S., Kubelkova, L., Haag, W. O., Rooney, J. J., Tsyganenko, A. A., Hall, W. K., ... Kolesnikov, I. M. (n.d.). Catalysis in the oil and gas industry. *Gubkin Russian State University of Oil and Gas.*

Komarewsky, V. I., & Ruther, W. E. (1950). Reactions of Paraffin Hydrocarbons in the Presence of Sulfuric Acid. *Journal of the American Chemical Society*, *72*(12), 5501–5503. doi:10.1021/ja01168a034

Lalancette, J. M., Fournier-Breault, M. J., & Thiffault, R. (1974). Metals Intercalated in Graphite. II. The Friedel–Crafts Reactions with AlCl3–Graphite. *Canadian Journal of Chemistry*, *52*(4), 589–591. doi:10.1139/v74-091

Lalancette, J. M., & Lafontaine, J. (1973). Intercalation of antimony pentafluoride in the lattice of graphite. *Journal of the Chemical Society. Chemical Communications*, (21): 815. doi:10.1039/c39730000815

Lei, Z. G., Dai, C. N., Wang, Y. L., & Chen, B. H. (2009). Process optimization on alkylation of benzene with propylene. *Energy & Fuels*, *23*(6), 3159–3166. doi:10.1021/ef900052j

Li, P., Zhang, W., Han, X., & Bao, X. (2010). Conversion of Methanol to Hydrocarbons over Phosphorus -modified ZSM-5/ZSM-11 Intergrowth Zeolites. *Catalysis Letters*, *134*(1-2), 124–130. doi:10.100710562-009-0214-6

Lu, B., Yakushi, Y., Oumi, Y., Itabashi, K., & Sano, T. (2006). Control of crystal size of high-silica mordenite by quenching in the course of crystallization process. *Microporous and Mesoporous Materials*, *95*(1-3), 141–145. doi:10.1016/j.micromeso.2006.05.017

Maier, W. H., & Ward, J. D. (1969). United States Patent No. US3478119A. US Patent Office.

Mattox, W. J. (1945). Conversion of Hyorocarbons. US Patent 2,385,524.

McClure, J. D. (1977). Ethylbenzene Process using an Unsupported Perfluorinated Polymer Catalyst. US Patent 4,041,090.

McClure, J. D. (1977). Hydrocarbon Conversion Process Usingan Unsupported Perfluornated Polymer Catalyst. US Patent 4,022, 847.

Meima, G. R. (1998). Advances in cumene production. *CATTech*, *2*(1), 5–12.

Meima, G. R., Van der Aalst, M. J. M., Samson, M. S. U., Garces, J. M., & Lee, J. G. (1996). Recent developments in the production of cumene with zeolite catalysts. *EdrolErdgasKohle*, *112*(7), 3156–3318.

Meyers, R. A. (2005). *Handbook of Heterogeneous Catalysis*. McGraw-Hill.

Michael Frenkel, G. J. K., Marsh, K. N., Roganov, G. N., & Wilhoit, R. C. (1994). *Thermodynamics of Organic Compounds in the Gas State* (Vol. 2). College Station, TX: Academic Press.

Ming-Yuan, H., Zhonghui, L., & Enze, M. (1988). Acidic and hydrocarbon catalytic properties of pillared clay. *Catalysis Today*, *2*(2-3), 321–338. doi:10.1016/0920-5861(88)85013-2

Mitchell, S., Michels, N.-L., Kunze, K., & Ramirez-Perez, J. (2012). Visualization of hierarchically structured zeolite bodies from macro to nano length scales. [PubMed]. *Nature Chemistry*, *4*(10), 825–831. doi:10.1038/nchem.1403

Mitsutani, A. (2002). Future possibilities of recently commercialized acid/base-catalyzed chemical processes. *Catalysis Today*, *73*(1-2), 57–63. doi:10.1016/S0920-5861(01)00518-1

Moreno, A., Cohen, L., & Berna, J. L. (1988). Influence of unsulfonated material and its sulfone content on the physical properties of linear alkylbenzenesulfonates. *Surfactant Detergent*, *25*, 216.

Namuangruk, S., Pantu, P., & Limtrakul, J. (2004). Alkylation of benzene with ethylene over faujasite zeolite investigated by the ONIOM method. *Journal of Catalysis*, *225*(2), 523–530. doi:10.1016/j.jcat.2004.04.016

O'Kelly, A. A., Kellet, J., & Plucker, J. (1947). Monoalkyibenzenes by Vapor-Phase Alkylation with Silica-alumina Catalyst. *Journal of Industrial and Engineering Chemistry*, *39*(2), 154–158. doi:10.1021/ie50446a016

Odedairo, T., & Al-Khattaf, S. (2011). Alkylation and transalkylation of alkylbenzenes in cymene production over zeolite catalysts. *Chemical Engineering Journal*, *167*(1), 240–254. doi:10.1016/j.cej.2010.12.043

Okuhara, T., Mizuno, N., & Misono, M. (1996). Catalytic Chemistry of Heteropoly Compounds. *Advances in Catalysis*, *41*, 113–252. doi:10.1016/S0360-0564(08)60041-3

Olah, G. A. (1973). *Friedel-Crafts Chemistry*. Wiley-Interscience.

Olah, G. A., Kaspi, J., & Bukala, J. (1977). Heterogeneous catalysis by solid superacids. 3. Alkylation of benzene and transalkylation of alkylbenzenes over graphite-intercalated Lewis acid halide and perfluorinated resin sulfonic acid (Nafion-H) catalysts. *The Journal of Organic Chemistry*, *42*(26), 4187–4191. doi:10.1021/jo00862a003

Olah, G. A., Prakash, G. K. S., & Sommer, J. (1979). Superacids. [PubMed]. *Science*, *206*(4414), 13–20. doi:10.1126cience.206.4414.13

Olah, G. A., & Ruff, J. K. (1964). Friedel-Crafts and related reactions. Volume I: General Aspects. *Inorganic Chemistry*, *3*(8), 1205–1206. doi:10.1021/ic50018a043

Perego, C., Amarilli, S., Carati, A., Flego, C., Pazzuconi, G., Rizzo, C., & Bellusi, G. (1999). Mesoporous silica-aluminas as catalysts for the alkylation of aromatic hydrocarbons with olefins. *Microporous and Mesoporous Materials*, *27*(2-3), 345–354. doi:10.1016/S1387-1811(98)00267-4

Perego, C., & Ingallina, P. (2002). Recent advances in the industrial alkylation of aromatics: New catalysts and new processes. *Catalysis Today*, *73*(1-2), 3–22. doi:10.1016/S0920-5861(01)00511-9

Progue, R. F. (1996). World Patent 96/34843. Dow Chemical Company.

Pujado, P. R., Salazar, J. R., & Berger, C. V. (1976). Cheapest Route to Phenol. *Hydrocarbon Processing*, *55*(3), 91–96.

Qn, L., & Hill, C. (1992). US Patent 5 118 894. US Patent Office.

Rodewald, P. G. (1975). Hydrocarbon Conversion in the Presence of a Catalyst of a Lewis acid Intercalated in Graphite. US Patent 3,925,495.

Satterfield, C. N. (1991). *Heterogeneous Catalysis in Industrial Practice* (2nd ed.). New York: McGraw-Hill.

Schmerling, L. (1946). Alkylation of Aromatic Compounds. US Patent 2,385,303.

Sebulsky, R. T., & Henke, A. M. (1971). Alkylation of Benzene with Dodecene-1 Catalyzed by Supported Silicotungstic Acid. *Industrial & Engineering Chemistry Process Design and Development*, *10*(2), 272–279. doi:10.1021/i260038a022

Shi, Y., Xing, E., Xie, W., Zhang, F., Mu, X., & Shu, X. (2015). Size-controlled synthesis of MCM-49 zeolites and their application in liquid-phase alkylation of benzene with ethylene. *The Royal Society of Chemistry's*, *5*, 13420–13429.

Slaughter, J. I., & McCaulay, D. A. (1959). United States Patent No.US2908728A. US Patent Office.

Smith, J., Sy, A., Chen, J., & Dautzenberg, F. (1993, January). Catalytic Distillation Route for Cumene. DeWitt Petrochemical Review.

Socha, R. F., Chang, C. D., Gould, R. M., Kane, S. E., & Avidan, A. A. (1987). Fluid-Bed Studies of Olefin Production from Methanol. Industrial Chemicals via C Processes: American Chemical Society, 328, 34-41. doi:10.1021/bk-1987-0328.ch003

Srirattnai, K., Damronglerd, S., Omi, S., Roengsumran, S., Petsom, A., & Ma, G. H. (2002). Encapsulated AlCl3: A convenient catalyst for the alkylation of benzene with dodecene. *Tetrahedron Letters*, *43*(25), 4555–4557. doi:10.1016/S0040-4039(02)00702-5

Stocker, M. (1999). Methanol-to-hydrocarbons: Catalytic materials and their behavior1. *Microporous and Mesoporous Materials*, *29*(1-2), 3–48. doi:10.1016/S1387-1811(98)00319-9

Svelle, S., Joensen, F., Nerlov, J., Olsbye, U., Lillerud, K. P., Kolboe, S., & Bjorgen, M. (2006). Conversion of methanol into hydrocarbons over zeolite H-ZSM-5: Ethene formation is mechanistically separated from the formation of higher alkenes. [PubMed]. *Journal of the American Chemical Society*, *128*(46), 14770–14771. doi:10.1021/ja065810a

Tecza, W., & Brzozowski, R. (2003). Modern industrial processes for making ethylbenzene and cumene. *Przemysl Chemiczny*, *82*(11), 1428–1434.

Teketel, S. Skistad, W., Benard, S., Olsbye, U., Lillcrud, K. P., Beato, P., & Svelle, S. (2012). Shape selectivity in the conversion of methanol to hydrocarbons: the catalytic performance of one-dimensional 10-ring zeolite: ZSM-22, MZSM-23,ZSM-48 and EU-1. American Chemical Society, 2(1), 26-37.

Teketel, S., Svelle, S., Lillerud, K. P., & Olsbye, U. (2009). Shape-Selective Conversion of Methanol to Hydrocarbons Over 10-Ring Unidirectional-Channel Acidic H-ZSM-22. *ChemCatChem*, *1*(1), 78–81. doi:10.1002/cctc.200900057

Twigg, M. V. (1996). *The Catalyst Handbook* (2nd ed.). London: Manson Publishing.

Uitti, K. D. (1977). Process for Production and Dehydrogenation of Ethylbenzene. US Patent 4,009,217.

Universal Oil Products Co. (1971). Preparation of monoalkylated aromatic compounds. Patent 2,032,048.

Ventuto, P. B., & Hamilton, L. A. (1966). Organic Reaction Catalized by crystalline aluminosilicates:II. Alkylation Reactions: Mechanistic and Aging Consideration. *Journal of Catalysis*, *5*(3), 484–493. doi:10.1016/S0021-9517(66)80067-2

Vincent, W. A., Kevin, F. J., & Gelbke, H.-P. (2005). Ethylbenzene. In *Ullman's Encyclopedia of Industrial Chemistry*. Weinheim: Wiley.

Vogt, E. T., Whiting, C., Dutta, G. T., Chowdhury, A., & Weckhuysen, B. M. (2015). Zeolites and zeotypes for oil and gas conversion. *Advances in Catalysis*, *58*, 143–314. doi:10.1016/bs.acat.2015.10.001

Vora, B. V., Marker, T. L., Barger, P. T., Nilsen, H. R., Kvisle, S., & Fuglerud, T. (1997). Economic route for natural gas conversion to ethylene and propylene. *Studies in Surface Science and Catalysis*, *107*, 87–98. doi:10.1016/S0167-2991(97)80321-7

Wang, S. H. (1993). Styrene, PEB Report 33C, Supplement C, Process Economic Program. *Papers presented at SRI International Conference*.

Wang, T., Tang, X. P., Huang, X. F., Qian, W., Cui, Y., Hui, X., ... Wei, F. (2014). Conversion of methanol to aromatics in fluidized bed reactor. *Catalysis Today*, *233*, 8–13. doi:10.1016/j.cattod.2014.02.007

Ward, J. D. (1971). Alkylation-transalkylation process. Patent 1,246,728/19.

Weber, M., & Klenine-Boymann, M. (2004) Phenol. In Ullmann's Encylopedia of Industrial Chemistry. Wiley-VCH Verlag GmbH & Co.

Welch, V. A., Fallon, K. J., & Gelbke, H.-P. (2005). Ethylbenzene. In *Ullman's Encyclopedia of Industrial Chemistry*. Wiley-VCH.

Wight, C. G. (1979). US Patent No. 4,169,111. US Patent Office.

Wunderly, H. L., Sowa, J. F., & Nieuwland, J. A. (1936). The Olefin—Benzene Condensation with the Sulfuric Acid Catalyst. *Journal of the American Chemical Society*, *58*, 1007. doi:10.1021/ja01297a050

Xing, E., Shi, Y., Xie, W., Zhang, F., Mu, X., & Shu, X. (2016). Size-controlled synthesis of MCM-49 zeolite from NaX for liquid-phase alkylation of benzene with ethylene. *Microporous and Mesoporous Materials*, *236*, 54–62. doi:10.1016/j.micromeso.2016.08.026

Yadav, G. D., & Asthana, N. S. (2003). Selective decomposition of cumene hydroperoxide into phenol and acetone by a novel cesium substituted heteropolyacid on clay. *Applied Catalysis A, General*, *244*(2), 341–357. doi:10.1016/S0926-860X(02)00605-1

Yang, W., Wang, Z., Sun, H., & Zhang, B. (2016). Advances in development and industrial applications of ethylbenzene processes. *Chinese Journal of Catalysis*, *37*(1), 16–26. doi:10.1016/S1872-2067(15)60965-2

Young, L. B., & Skillman, N. J. (1979). US Patent 4 301 316. US Patent Office.

Young, L. B., & Skillman, N. J. (1981). US Patent 4 301 317. US Patent Office.

Yurchak, S., Voltz, S. E., & Warner, J. P. (1979). Process Aging Studies in the Conversion of Methanol to Gasoline in a Fixed Bed Reactor. *Industrial & Engineering Chemistry Process Design and Development*, *18*(3), 527–534. doi:10.1021/i260071a030

Zaidi, H. A., & Pant, K. K. (2008). Activity of Oxalic Acid Treated ZnO/CuO/HZSM-5 Catalyst for the Transformation of Methanol to Gasoline Range Hydrocarbons. *Industrial & Engineering Chemistry Research*, *47*(9), 2970–2975. doi:10.1021/ie071339y

Zhu, X. X., Chen, C. F., An, J., Zeng, P., & Xu, Y. (2011). Development and Industrialization of the Ethylbenzene Production Technologies from Dilute Ethylene in FCC Dry Gas. *Scientific Net Publisher in Materials Science & Engineering, 233*(235), 1708–1713.

Chapter 2
Conversion of CO_2 to High Value Products

Nibedita Nath
G. M. University, India

ABSTRACT

Due to human activities and rapid industrialization, the amount of CO_2 emitted into the atmosphere increases by the day. It is an environmental pollutant and is indirectly responsible for causing climate change and global warming. Thus, research has been carried out for the conversion of CO_2 into value added chemicals. However, CO_2 is a chemically-inert and thermodynamically-stable molecule; thus, external energy should be supplied or a suitable catalyst should be designed for their conversion into useful chemical. Methanol, dimethyl ether, higher alcohols, methyl, formic acid, formaldehyde, organic carbonates, etc. are the different chemicals that are prepared from CO_2. CO_2 is an environmentally friendly raw material as it is nontoxic, abundant, and economical. A lot of research has been carried out on the reaction using CO_2 as a raw material. This chapter mainly focused on synthesis of various chemicals from CO_2 as a raw material.

INTRODUCTION

CO_2 is the main greenhouse gases (GHGs) which is responsible for global warming and climate change. With increasing emission of CO_2 climate change is one of the serious issues for the development of future energy systems. The concentration of atmospheric CO_2 have been increasing from 280 parts per million (ppm) to 400 ppm in 2014 due to population growth, deforestation, rapid industrialization and intensive agriculture (the guardian.com). From the Greenhouse Gas Bulletin & the

DOI: 10.4018/978-1-5225-8033-1.ch002

UN weather agency's annual flagship report. "Globally averaged concentrations of CO_2 is 400.00 ppm in 2015 while it is 403.3 ppm in 2016 due to this the average global temperature is increasing (ElMekawy, et.al.,2016). The average temperature of land and ocean on February 2017 is 0.98°C above the temperature of February 2016. It was reported that 30 Gton of CO_2 is emitted per year which is responsible for increase in atmospheric concentration of CO_2 and global warming. CO_2 emission drastically increased from 15 Gton in 1971 to 30 Gton in 2007. According to United States Energy Information Agency (EIA 2014) that till 2030, the energy consumption is increased by 57%, and if the primary source of energy is coal then till 2030, the expected emission of CO_2 is 40.2 Gton (Johnson, et.al.,2014).

This increase in the concentration of CO_2 in the atmosphere has negative impact on our climate, environment, society and future generations (Johnson, et.al., 2014, Field, et.al., 2014, Lagarde, 2013, Nichols Roth, 2015). Therefore, the conversion of CO_2 into value-added chemicals is one of the main challenges for the 21st century. On increasing population, the demand of carbon based plastics to medicines will also be increasing and to fulfill the demand CO_2 will have to be recycle. Carbon capture and recycling will capture CO_2 from source, and using any alternative energy source converted it to useful materials and fuels.

Conversion of CO_2 An Overview

The utilization of CO_2 in Industrial processes was developed earlier in 1880–1893s, for the synthesis of urea (Bazarov, 1870), hydrogen carbonates (Steinhauser, 2008), and the production of salicylic acid (Kolbe, 1874). Since the late 1800s the chemistry of CO from fossil carbon has been developed for the synthesis of chemicals and fuels that had its golden era in the 1900–1970s, and still it plays an important role in the production of value added chemicals and fuels. In 1970s CO_2 was used as additive for synthesis of methanol with CO (Liu et. al., 2003) and used as a co-monomer for the synthesis of organic carbonates from epoxides (Tsuda, et.al., 1976). In 2013 Aresta and his co-worker describe the utilizaton of CO_2 in the synthesis of value added chemicals and its perspective use at 2030, on considering the market growth for the chemicals and the new technologies (Aresta, et. al., 2013).

CO_2 is one of the easily available feedstock used as a raw material for the production of many value-added chemicals and fuels, which provides a solution to reducing CO_2 emission and energy supply challenges (Markewitz, et.al. 2012). The CO_2 capture, utilization, and sequestration (CCUS) is a promising approach to reduce the emission of CO_2. The Carbon capture and storage (CCS) is a technology which involves in reducing CO_2 and fixing into a long-lived compound (Boot-Handford, et. al., 2014). However, the technical and economic barriers of CCS should be overcome before it can be utilize on a large scale. The main economic obstacles is the high cost of CCS that comes primarily from capture and compression which is about 75% of total cost of CCS (Styring, et. al., 2011).

Carbon capture and utilization (CCU) have recently started to attract great deal of attention worldwide because it can convert CO_2 into valuable products, instead of permanently sequestering it. The advantages of CCU as compared to CCS are that utilization of CO_2 is normally a profitable activity as the products have various applications. (Styring,et.al., 2011). CCS and CCU technology used to capture CO_2 emissions from the sources such as power plants and different industrial processes, to reduce the CO_2 emission. (Markewitz, et. al., 2012). In CCS, CO_2 is capture and storage in a suitable place for a long-time (Metz, et.al., 2005, Weisser,2007, Hertwich, et.al.,2008, Cooper,2009, Nagashima, et.al., 2011, Zapp, et.al., 2012), while in CCU, CO_2 is capture and converted into various useful products (Styring, et.al., 2011, Markewitz, et.al.,2012). Different CCS and CCU options are summarized in Figure 1 and described below.

Figure 1. Different technology for CO_2 storage and utilization

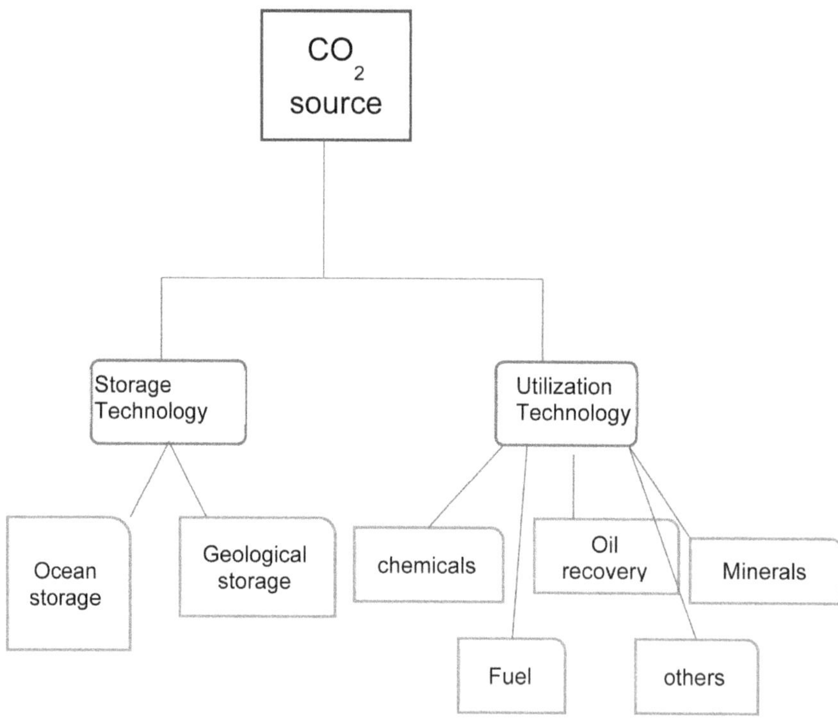

CONVERSION OF CO_2 INTO VALUE ADDED PRODUCT

CO_2 is a nontoxic, abundant and economical raw material for different industrial process. However, few industries can utilize CO_2 as a raw material. Because in CO_2 carbon atom is in its most oxidized form which is the biggest problem for using CO_2 as a raw material in industrial processes due to its low energy level. So large quantity of energy is needed for conversion of CO_2 into useful products.

There are four methodologies of altering this:

1. By using high energy starting materials i.e. hydrogen and organometallics.
2. Choosing low energy synthetic targets.
3. Removing a compound on the product side, and forcing the equilibrium to the right.
4. Supplying external energy, i.e., light or electricity.

Choosing the suitable conditions for the CO_2 conversion to achieve a negative Gibbs energy for the reaction. (Sakakura, et. al., 2007) Therefore, active substrates having high free-energy are used for the preparation of the target products like unsaturated compounds such as alkene, alkyne etc. (Yu, et.al.,2013, Wu, et.al.,2014), three membered ring compounds like aziridines (Song, et.al.,2012), epoxides (Lu, et.al., 2012) *etc.*, organometallic compounds like organozinc (Yeung, et. al.,2008) *etc.* with relevant low energy state e.g. organic carbonates, carbamates. The working principle of catalysts for the preparation of different target product from CO_2 has been explained by (Muller et al., 2011, 2012).

CO_2 is a non-polar & chemically unreactive molecule under standard conditions. Due to its low reactivity, it is converted into different valuable products by reducing activation energy of the reaction using catalysts.

Conversion of CO_2 into Organic Carbonate

Organic carbonate is the most promising target chemicals in terms of green and sustainable chemistry due to its low toxicity, non-corrosiveness and biodegradability (Peng, et.al, 2012). Organic carbonates have been widely used as starting materials (Kathalewar, et. al., 2013, Grego, et. al., (2012) to produce polycarbonate resins, polyurethane resins, electrolytes in lithium ion batteries, alkylating and carbonylating reagents and inert solvents. Moreover, in the future, they are expected to be a fuel additive (Arteconi, et. al., 2011).

Organic carbonates are further classified into cyclic and linear carbonates (Shaikh & Sivaram, 1996) because both compounds have three oxygen atoms in each molecule; according to thermodynamic point of they are suitable synthetic targets from CO_2. From the industrial point of view the important organic carbonates are dimethyl carbonate (DMC), diphenyl carbonate (DPC), ethylene carbonate (EC) and propylene carbonate (PC). DMC and DPC have open chain structures, while EC and PC are cyclic.

Cyclic carbonates and polycarbonates are synthesized by the addition of CO_2 to epoxides it was discovered by Inoue et al. 50 years ago (Inoue et. al., 1969). Before this discovery, phosgene is used as starting material for organic carbonates synthesis which is a very toxic compound. Thus, the utilization of non-toxic CO_2 as a raw material offers a safer and greener copolymerization process. Various catalyst systems have been developed for CO_2/epoxides copolymerization.

In 2017 Xiaohe, et.al., design and synthesized Co(II) metal-organic framework (MOF) {[Co(μ3-L) (H_2O)]·0.5H_2O}n (H_2L = thiazolidine 2,4-dicarboxlic acid) catalyst for the conversion of CO_2 and propylene oxide into propylene carbonate with a yield up to 98% under 50°C and 1 atmosphere (Xiaohe,et.al.,2017) Scheme 1.

Scheme 1

CO_2 were directly converted into propylene carbonate using ionic liquid (IL) 1-methyl-3-(3-trimethoxysilylpropyl)imidazolium chloride supported pristine sodium titanate nanotubes (TNT) and nanowires (TNW) (Monteiro et.al., 2017) (scheme 2). The catalyst (IL-TNTs and IL-TNWs) showed good catalytic activity for cycloaddition of CO_2 to propylene oxide (Turnover frequencies TOF = 46 and 49 h-1, respectively) and 100% selectivity in propylene carbonate (propylene/catalyst molar ratio = 580, cat/cocat molar ratio = 2.4, 4.0 MPa of CO_2, 383.15 K, 6 h).

Scheme 2

Molten tetralkylammonium salts having halides as counterions catalyzed synthesis of cyclic carbonates from epoxide and CO_2 has been reported (Vincenzo, et. al., 2002) Scheme 3. A highly efficient binary catalytic system has been developed based on tannic acid/NBu_4X (X=Br, I) for coupling of CO_2 and oxiranes to organic carbonates, Scheme 4 in good yields at very low catalyst loading (Sopena, et. al., 2015). The presence of multiple polyphenol fragments in tannic acid is beneficial

Scheme 1

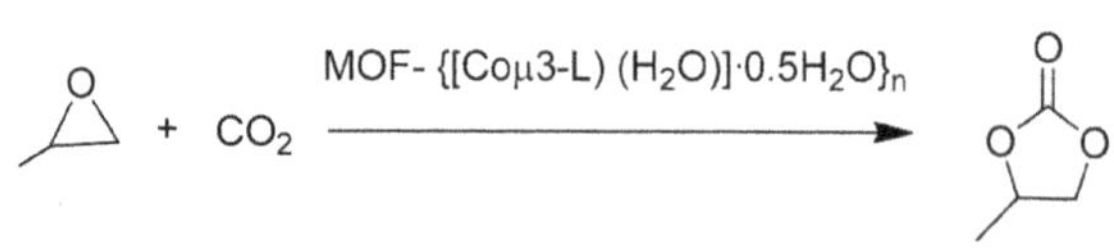

Scheme 2

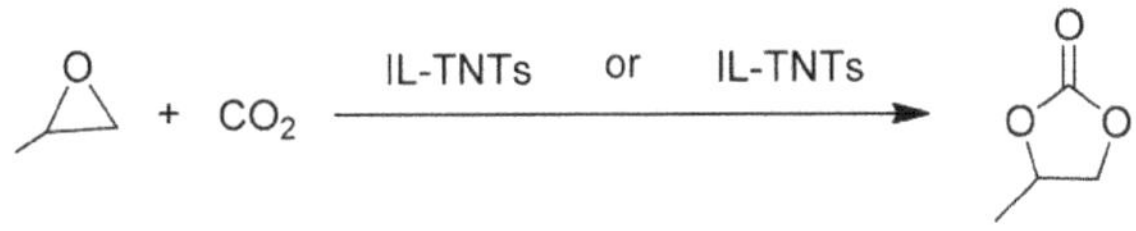

for synergistic effects that increase the stability. The TOF above 200 h^{-1} which is highest among the organocatalysts reported to date for CO_2 conversion.

Scheme 3

Scheme 4

Quaternary phosphonium salts and PEG-supported phosphonium salt were found to be efficient catalysts for synthesis cyclic carbonates from carbon dioxide and epoxides under solvent free condition (Tian, et.al.,2008) Scheme 5.

Scheme 3

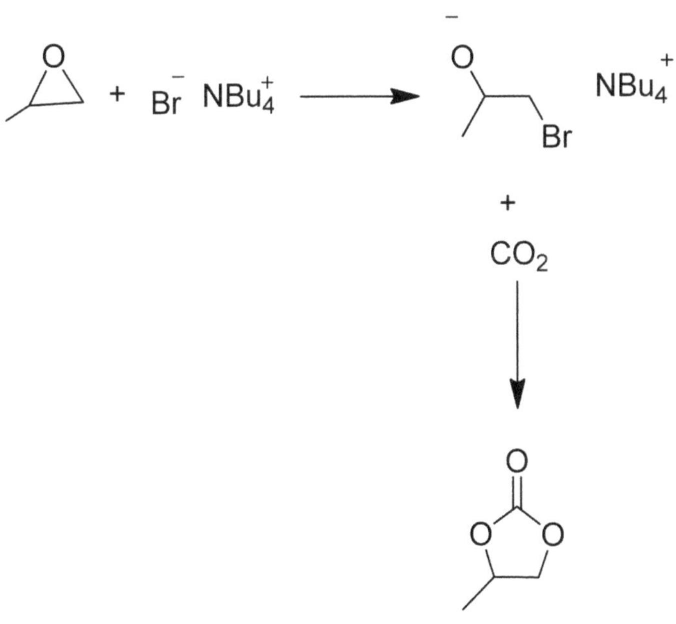

Scheme 4

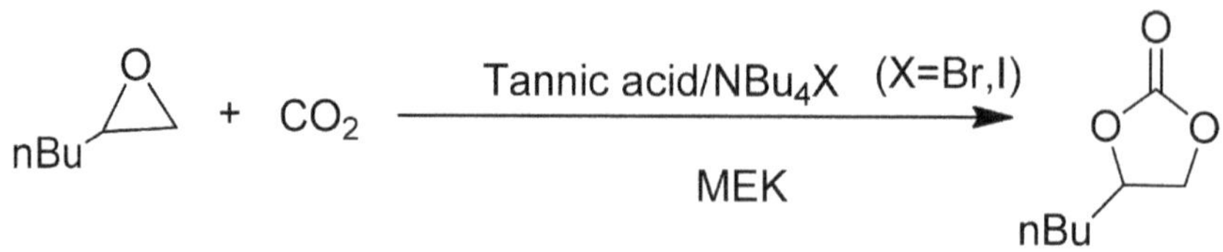

Scheme 5

Boronic acids with onium salts used as an efficient organocatalysts for the conversion of CO_2 with epoxides to cyclic carbonates in H_2O under mild conditions (Wang & Zhang, 2016) scheme 6.

Scheme 6

An amino triphenolate ligand scaffold based aluminum catalyst shows high catalytic activity (initial TOF up to 36 000 h^{-1}), broad substrate scope, and functional group tolerance in the formation of organic carbonates from epoxides and CO_2 (Whiteoak, et. al., 2013) Scheme 7.

Scheme 7

4-(dimethylamino)pyridine hydrobromide ([DMAPH]Br) as a highly efficient and recyclable catalyst for the formation of cyclic carbonates from atmospheric CO_2 and epoxides in the presence of 1 mol % of [DMAPH]Br under solvent-free conditions (Zhang, et.al,2017) Scheme 8.

Scheme 5

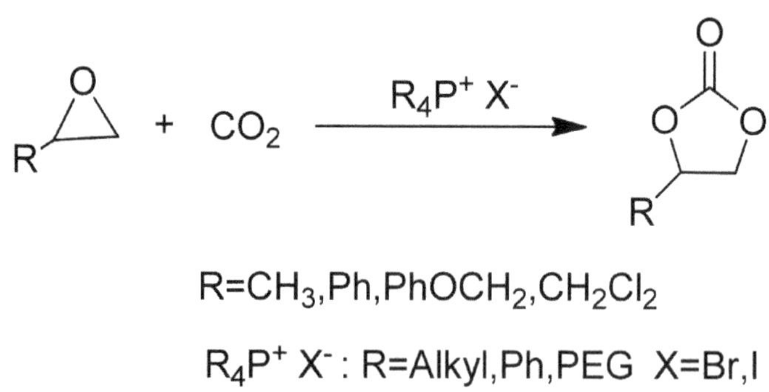

Scheme 6

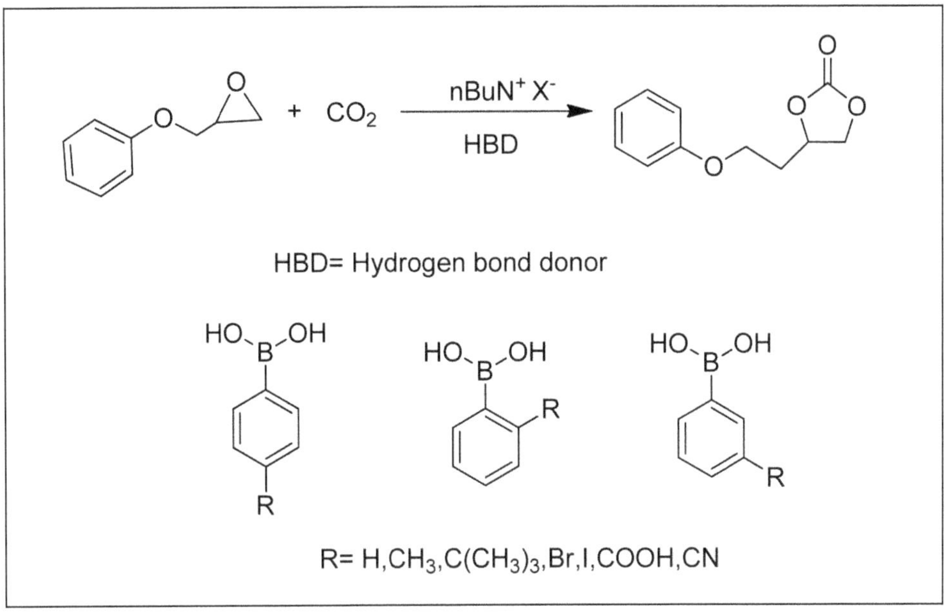

Scheme 7

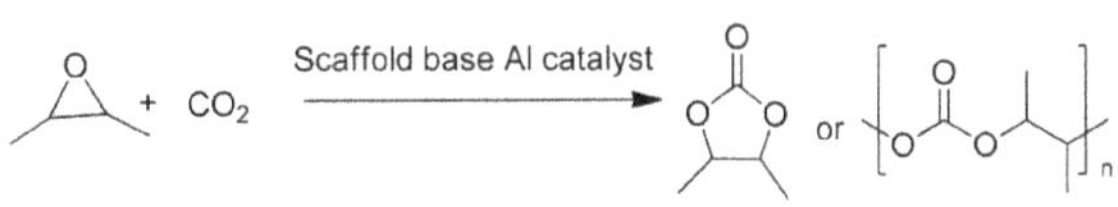

Scheme 8

Cyclic carbonate was synthesized from 4-vinyl-1-cyclohexene-1,2-epoxide (VCHO) and carbon dioxide under solvent free condition using ionic liquid as a catalyst and zinc halide as co catalyst. Ionic liquids based on 1-alkylmethylimidazolium salts of different alkyl groups (ethyl, butyl, hexyl, octyl) and different anions (Cl^-, BF_4^-, PF_6^-) were used as catalysts (Lee, et. al., 2008) Scheme 9.

Scheme 8

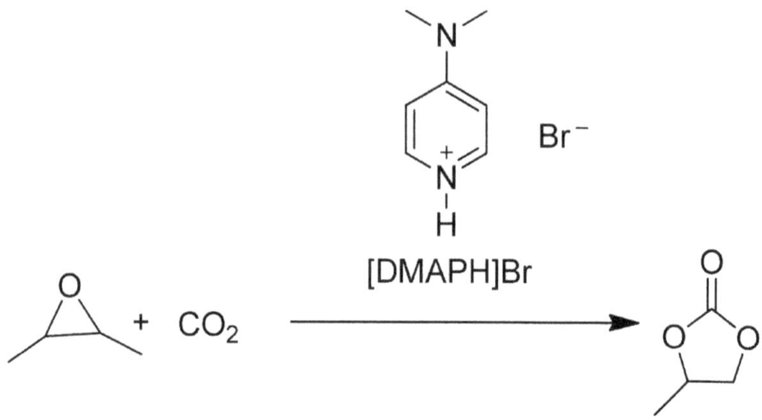

Scheme 9

Conversion of CO_2 into Methanol

Methanol is an important source as the feedstock to produce varieties of product (Clary, 2013). Because the energy efficiency of methanol is higher than that of its derivatives, hence more research has been focused on the synthesis of methanol. Worldwide, production of methanol is about 75 million metric tons, and daily more than 100 000 tons of methanols are used as a chemical feedstock or as a fuel. Methanol has been widely used in medicals, pharmaceuticals, dyes, paints, plastics, rubbers

Scheme 9

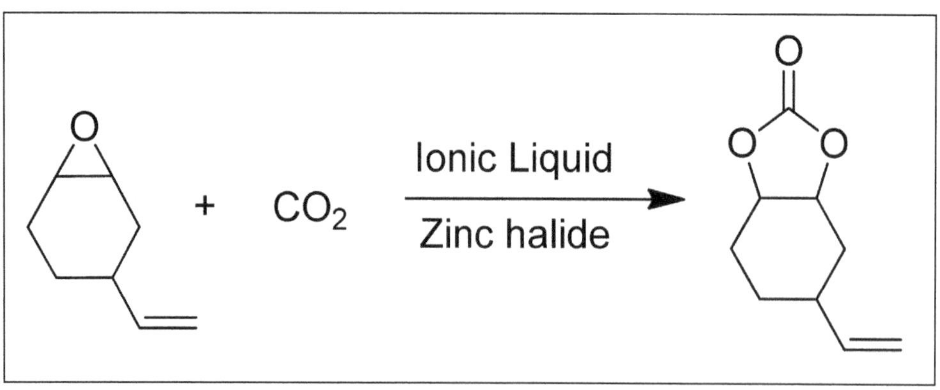

fibers, etc. In 1920s, the synthesis of methanol from syngas on an industrial scale was introduced by BASF, Germany. The $Cu/ZnO/Al_2O_3$ catalyst system is employed in the industrial synthesis of methanol (Behrens, et. al., 2013). As an alternative feedstock, CO_2 replace CO in the production of methanol (Ma, et. al., 2009). Direct hydrogenation of CO_2 formed water vapour that inhibits the reaction strongly and result serious catalyst deactivation (Li, et. al., 2014, Gao, et. al., 2015). In addition, to water vapour other by-products like CO and hydrocarbons are formed during the reaction. Therefore, an efficient catalytic system is to be needed to improve catalytic stability and avoid the formation of undesired products during methanol synthesis.

Methanol (CH_3OH) is most attractive hydrogenation products of CO_2, because it is directly used as a transportation fuel in internal combustion engines and direct methanol fuel cells (DMFC). CH_3OH is also a hydrogen storage medium (12.5 wt% H_2) and an important raw material for the production of a number of chemicals and products including ethylene and propylene (Goeppert, et. al., 2014). CH_3OH is already one of the most important building blocks in the chemical industry with an annual production in excess of 70 million tons.

Pentaethylene hexamine and Ru-Macho-BH (Pincer catalyst) was found to be highly efficient homogeneous catalyst system for the hydrogenation of CO_2 to CH_3OH at 125– 165 °C in an ethereal solvent, initial TOF = 70 h^{-1} at 145 °C (Kothandaraman, et.al., 2016) Scheme 10.

Scheme 10

Photoelectrochemical (PEC) conversion of CO_2 using photo- cathodes based on Cu_2O nanowires (NWs) overcoated with Cu^+ incorporated crystalline TiO_2 (TiO_2–Cu^+) shell. Cu_2O NW photocathodes show remanent photocurrent of 5.3% after 30 min of PEC reduction of CO_2. After coating Cu_2O with TiO_2 –Cu^+ overlayer, the remanent photocurrent is 27.6%, which is an increase by 5.2 fold. The charge transfer resistance of Cu_2O/TiO_2–Cu + is 0.423 k Ω/cm^2, whereas Cu_2O photocathode shows resistivity of 0.781 kΩ /cm ^{2} under irradiation (Lee, et, al., 2017).

Water splitting technique is used for reduction of CO_2 into methanol was developed using Al and Cu powder as a reductant and catalyst, respectively. It is a simple and green technique, and also provides an example of a multitude of possible chemical reactions for CO_2 conversion (Lyu, et. al., 2015) Scheme 11.

Scheme 11

Photo conversion of CO_2 is a promising method for the reduction of CO_2 concentration in atmospheric and also mitigating energy problems. For this purpose various efficient and stable semiconductor photocatalysts were used among them layered double

Scheme 10

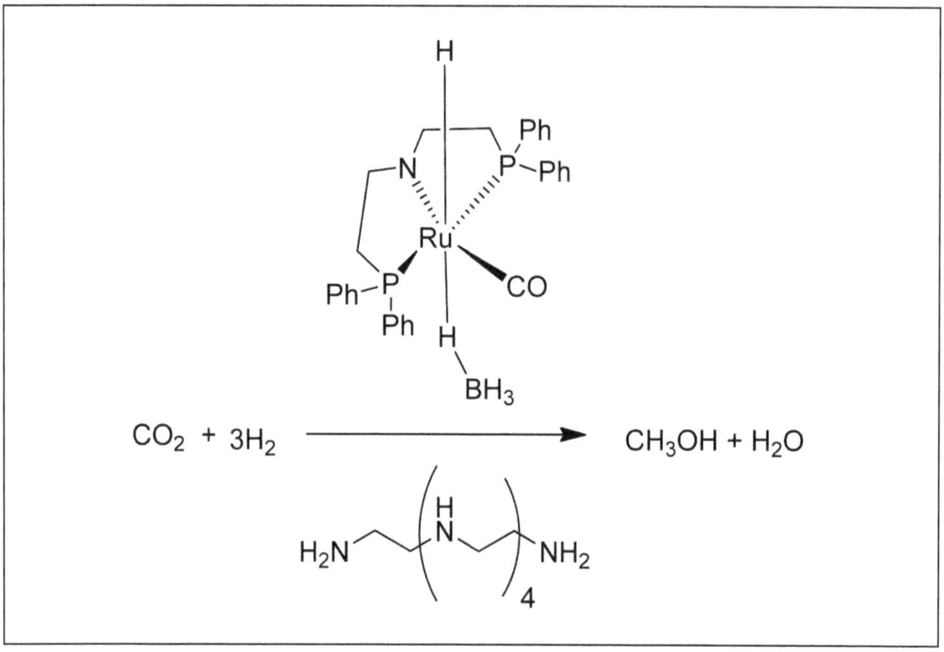

Scheme 11

$$2\,Al + 4H_2O \longrightarrow 2AlO(OH) + 3\,H_2 + CO_2 \longrightarrow CH_3OH + H_2O$$

hydroxides (LDHs) have gain more attention as catalysts for photoconversion of CO_2 into methanol. Various LDHs of the formula $[M^{II}_3 Ga^{III}(OH)_8]_2A \cdot mH_2O$ $\{M^{II} = Zn^{II}$, Cu^{II}; $A^{2-} = CO_3{}^{2-}$, $[Cu(OH)_4]^{2-}$ were synthesized and used for photoconversion of CO_2 into methanol at a reaction pressure of 0.40 MPa in the presence of H_2 (Miyano, et.al., 2017) Scheme 12.

Scheme 12

The highly efficient iron (II) scorpionate catalyst $[FeCl_2\{\kappa 3\text{-}HC(pz)_3\}]$ (pz=pyrazol-1-yl) was developed for eco-friendly synthesis of methanol (yield= 44%, TON up to 2.3×10^3) directly from CO_2 and H_2 in a solvent or amine free condition (Ribeiro, et.al., 2013) scheme 13.

Scheme 13

Cu-loaded carbon modified titanium oxide (Cu-C/TiO2) nanoparticles catalyzed photocatalytic reduction of CO_2 into methanol has been investigated under irradiation of UV and natural sunlight (Scheme). Cu-C/TiO2 nanoparticles were prepared by sonicated sol–gel method (Kavil, et.al., 2017) Scheme 14.

Scheme 14

Methanol is also synthesized from epoxide and CO_2 via cyclic carbonate by using a very efficient ruthenium catalyst as shown in Scheme 15 where a TON greater than 87000 (Li, et. al., 2013).

Scheme 15

Conversion of CO_2 into Carboxylic Acid

Like carbonates, carboxylic acids also a suitable raw material producing from CO_2. The potassium or sodium salt of phenols react with CO_2 to give carboxylic acids.

Scheme 12

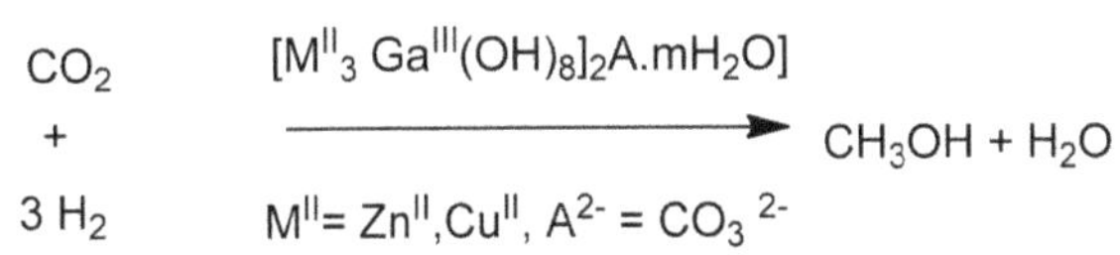

Scheme 13

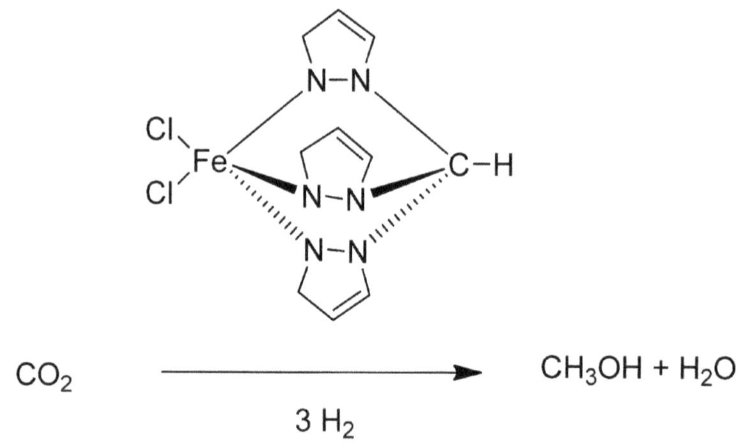

Scheme 14

$$CO_2 \xrightarrow[3\ H_2]{Cu\text{-}C/TiO_2\ \ UV,h\nu} CH_3OH + H_2O$$

Scheme 15

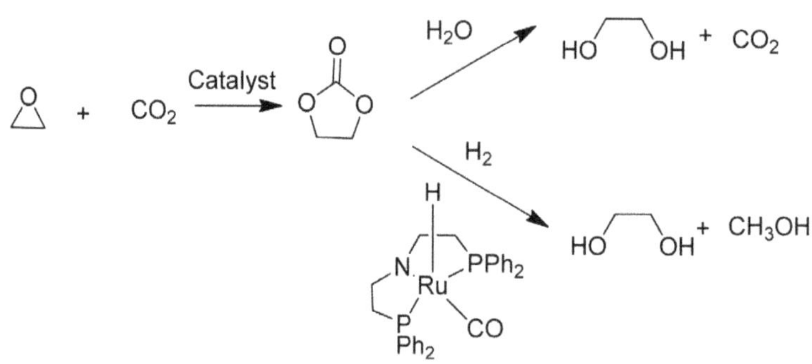

It is used for the preparation of numerous pharmaceutical products and dyestuff intermediates. Practically it is synthesized by Kolbe–Schmitt process in which sodium salt is heated under pressure in an autoclave and the acid is recovered by acidification with H_2SO_4 (Sneeden, et.al., 1982). One of the most important acids prepared by this process is salicylic acid Scheme 16 which is used as intermediate in the synthesis of aspirin.

Scheme 16

Carboxylation of carbon nucleophile with CO_2 was traditionally achieved by the use of highly reactive organolithium or Grignard reagents (Ebert, et. al., 2005, M. Aoki, et. al., 2006).

Scheme 17

In 1986 Hoberg and his co-worker developed the carboxylation of alkene by using by using nickel/DBU catalyst with the formation of a key five-membered metallactone intermediate (Hoberg, et.al., 1986) Scheme 18.

Scheme 18

Styrene derivatives undergo reductive carboxylation with CO_2 using organometallic catalyst under mild conditions to produce saturated phenylacetic acid derivatives (Williams, et. al., 2008) Scheme 19.

Scheme 19

Hydrocarbon converted to carboxylic acids with metal carbonates under a high temperature and pressure (Henkel reaction) (Thomas,et.al). Also carboxylic acid is

Scheme 16

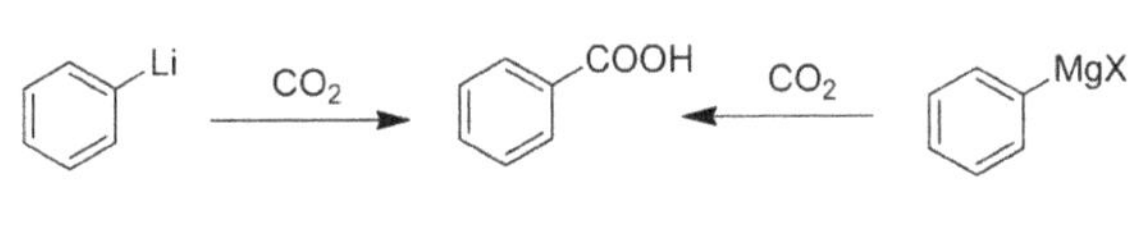

Scheme 17

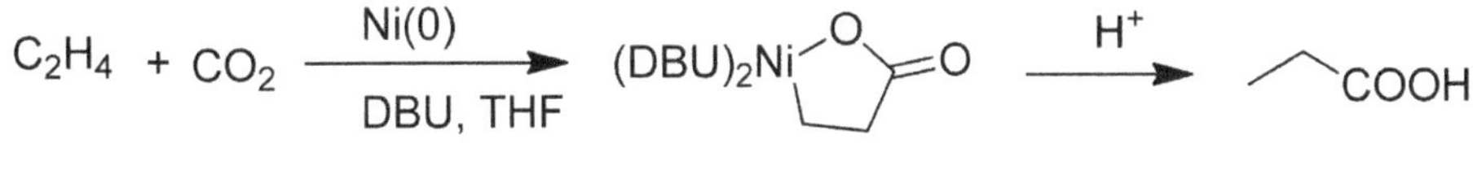

Scheme 18

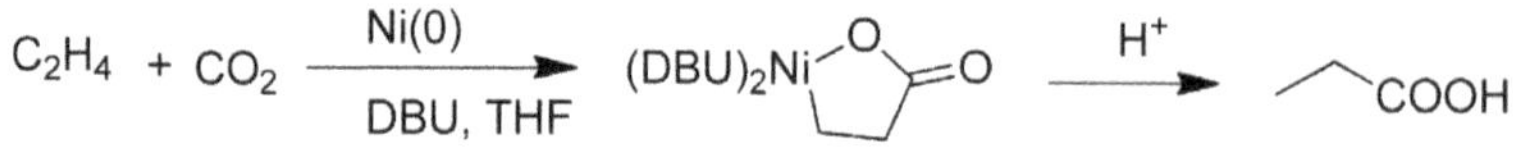

Scheme 19

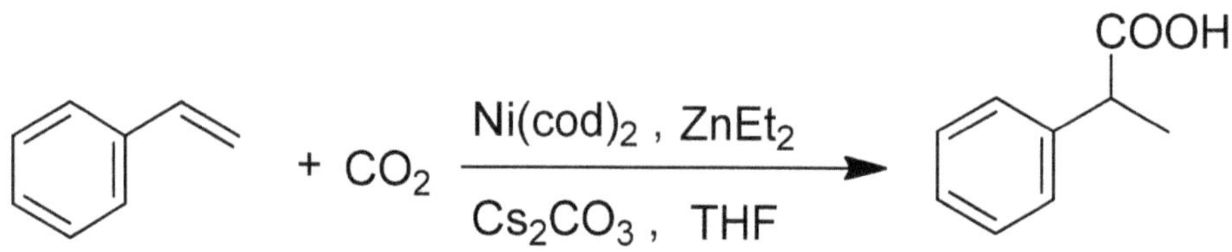

also prepared by the reaction of olefin with CO_2 under acidic condition as shown in Scheme 20.

Scheme 20

A simple and straight forward method was developed by Zhang's group for direct carboxylation of terminal alkynes Cs_2CO_3 as base and in the absence of catalyst, Scheme 21 (Yu, et. al., 2011). However Phenanthroline backbone and phosphineligand based copper catalyst was also used as an efficient catalyst for carboxylation of terminal alkynes, Scheme 22 (Gooßen, et. al., 2010).

Scheme 20

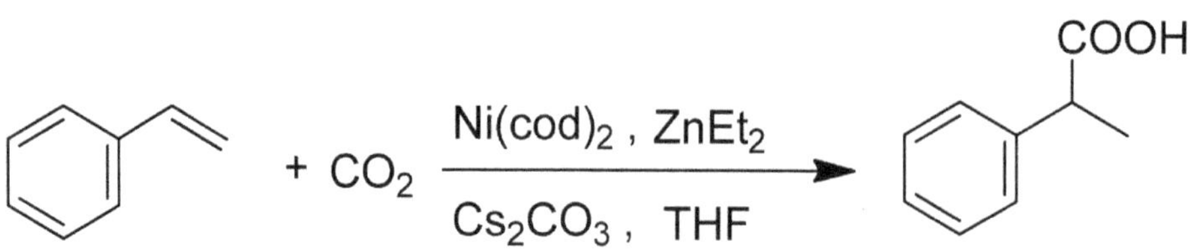

Scheme 21

Scheme 22

The carboxylation of a wide range of α- substituted α-siloxy silanes in the presence of CsF under a CO_2 atmosphere via Brook rearrangement has been reported Scheme 23 (Mita, et. al., 2014).

Scheme 23

Alkylboron compounds (alkyl-9-BBN) undergo Carboxylation with CO_2 in the presence of catalytic amounts of CuOAc/1, 10-phenanthroline and KO^{t}Bu as shown in Scheme 24 (Ohmiya, et. al., 2011).

Scheme 21

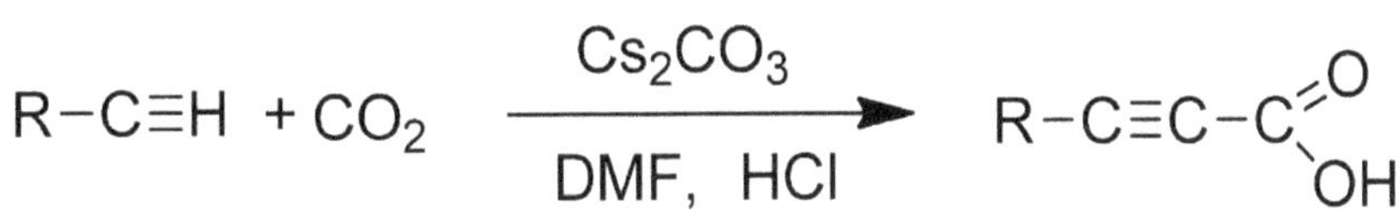

Scheme 22

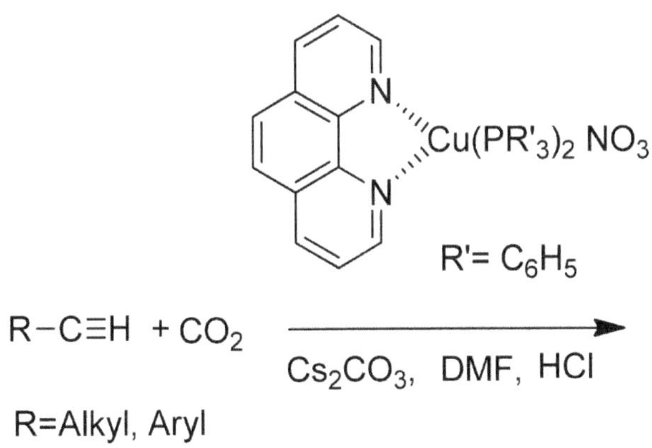

Scheme 23

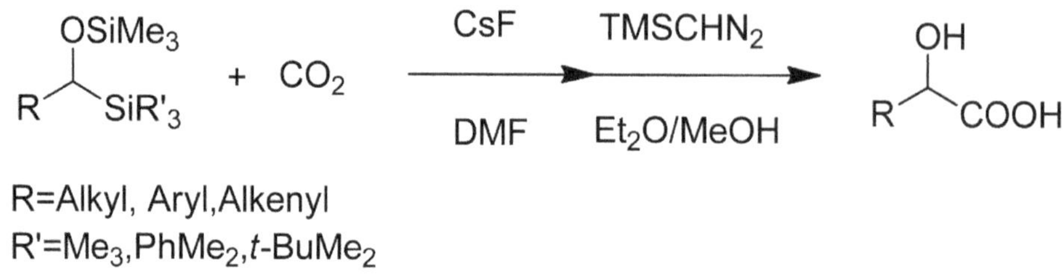

Scheme 24

Synthesis of α,β-unsaturated carboxylic acids from carboxylation of symmetrical and unsymmetrical alkynes with CO_2 catalyzed by iron has been reported Scheme 25 (Santhoshkumar, et.al.2016).

Scheme 25

A simple and straightforward method was developed by Yoo's group for direct carboxylation of indoles (Yoo, et.al., 2012). Here indole reacts with CO_2 under basic condition to give indole-3-carboxylic acids as shown in Scheme 26.

Scheme 24

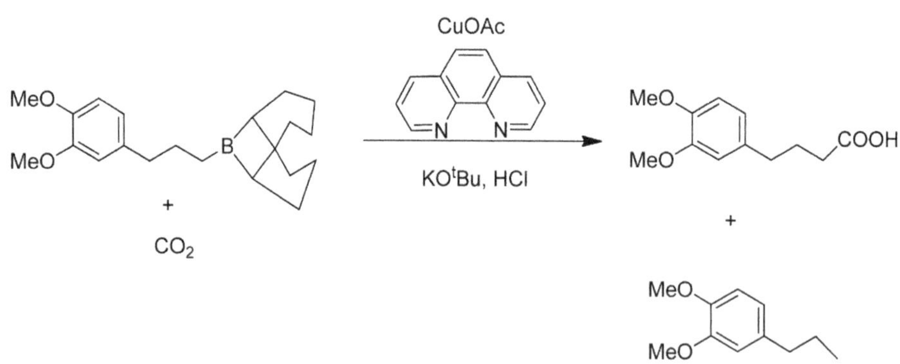

Scheme 25

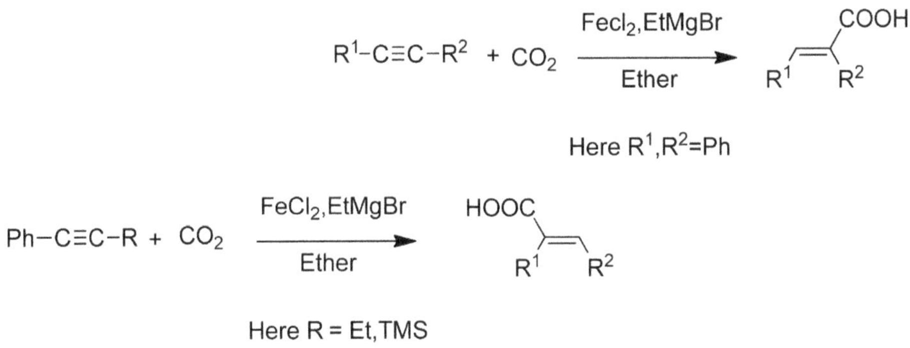

Scheme 26

A N-heterocyclic carbene NHC/ Ag_2O catalytic system has been developed by Yuan and co-workers.Terminal alkynes were converted to corresponding acid with CO_2 via carboxylation (Yuan, et.al., 2017) Scheme 27.

Scheme 27

Conversion of CO_2 into Dimethyl Ether

Dimethyl ether (DME) is also called wood ether or methyl oxide. It is the simplest ether and mainly used as a pressurizing agent in consumer products, in paints and for insect control. DME is an environmentally friendly fuel having advantages over methanol, as it is considered to be an excellent diesel and LPG substitute (Semelsberger,et.al., 2006, Grove, et.al., 2017). In addition, it is a suitable H_2 vector for fuel cells (Erena, et. al., 2013, Oar-Arteta, et. al., 2016, Gonzalez-Gil, et. al.,

Scheme 26

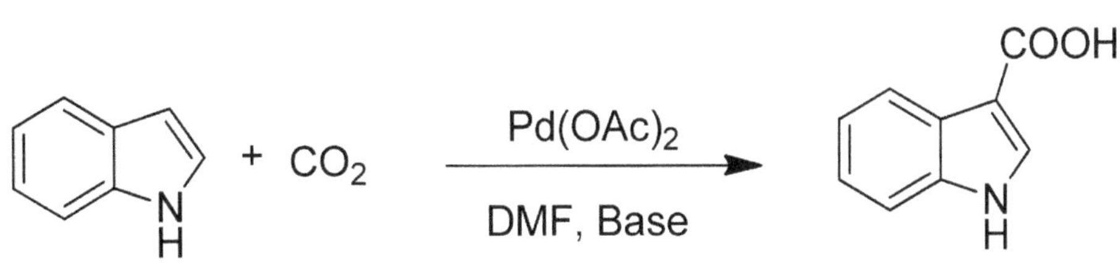

Scheme 27

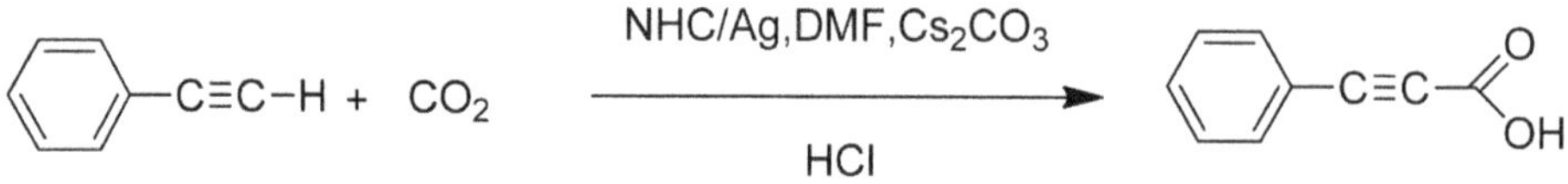

2016) and for H_2 production on a large-scale (Elewuwa, et.al., 2015, Elewuwa, et.al., 2016). It is also used as intermediate for the production of olefins in place of methanol (Perez-Uriarte, et. al., 2016). DME is synthesized from methanol and syngas by dehydration using variety of catalysts.

DME can be synthesized from CO_2 or syngas by hydrogenation via two different steps. The first step involves the synthesis of methanol from CO_2 or syngas using dehydration catalyst. In the second step process methanol is dehydrated to DME using dehydration catalyst. On the other hand DME is also directly prepared from dehydration of CO_2 by reverse water gas shift reaction (Catizzone, et. al., 2018).

Scheme 28

In 2011 Naik and co-worker has been developed different bifunctional hybrid catalyst ($6CuO–3ZnO–Al_2O_3/\gamma$-Al_2O_3 and $6CuO–3ZnO–Al_2O_3$/HZSM-5) for synthesis of DME from a CO_2/H_2 gas mixture (Naik, et. al., 2011). Tao et. al reported ZSM-5 based catalyst which was found to be active and stable catalyst for the production of DME from CO_2 and H_2 (Tao, et. al., 2001). Also the $CuO–Fe_2O_3–ZrO_2$/HZSM-5 bifunctional catalyst has been developed and used for the direct synthesis of DME from CO_2 and H_2 (Liu, et. al., 2013).

The Cu-Zn-Zr/zeolite hybrid catalytic systems, has been exploited by Frusteri and co-workers. It contains both metallic and acidic functionalities which are successfully produced DME by CO_2 hydrogenation (Frusteri, et. al., 2017). Bonura et. al, developed CuZnZr-ferrierite hybrids catalyst combining in a single pellet for dehydration of methanol to DME. Here the metal-oxide sites facilate the CO_2 activation or hydrogenation and the acidic sites facilate the dehydration of methanol to DME (Bonura, et. al., 2017). Recently Kornas has reported the use of CuO/ZrO_2 + montmorillonite K10 hybrid catalysts for the direct synthesis of DME from CO_2 (Kornas, et. al., 2017).

Scheme 28

Step-1 Synthesis of Methanol

$$CO_2 + 3H_2 \rightleftharpoons CH_3OH + H_2O$$

Step-2 Dehydration of Methanol

$$2CH_3OH \rightleftharpoons CH_3OCH_3 + H_2O$$

Step-3 Reverse-water gas shift reaction

$$CO_2 + H_2 \rightleftharpoons CO + H_2O$$

Conversion of CO_2 into Formic Acid

Formic acid (FA) is a very important chemical widely used in leather, textile and pharmaceuticals industry. It is also useful for coagulation of rubber latex, crop protection, and additive for cleaning agent (Aresta, 2010). Formic acid finds its applications in textiles, and food chemicals, due to its strong acidic nature and reducing properties. Usually, the biggest consumer of FA the leather Industry (http://www.icis.com). Since the European Union ban on non-prescribed feed antibiotics in 2006 it is used mainly as a preservative and antibacterial agent in livestock feed (http://eur-lex.europa.eu, https://en.wikipedia.org). FA can be obtained renewably and therefore excellent chemical hydrogen storage (CHS) material. Synthesis of FA process from CO_2 and H_2 using various catalytic systems are summarized below.

Presently, FA is produced in two steps. In the first step, CO reacts with methanol to form methyl formate (MF), which on hydrolysis gives FA, either it react with ammonia to give formamide, which on hydrolyzed with H_2SO_4.

Scheme 29

These processes do not use renewable carbon sources, and all are atom inefficient. So there is a large scale development on renewable syntheses of FA from CO_2. CO_2 on hydrogenation give FA. Consequently, a lot of research has been carried out in the catalytic hydrogenation of CO_2 to FA Scheme 30.

Scheme 30

Recently Kumar and co-worker developed graphene oxide modified cobalt metallated aminoporphyrin (GO-Co-ATPP) photocatalyst for the conversion of CO_2 to formic acid using visible light (Kumar, et. al., 2018). Electro-biocatalytic conversion of CO_2 into formic acid using whole-cell biocatalysts was reported. Shewanella oneidensis MR-1 (S. oneidensis MR-1), a facultative aerobic bacterium that has been widely used as biofuel cells and detoxification of heavy metal oxides is taken as whole-cell biocatalyst for synthesis of formic acid from CO_2 and electrons supplied from an electrode (Tuan Le, et. al., 2018). In 2016 Lu, et. al., 2016 investigated the direct hydrogenation of CO_2 to formic acid in water using highly efficient iridium catalyst, [Cp*Ir(N,N')Cl]Cl (N,N'= 2,2'-bi-1,4,5,6-tetrahydropyrimidine) in the absence of a base Scheme 31 (Lu, et. al., 2016).

Scheme 29

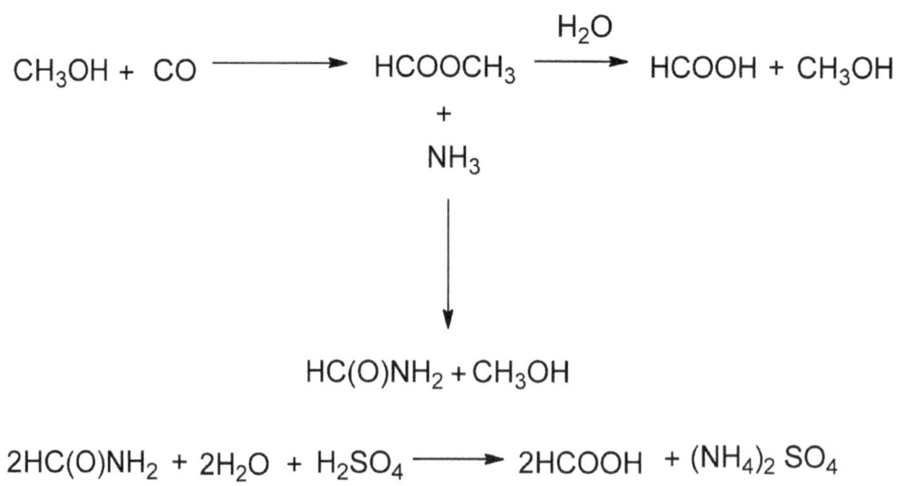

Scheme 30

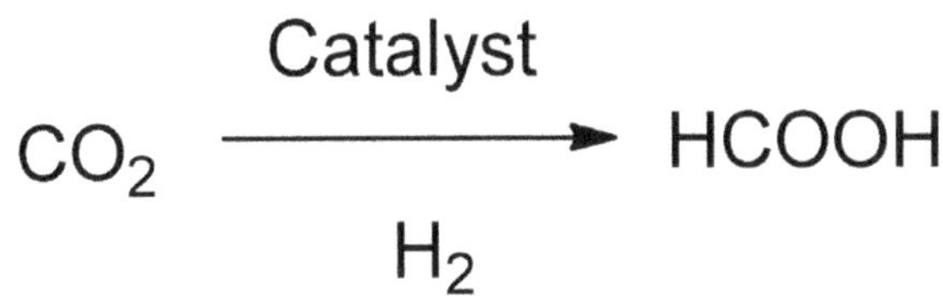

Scheme 31

Maihom and co-worker theoretically investigate the hydrogenation of CO_2 to formic acid on Cu-alkoxide-functionalized MOF-5 (Cu-MOF-5) by using DFT calculations (Maihom, et. al., 2013). On the other hand a copper hydride complex having a 1,2-bis(diphenylphosphino) benzene ligand has been reported for formic acid synthesis from CO_2 with hydrosilanes Scheme 32 (Motokura, et.al., 2012). A palladium catalyst on chitin catalyzed hydrogenation of CO_2 has also been proposed (Songa, et. al., 2017).

Scheme 31

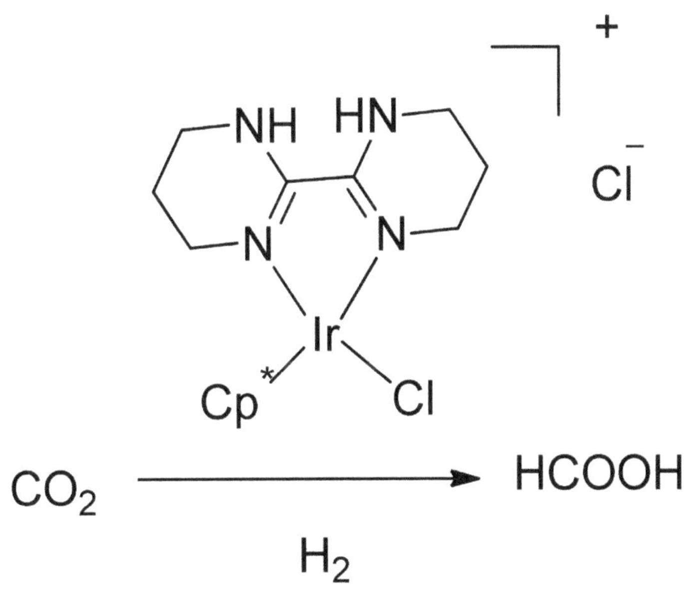

Scheme 32

More recently, Weilhard, et.al., 2018 investigated an active and selective catalytic system, where the Ionic liquid (ILs) plays multiple roles in the direct hydrogenation of CO_2 to formic acid using a Ru-catalyst (Weilhard, et.al., 2018) and Zhang and co-worker found that Ru-PPh_3/Al_2O_3 catalysts proved to be a novel heterogeneous catalyst for synthesis of formic acid from hydrogenation of CO_2(Zhang, et.al., 2018). The research group of Bai showed the exclusive conversion of CO_2 into formic acid using Pd-Sn alloy electrocatalyst in an aqueous solution (Wei Chen, et. al., 2017).

Conversion of CO_2 into Ester and Lactones

CO_2 react with various unsaturated compounds in presence of transition metal catalyst to give esters and lactones. For example 1,3-Butadiene reacts with CO_2 in the presence of a palladium catalyst as shown in Scheme 33 (Behr &Heite, 2000, Behr & Brehme, 2002).

Scheme 32

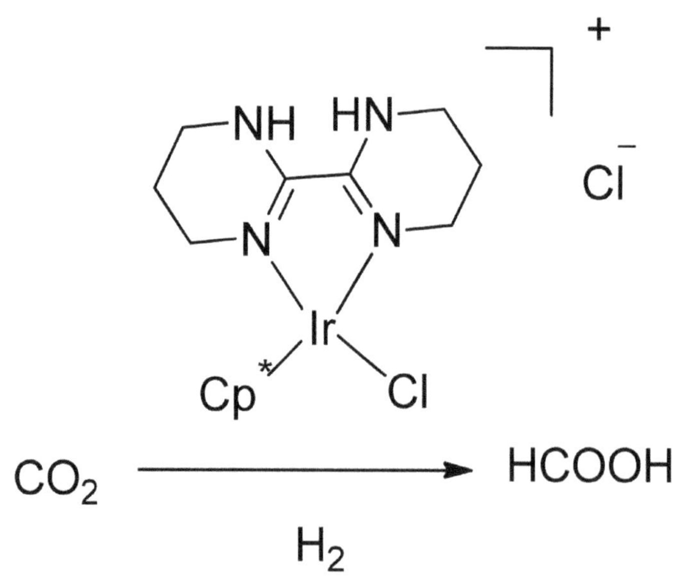

Scheme 33

Different unsaturated compound used for synthesis of lactones are acetylene Scheme34, allene Scheme 35 and diynes Scheme 36 (Tsuda, et.al., 1992, 1995) are react with CO_2 in the presence of a low valent transition metal catalyst(Nickel(0)) .

Scheme 34

Scheme 35

Scheme 36

Recently Song and co-worker developed palladium nanoparticles as the catalyst which is an efficient catalyst for the synthesis of δ -lactone from $CO_2$and 1,3- butadiene (Song, et.al., 2016) Scheme 37.

Scheme 37

The group of Frogneux reported the one-step conversion of CO_2 into heteroaromatic esters using fluoride anions as promoters for the C-Si bond activation, organosilanes are successfully carboxylated with CO_2 in the presence of an electrophile (Frogneux, et. al., 2016) Scheme 38, and Kondo group developed the carboxylation of alkynylsilanes (Yonemoto-Kobayashi, et. al., 2013) Scheme 39.

Scheme 33

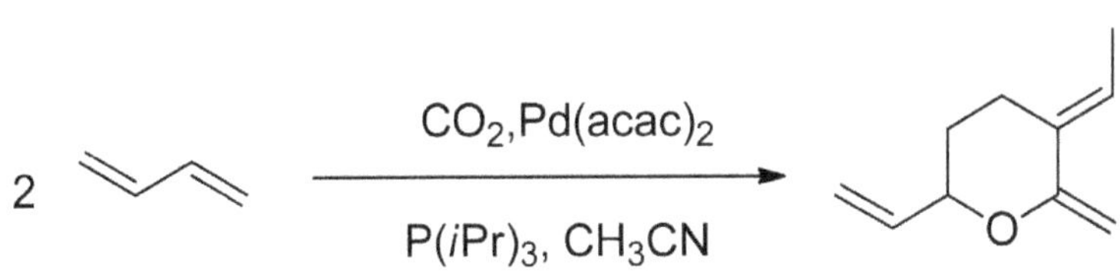

Scheme 34

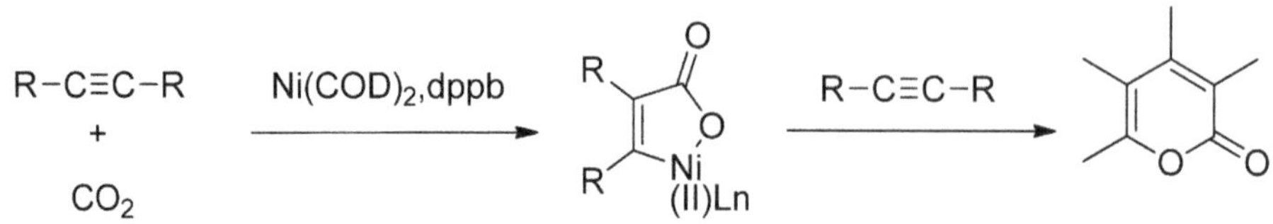

Scheme 35

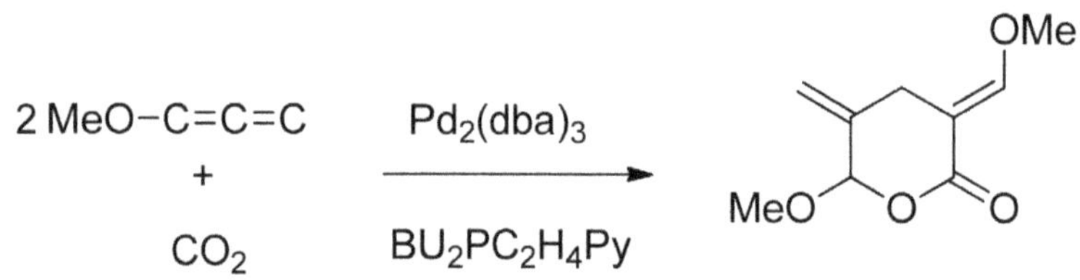

Scheme 36

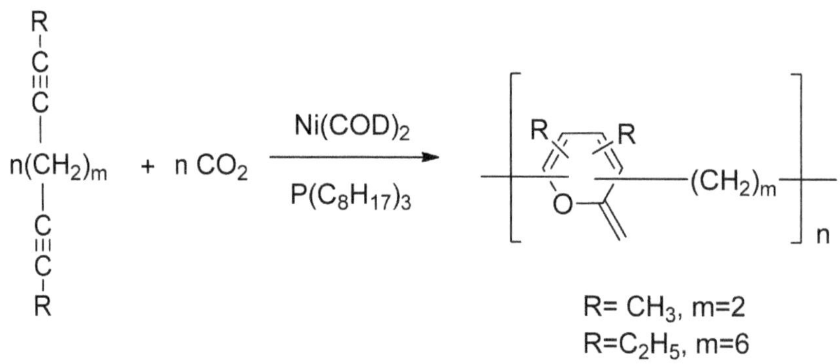

Scheme 37

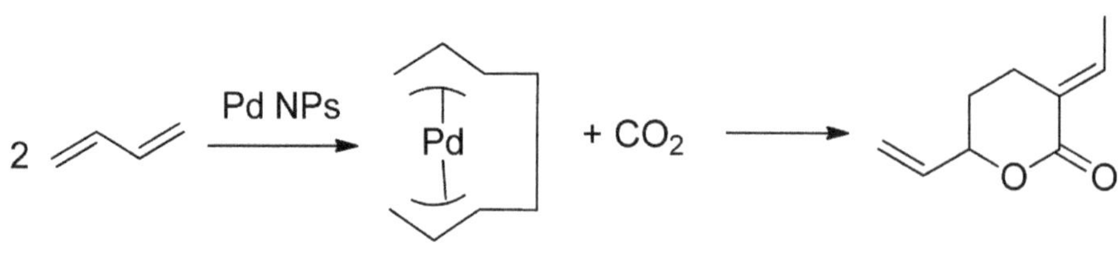

Scheme 38

Scheme 39

Conversion of CO_2 into Isocyanate

Isocyanates are highly valuable intermediates in organic synthesis. They are readily converted into the ureas and carbamates which are found in many biologically active compounds and polymers like polyurethanes (Ghosh & Brindisi, 2015). The diisocyanates derived from arylamines are used as raw material in the pharmaceutical industry and in the manufacturing of polyurethanes. Isocyanate is synthesized usually from the reaction of toxic phosgene or its equivalents with amines Scheme 38 (Engels, et. al., 2013, Six & Richter, 2003, Storace, et. al., 2002). But phosgene is a toxic material and there are also major problem associated with its production and storage(https://www.opcw.org). To avoid this problem CO_2 is used in place of phosgene for the synthesis of isocyanate because it is a non-toxic, abundant, renewable and economical(Yuan, et. al., 2017, Quaranta & Aresta, 2010). The reaction is carried out under high pressures and temperatures due to the thermodynamic stability of CO_2 (Heyn, et. al., 2014, Shi, et. al., 2003).

Scheme 38

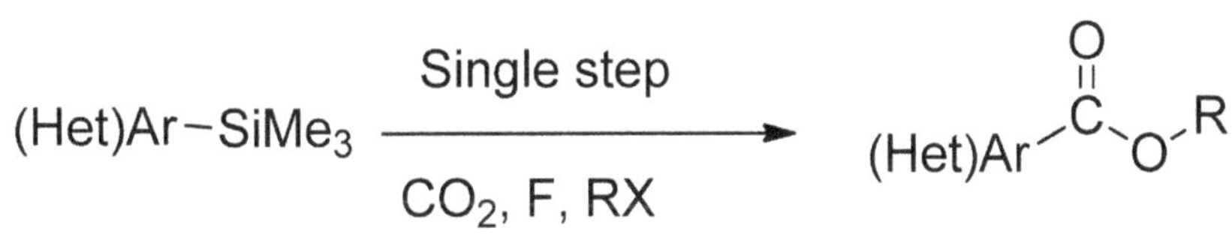

Scheme 39

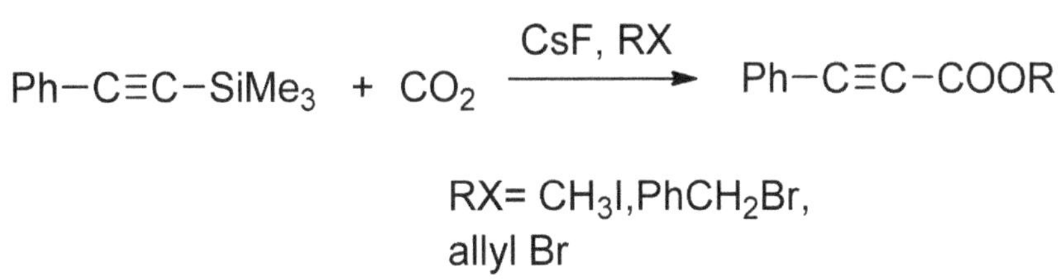

Scheme 40

More recently Ren group reported a metal-free synthesis of aryl isocyanates from arylamines and CO_2 under mild condition (Ren &Sophie, 2018) Scheme 40.

Scheme 41

Camp and co-worker reported the reaction of CO_2 with sterically saturated uranium(III) tetrasilylamido complex $[K(18c6)][U(N(SiMe_3)_2)_4]$ which leads to insertion of CO_2 into the U–N bond produce a stable U(IV) isocyanate complex $[K(18c6)][U(N(SiMe_3)_2)_3(NCO)_2]_n$ as shown in (Camp, et. al., 2015) Scheme 42.

Scheme 40

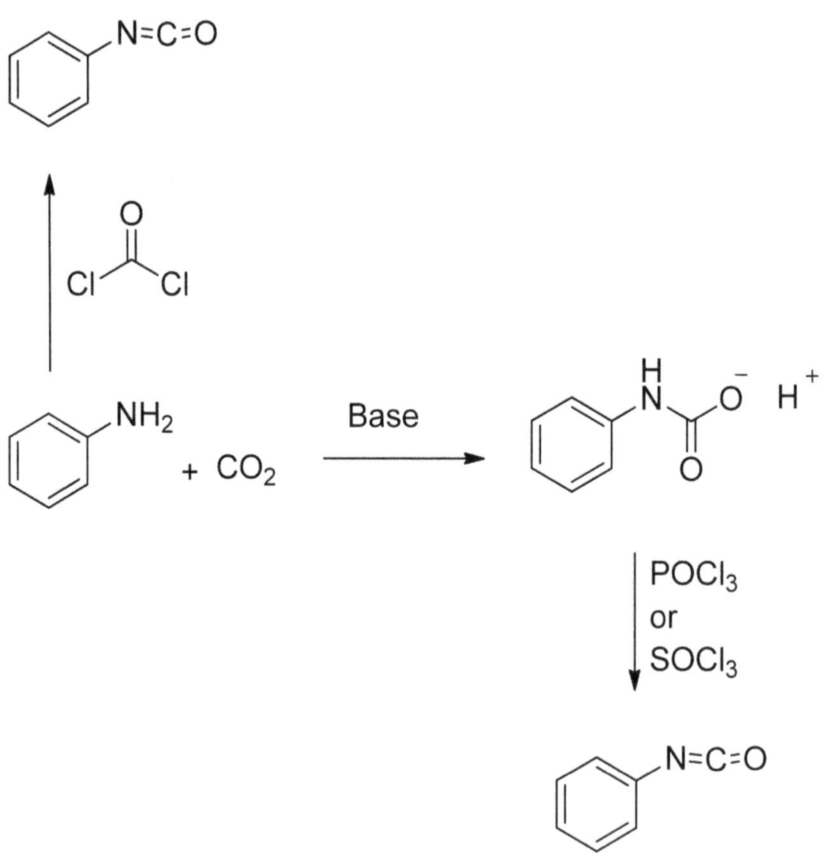

Scheme 41

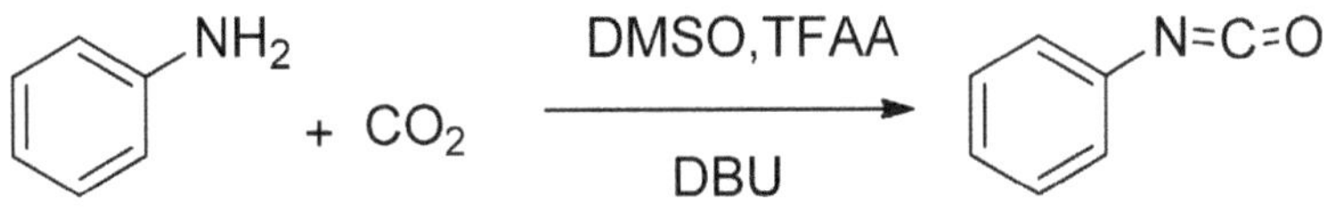

Scheme 42

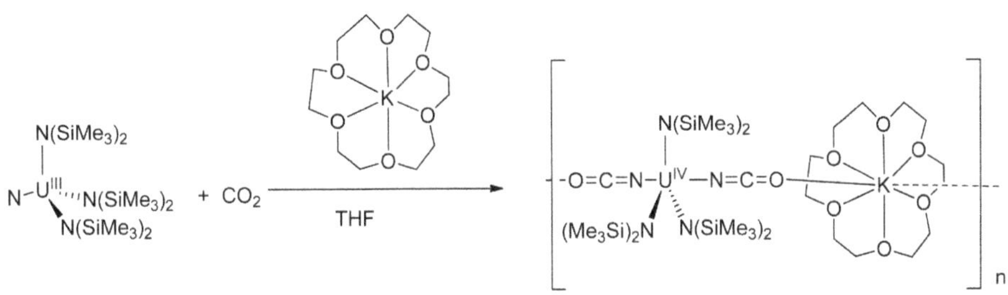

Scheme 42

Isocyanates can also be synthesized by dehydration of Amine with $POCl_3$ or P_4O_{10} in the presence of tertiary amines (Waldman & McGhee, 1994, Saylik, et.al., 1999) Scheme 41. But this reaction does not carried out on a large industrial scale because it requires an equimolar amount of dehydrating agents, bases, and produces a large amount of waste. As per related technology, urethane on thermolysis, gives isocyanates (Valli & Alper, 1995) Scheme 43.

Scheme 43

Scheme 44

Conversion of CO_2 into Urea and Urethane

Recently urea synthesis is the largest use of CO_2 in organic synthesis. Urea is mainly used as a Nitrogen fertilizer, resins and as an animal feed additive. Industrially urea is synthesized by two step process. The first step involves the reaction of ammonia and CO_2 to form ammonium carbamate which is then dehydrated to urea in the

Scheme 43

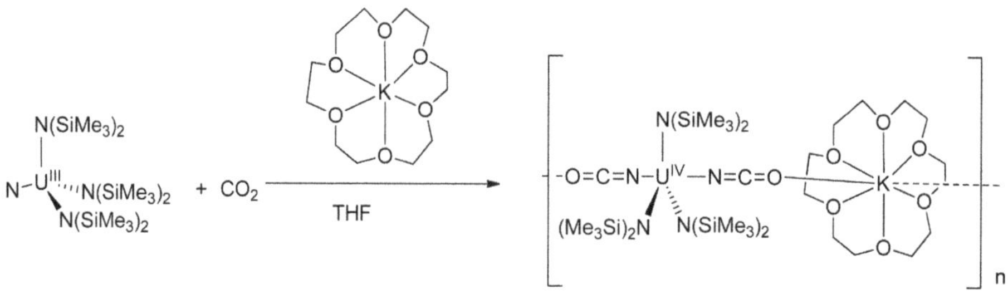

Scheme 44

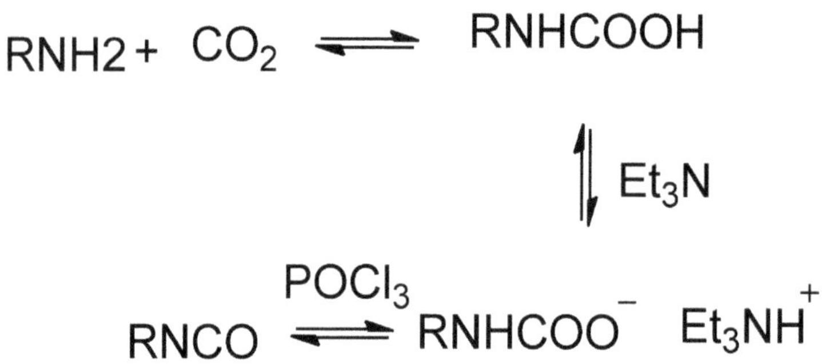

second step Scheme 44. CO_2 readily reacts with nucleophiles to give urea (Shi,et. al., 2003, Tai, et.al., 2002, Nomura, et.al., 1992) Scheme 44. Dehydration is the key reaction to get substituted urea. Substituted urea can be formed by using a dehydrating agent. Recently, The synthesis of N,N-dialkylurea has been achieved by using hydrophilic ionic liquids as reaction media and dehydrating agent as shown in Scheme (45 and 46)))).

Scheme 45

Scheme 46

Scheme 47

Wu and co-worker developed an effective route for the synthesis of urea derivatives from amines and CO_2 in the absence of any catalyst, organic solvent or other additives. They react *n* -butylamine (BA) with CO_2 to formed N, N-dibutylurea (DBU) as the main-product and no by-products except water as shown in (Wu, et. al., 2010) Scheme 48.

Scheme 48

Xiang group reported for the first time the formation of urea from CO_2 at ambient temperatures by negative corona discharge in gas phase Xiang, et. al., 2012).

Scheme 45

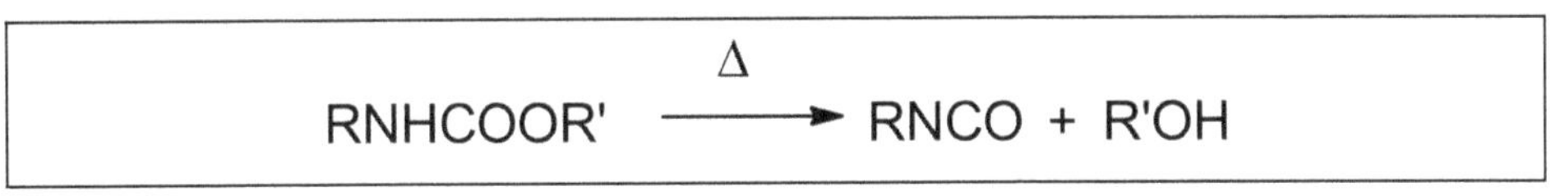

Scheme 46

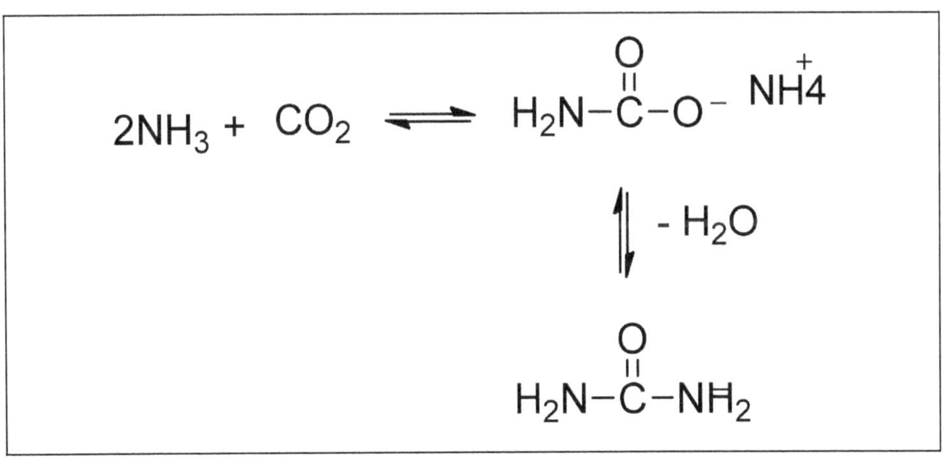

Scheme 47

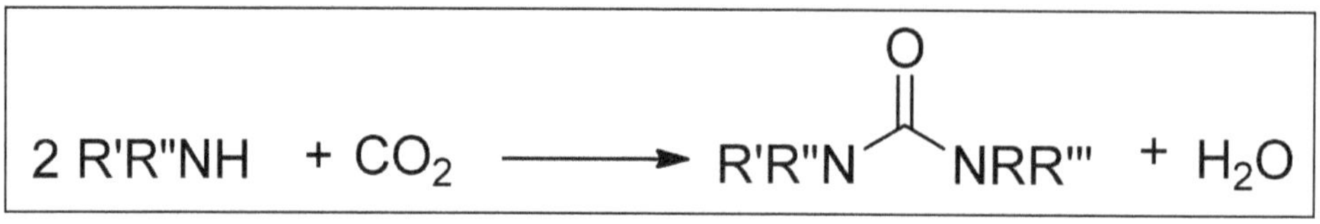

Scheme 48

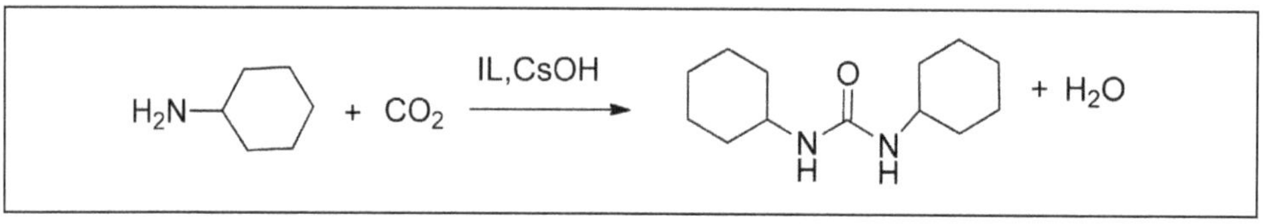

Carbamates (also known as urethane) are a stable class of compounds derived from the unstable carbamic acid. Carbamic acids are obtained when primary or secondary amines react with CO_2 as shown in Scheme 47 (Dell'Amico, et. al., 2003). Then *insitu* generated Carbamic acids with electrophile like organic halide to give urethane Scheme 48. Different types of homogeneous and heterogeneous catalysts are used for synthesis urethane (Salvatore, et.al., 2001, Yoshida, et.al., 2000, McGhee, et.al., 1995).

Scheme 49

Scheme 50

Carbamates can also be formed from primary amines and alcohols in the presence of an organotin catalyst and an acetal as a dehydrating agent under high CO_2 pressure, thereby replacing the organic halide with an alcohol as shown in Scheme 51(Alba, et. al., 2001).

Scheme 51

Nucleophilic addition of carbamate ion to alkenes in presence of allylic chloride and palladium/phosphine catalysis gives O-Allyl carbamates(Yang, et.al., 2008, McGhee, et.al., 1993). The carbamate ion used is obtained from primary or secondary amines with CO_2 and base illustrate in Scheme 52.

Scheme 49

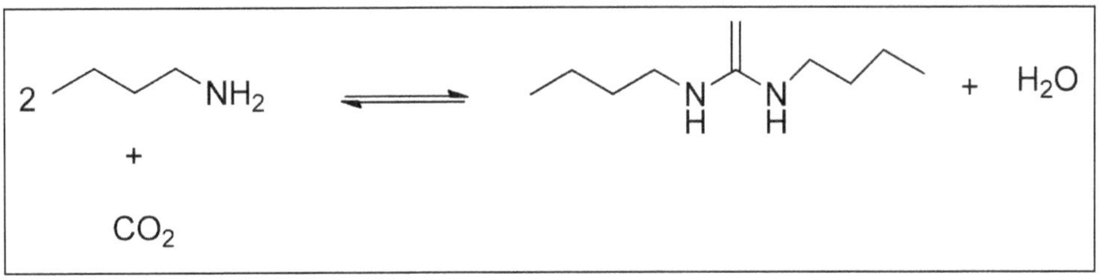

Scheme 50

$$2\ R'R''NH + CO_2 \rightleftharpoons 2\ R'R''NCOOH \xrightleftharpoons{\text{Base}} 2\ R'R''NCOO^{-}\ H^{+}$$

Scheme 51

$$RNH2 + CO_2 + R'X + \text{Base} \longrightarrow RNHCOOR' + HX$$

Scheme 52

Ion group developed different carbamates prepared from cheap and readily available reagents such as amines, alcohols and CO_2 in halogen-free way. Basic catalysts can convert both linear and branched aliphatic amines to the corresponding carbamates, under mild conditions in the absence of dehydrating agents illustrate in Scheme 53 (Ion, et. al., 2008).

Scheme 53

Mesoporous silica containing ammonium salts and adenine-modified Ti-incorporated mesoporous silica used as an effective catalyst for synthesis of urethane from CO_2 without any base Scheme 54 (Srivastava, et.al., 2005,2006).

Scheme 52

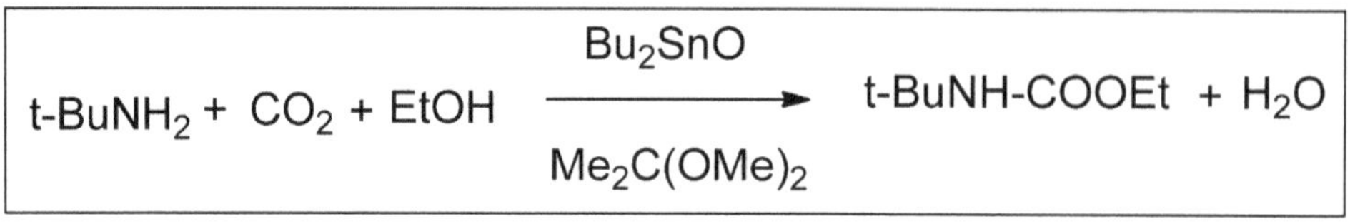

Scheme 53

Scheme 54

Carbamates can also be synthesized through nucleophilic addition of carbamic acids to alkenes Scheme 55 (McGhee & Riley, 1995).

Scheme 55

Market Size

European Union (EU) has decided to reduce greenhouse gas emissions by 40% below the 1990 levels by 2030, and down to 80-95% by 2050. The technology used for this context is carbon capture and storage (CCS) and carbon capture and utilization (CCU) (Impact assessment and scenario analysis by European Commission, 2011). Utilization of CO_2 into various useful products has become a promising field of

Scheme 54

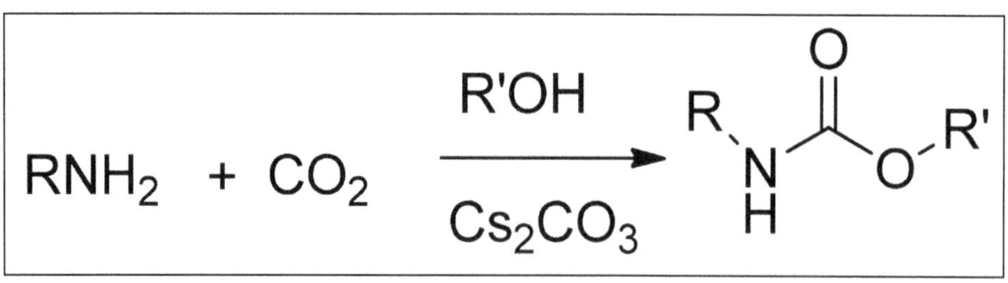

Scheme 55

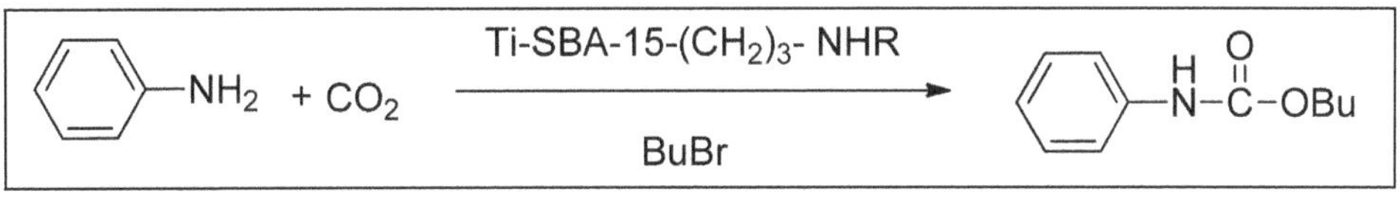

research in recent years. CO_2 used as a raw material for the preparation of urea, salicylic acid, organic carbonates, methanol and polycarbonates etc. Large scale preparation of those chemicals would be useful for industrial application, resulting consumption of approximately 116 MT of CO_2 per annum, 94% of this is used for the production of urea (Otto, et. al., 2015). In China about 0.6 MT of dimethyl carbonate per annum is produced from CO_2 in China. Even though only a small proportion (~0.1%) of the total abundance of CO_2 is currently used in chemical industry (https://hub.globalccsinstitute.com). If CO_2 used as a source for carbon instead of other fossil hydrocarbons then CO_2 emissions should be reduced, that contribute to the sustainable economy.

The funding support for CO_2 utilisation will be increase day by day. The Paris Agreement, 2015 have budget of at least 100 billion USD per year also the G7 countries agreed on the decarbonization of the world economy until 2100. Overall both declarations shape a strong future demand of low emission technologies and circular economy concepts worldwide (https://hub.globalccsinstitute.com).

CONCLUSION AND FUTURE PROSPECTS

Utilization of CO_2 as a raw material is a technique for the reduction of CO_2 in the atmosphere. CO_2 is recycle from the sources and used as a renewable carbon source. CO_2 is a major greenhouse gas and its increased concentration in the atmosphere is responsible for the global warming and climate changes. Therefore, the reduction of CO_2 concentration becomes the global challenge. As CO_2 is a renewable source of carbon, conversions of CO_2 into fuels and chemicals provide opportunities to reduce the concentration of CO_2 in atmosphere.

The utilization of CO_2 into useful chemicals, fuels, materials, Plastic etc. would be particularly promising area of research worldwide. Other technologies, such as conversion of CO_2 to syngas and novel materials have also been investigated. However, due to the thermodynamic stability and kinetic inertness, few industries can use CO_2 as a raw material. However there are still many years of research ahead before the utilization of CO_2 as a raw material. Therefore efforts must be focused on the fast and low energy pathways for capture and the utilization of CO_2.

REFERENCES

Alba, M., Choi, J.-C., & Sakakura, T. (2001). Halogen-free process for the conversion of carbon dioxide to urethanes by homogeneous catalysis. *Chemical Communications (Cambridge)*, 2238–2239. PMID:12240129

Aoki, U., Takaya, J., & Iwasawa, N. (2006). Rhodium(I)-Catalyzed Carboxylation of Aryl- and Alkenylboronic Esters with CO_2. *Journal of American Chemical Society, 128*, 8706-8707.

Aresta, M. (2010). *Carbon dioxide Recovery and Utilization*. Boston, MA: Kluwer Academic Publishers.

Aresta, M., Dibenedetto, A., & He, L.N. (2013). *Analysis of Demand for Captured CO_2 and Products from CO_2 Conversion*. TCGR Report.

Arteconi, A., Mazzarini, A., & Nicola, G. D. (2011). Emissions from Ethers and organic carbonate Fuel Additives: A Review. *Water, Air, and Soil Pollution, 221*(1-4), 405–423. doi:10.100711270-011-0804-y

Bazarov, A. I. (1870). Obtaining urea directly from Carbon dioxide and Ammonia. *Journal für Praktische Chemie, 2*, 283–312.

Behr, A., & Brehme, V. A. (2002). Homogeneous and heterogeneous catalyzed three-step synthesis of 2-ethylheptanoic acid from carbon dioxide, butadiene and hydrogen. *Journal of Molecular Catalysis A Chemical, 187*(1), 69–80. doi:10.1016/S1381-1169(02)00190-5

Behr, A., & Heite, M. (2000). Telomerisation von Kohlendioxid und 1,3-Butadien: Verfahrensentwicklung via Miniplant-Technik. *Chemieingenieurtechnik (Weinheim), 72*(1-2), 58–61. doi:10.1002/1522-2640(200001)72:1/2<58::AID-CITE58>3.0.CO;2-F

Behrens, M., Zander, S., Kurr, P., Jacobsen, N., Senker, J., Koch, G., ... Schlogl, R. (2013). Methanol Steam Reforming, in Catalysis for alternative energy generation. *Journal of the American Chemical Society, 135*, 6061–6068. doi:10.1021/ja310456f PMID:23488720

Bonura, G., Cannilla, C., Frusteri, L., Mezzapica, A., & Frusteri, F. (2017). DME production by CO_2 hydrogenation: Key factors affecting the behaviour of CuZnZr/ferrierite catalysts. *Catalysis Today, 81*, 337–344. doi:10.1016/j.cattod.2016.05.057

Boot-Handford, M. E., Abanades, J. C., Anthony, E. J., Blunt, M. J., Brandani, S., Mac Dowell, N., ... Fennell, P. S. (2014). Carbon capture and storage update. *Energy & Environmental Science, 7*(1), 130–189. doi:10.1039/C3EE42350F

Camp, C., Chatelain, L., Kefalidis, C. E., Pecaut, J., Maron, L., & Mazzanti, M. (2015). CO_2 conversion to isocyanate via multiple N–Si bond cleavage at a bulky uranium(III) complex.(2015) ChemComm. *Chemical Communications (Cambridge)*, *51*(84), 15454–15457. doi:10.1039/C5CC06707C PMID:26346380

Catizzone, E., Bonura, G., Migliori, M., Frusteri, F., & Giordano, G. (2018). CO_2 Recycling to Dimethyl Ether: State-of-the-Art and Perspectives. *Molecules (Basel, Switzerland)*, *23*, 1–28.

ICIS Chemical Business. (2006). *Kemira to expand Finnish formic acid capacity*. Author.

Clary, J. J. (2013). *The Toxicology of Methanol* (1st ed.). John Wiley & Sons,Inc. doi:10.1002/9781118353110

Coates, G. W., & Moore, D. R. (2004). Discrete metal-based catalysts for the copolymerization of CO_2 and epoxides: Discovery, reactivity, optimization, and mechanism. *Angewandte Chemie International Edition*, *43*(48), 6618–6639. doi:10.1002/anie.200460442 PMID:15558659

Cooper, C. (2009). *A Technical Basis for Carbon Dioxide Storage, CO_2 Capture Project*. available at: http://unfccc.int/resource/docs/2011/smsn/ngo/276.pdf

Dell'Amico, D. B., Calderazzo, F., Labella, L., Marchetti, F., & Pampaloni, G. (2003). Converting carbon dioxide into carbamato derivatives. *Chemical Reviews*, *103*(10), 3857–3898. doi:10.1021/cr940266m PMID:14531715

Ebert, G. W., Juda, W. L., Kosakowski, R. H., Ma, B., Dong, L., Cummings, K. E., ... Luo, J. (2005). Carboxylation and Esterification of Functionalized Arylcopper Reagents. *The Journal of Organic Chemistry*, *70*(11), 4314–4317. doi:10.1021/jo047731s PMID:15903305

Elewuwa, F. A., & Makkawi, Y. T. (2015). Hydrogen production by steam reforming of DME in a large scale CFB reactor. Part I: Computational model and predictions. *International Journal of Hydrogen Energy*, *40*(46), 15865–15876. doi:10.1016/j.ijhydene.2015.10.050

Elewuwa, F. A., & Makkawi, Y. T. (2016). A computational model of hydrogen production by steam reforming of dimethyl ether in a large scale CFB reactor. Part II: Parametric analysis. *International Journal of Hydrogen Energy*, *41*(44), 19819–19828. doi:10.1016/j.ijhydene.2016.08.072

ElMekawy, A., Hegab, H. M., Mohanakrishna, G., Elbaz, A. F., Bulut, M., & Pant, D. (2016). Technological advances in CO_2. conversion electro-biorefinery: A step toward commercialization. *Bioresource Technology, 215*, 357–370. doi:10.1016/j.biortech.2016.03.023 PMID:27020396

Engels, H.-W., Pirkl, H.-G., Albers, R., Albach, R. W., Krause, J., Hoffmann, A., ... Dormish, J. (2013). Polyurethanes: Versatile materials and sustainable problem solvers for today's challenges. *Angewandte Chemie International Edition, 52*(36), 9422–9441. doi:10.1002/anie.201302766 PMID:23893938

Erena, J., Vicente, J., Aguayo, A. T., Gayubo, A. G., Olazar, M., & Bilbao, J. (2013). Effect of combining metallic and acid functions ib CZA/HZSM-5 desilicated zeolite catalysts on the DME steam reforming in a fluidized bed. *International Journal of Hydrogen Energy, 38*(24), 10019–10028. doi:10.1016/j.ijhydene.2013.05.134

European Commission. (2003). European Commission Regulation (EC) No 1831/2003 of the European Parliament and the Council of 22nd September 2003 on additives for use in animal nutrition. Off J Eur Union 2003. L268 18.10.

Field, C. B., Barros, V. R., Dokken, D. J., Mach, K. J., Mastrandrea, M. D., Bilir, T. E., ... White, L. L. (2014). *IPCC, 2014: Climate Change 2014 Impacts, Adaptation, and Vulnerability. Part A: Global and Sectoral Aspects. Contribution ofWorking Group II to the Fifth Assessment Report of the Intergovernmental Panel on Climate Change*. Cambridge University Press.

Frogneux, X., von Wolff, N., Thuery, P., Lefevre, G., & Cantat, T. (2016). CO_2 Conversion into Esters by Fluoride-Mediated Carboxylation of Organosilanes and Halide Derivatives. *Chemistry (Weinheim an der Bergstrasse, Germany), 22*(9), 2930–2934. doi:10.1002/chem.201505092 PMID:26712565

Frusteri, F., Migliori, M., Cannilla, C., Frusteri, L., Catizzone, E., Aloise, A., Giordano, G., & Bonura, G. (2017), Direct CO_2-to-DME hydrogenation reaction: New evidences of a superior behaviour of FER-based hybrid systems to obtain high DME yield. *Journal of CO_2 Utilization, 18,* 353–361.

Gao, P., Zhong, L. S., Zhang, L. N., Wang, H., Zhao, N., Wei, W., & Sun, Y. H. (2015). Yttrium oxide modified Cu/ZnO/Al_2O_3 catalysts *via*hydrotalcite-like precursors for CO_2hydrogenation to methanol. *Catalysis Science & Technology, 5*(9), 4365–4377. doi:10.1039/C5CY00372E

Ghosh, A. K., & Brindisi, M. (2015). Organic carbamates in drug design and medicinal chemistry. *Journal ofMedicinal Chemistry, 58*(7), 2895–2940. doi:10.1021/jm501371s PMID:25565044

Goeppert, A., Czaun, M., Jones, J.-P., Prakash, G. K. S., & Olah, G. A. (2014). Recycling of carbon dioxide to methanol and derived products - closing the loop. *Chemical Society Reviews*, *43*(23), 7995–8048. doi:10.1039/C4CS00122B PMID:24935751

Gonzalez-Gil, R., Herrera, C., Larrubioa, M. A., Kowalik, P., Pieta, I. S., & Alemany, L. J. (2016). Hydrogen production by steam reforming of DME over Ni-based catalysts modified with vanadium. *International Journal of Hydrogen Energy*, *41*(43), 19781–19788. doi:10.1016/j.ijhydene.2016.05.074

Gooßen, L. J., Rodríguez, N., Manjolinho, F., & Lange, P. P. (2010). Synthesis of propiolic acids via copper-catalyzed insertion of carbon dioxide into the C–H bond of terminal alkynes. *Advanced Synthesis & Catalysis*, *352*(17), 2913–2917. doi:10.1002/adsc.201000564

Grego, S., Aricò, F., & Tundo, P. (2012). Phosgene-free carbamoylation of aniline via dimethyl carbonate. *Pure and Applied Chemistry*, *84*(3), 695–705. doi:10.1351/PAC-CON-11-07-17

Grove, J., Greig, C. R., Smart, S., & Lant, P. A. (2017). Producing a CO_2-neutral clean cooking fuel in India- where and at what cost. *International Journal of Hydrogen Energy*, *42*(30), 19076–19078. doi:10.1016/j.ijhydene.2017.06.070

Hertwich, E. G., Aaberg, M., Singh, B., & Strømman, A. H. (2008). Life-cycle Assessment of Carbon Dioxide Capture for Enhanced Oil Recovery. *Chinese Journal of Chemical Engineering*, *16*(3), 343–353. doi:10.1016/S1004-9541(08)60085-3

Heyn, R. H., Jacobs, I., & Carr, R. H. (2014). *CO_2 Chemistry*. In M. Aresta & R. van Eldik (Eds.), *Advances in Inorganic Chemistry* (Vol. 66, pp. 83–115). Boca Raton, FL: Academic Press.

Hoberg, H., Peres, Y., & Milchereit, A. (1986). C-C-verknijpfung von alkenen mit co, an nickel(o); n-pentensauren aus ethen. *Journal of Organometallic Chemistry*, *307*, C38–C40. doi:10.1016/0022-328X(86)80487-9

Hoberg, H., Schefer, D., Burkhart, G., Kruger, C. M., & Romao, J. (1984). Nickel(0)-induzierte C-C-verknüpfung zwischen kohlendioxid und alkinen sowie alkenen. *Journal of Organometallic Chemistry*, *266*(2), 203–224. doi:10.1016/0022-328X(84)80129-1

Inoue, S., Koinuma, H., & Tsuruta, T. (1969). Copolymerization of carbon dioxide and epoxide. *Journal of Polymer Science. Part B: Polymer Letters*, *7*(4), 287–292. doi:10.1002/pol.1969.110070408

Ion, A., Van Doorslaer, C., Parvulescu, V., Jacobs, P., & De Vos, D. (2008). Green synthesis of carbamates from CO_2, amines and alcohols. *Green Chemistry, 10*(1), 111–116. doi:10.1039/B711197E

Johnson, L., Grant, J., & Low, P. L. (2014). Two degrees of separation: ambition and reality: Low Carbon Economy Index. *The Journal of Organic Chemistry, 67*, 9070–9072.

Kathalewar, M. S., Joshi, P. B., Sabnis, A. S., & Malshe, V. C. (2013). Non-isocyanate polyurethanes: From chemistry to applications. *RSC Advances, 3*(13), 4110–4129. doi:10.1039/c2ra21938g

Kavil, Y. N., Shaban, Y. A., Al Farawati, R. K., Orif, M. I., Zobidi, M., & Khan, S. U. M. (2017). Photocatalytic conversion of CO_2 into methanol over Cu-C/TiO_2 nanoparticles under UV light and natural sunlight. *Journal of Photochemistry and Photobiology A Chemistry, 347*, 244–253. doi:10.1016/j.jphotochem.2017.07.046

Kolbe, H. (1874). Ueber eine neue Darstellungsmethode und einige bemerkenswerte Eigenschaften der Salicylsäure. *Journal für Praktische Chemie, 10*(1), 89–107. doi:10.1002/prac.18740100106

Kornas, A., Grabowski, R., Sliwa, M., Samson, K., Ruggiero-Mikołajczyk, M., & Zelazny, A. (2017). Dimethyl ether synthesis from CO_2 hydrogenation over hybrid catalysts: Effects of preparation methods. *Reaction Kinetics, Mechanisms and Catalysis, 121*(1), 317–327. doi:10.100711144-017-1153-7

Kothandaraman, J., Goeppert, A., Czaun, M., Olah, G. A., & Surya Prakash, G. K. (2016). Conversion of CO_2 from Air into Methanol Using a Polyamine and a Homogeneous Ruthenium Catalyst. *Journal of the American Chemical Society, 138*(3), 778–781. doi:10.1021/jacs.5b12354 PMID:26713663

Kumar, S., Yadav, R.K., Ram, K., Aguiar, A., Koh, J., & Sobral, A.J.F.N., (2018), Graphene oxide modified cobalt metallated porphyrin photocatalyst for conversion of formic acid from carbon dioxide. *Journal of CO_2 Utilization, 27*,107-114.

Lagarde, C. (2013). *A New Global Economy for a New Generation*. International Monetary Fund.

Lapidus, A. L., Pirozhkov, S. D., & Zelinsky, N. D. (1988). *Proc. 9th Int. Congr. Catal., 3*, 1028.

Lee, E.-H., Ahn, J.-Y., Dharman, M. M., Park, D.-W., Park, S.-W., & Kim, I. (2008). Synthesis of cyclic carbonate from vinyl cyclohexene oxide and CO_2 using ionic liquids as catalysts. *Catalysis Today, 131*(1-4), 130–134. doi:10.1016/j.cattod.2007.10.012

Lee, K., Lee, S., Cho, H., Jeong, S., Kim, W., Lee, S., & Lee, D. C. (2017). Cu^+ -incorporated TiO_2 overlayer on Cu_2O nanowire photocathodes for enhanced photoelectrochemical conversion of CO_2 to methanol. *Journal of Energy Chemistry*.

Li, C., Yuan, X., & Fujimoto, K. (2014). Development of highly stable catalyst for. methanol synthesis from carbon dioxide. *Applied Catalysis A, General*, *469*, 306–311. doi:10.1016/j.apcata.2013.10.010

Li, Y., Junge, K., & Beller, M. (2013). Improving the efficiency of the hydrogenation of carbonates and carbon dioxide to methanol. *ChemCatChem*, *5*(5), 1072–1074. doi:10.1002/cctc.201300013

Liu, R., Qin, Z., Ji, H., & Su, T. (2013). Synthesis of Dimethyl Ether from CO_2 and H_2 Using a Cu−Fe−Zr/HZSM-5 Catalyst System. *Industrial & Engineering Chemistry Research*, *52*(47), 16648–16655. doi:10.1021/ie401763g

Liu, X.-M., Lu, G. Q., Yan, Z.-F., & Beltramini, J. (2003). Recent Advances in Catalysts for Methanol Synthesis via Hydrogenation of CO and CO_2. *Industrial & Engineering Chemistry Research*, *4225*(25), 6518–6530. doi:10.1021/ie020979s

Lu, S.-M., Wang, Z., Li, J., Xiao, J., & Li, C. (2016). Base-free hydrogenation of CO_2 to formic acid in water with an iridium complex bearing a *N,N′*-diimine ligand. *Green Chemistry*, *18*(16), 4553–4558. doi:10.1039/C6GC00856A

Lu, X.-B., & Darensbourg, D. J. (2012). Cobalt catalysts for the coupling of CO_2 and epoxides to provide polycarbonates and cyclic carbonates. *Chemical Society Reviews*, *41*(4), 1462–1484. doi:10.1039/C1CS15142H PMID:21858339

Lyu, L., Jin, F., Zhong, H., Chen, H., & Yao, G. (2015). A novel approach to reduction of CO_2 into methanol by water splitting with aluminum over a copper catalyst. *RSC Advances*, *5*(40), 31450–31453. doi:10.1039/C5RA02872H

Ma, J., Sun, N. N., Zhang, X. L., Zhao, N., Mao, F. K., Wei, W., & Sun, Y. H. (2009). A short review of catalysis for CO2 conversion. *Catalysis Today*, *148*(3-4), 221–231. doi:10.1016/j.cattod.2009.08.015

Maihom, T., Wannakao, S., Boekfa, B., & Limtrakul, J. (2013). Production of Formic Acid via Hydrogenation of CO_2 over a Copper- Alkoxide-Functionalized MOF: A Mechanistic Study. *The Journal of Physical Chemistry C*, *117*(34), 17650–17658. doi:10.1021/jp405178p

Markewitz, P., Kuckshinrichs, W., Leitner, W., Linssen, J., Zapp, P., Bongartz, R., ... Mgller, T. E. (2012). Worldwide innovations in the development of carbon capture technologies and the utilization of CO_2. *Energy & Environmental Science*, *5*(6), 7281–7305. doi:10.1039/c2ee03403d

McGhee, W., Riley, D., Christ, K., Pan, Y., & Parnas, B. (1995). Carbon Dioxide as a Phosgene Replacement: Synthesis and Mechanistic Studies of Urethanes from Amines, CO_2, and Alkyl Chlorides. *The Journal of Organic Chemistry*, *60*(9), 2820–2830. doi:10.1021/jo00114a035

McGhee, W. D., & Riley, D. P. (1992). Palladium-mediated synthesis of urethanes from amines, carbon dioxide, and cyclic diolefins. *Organometallics*, *11*(2), 900–907. doi:10.1021/om00038a060

Metz, B., Davidson, O., de Coninck, H., Loos, M., & Meyer, L. (2005). *IPCC Special Report on Carbon Dioxide Capture and Storage*. Intergovernmental Panel on Climate Change. Available at: http://www.ipcc.ch/pdf/special-reports/srccs/srccs_wholereport.pdf

Mita, T., Higuchi, Y., & Sato, Y. (2014). Carboxylation with CO_2 via Brook Rearrangement: Preparation of α-Hydroxy Acid Derivatives. *Organic Letters*, *16*(1), 7–14. doi:10.1021/ol403099f PMID:24313846

Miyano, M., Zhang, H., Yoshiba, M., & Izumi, Y. (2017). Selective Photoconversion of Carbon Dioxide into Methanol Using Layered Double Hydroxides at 0.40 MPa. *Energy Technology (Weinheim)*, *5*(6), 892–900. doi:10.1002/ente.201600578

Monteiro, W. F., Vieira, M. O., Aquino, A. S., de Souza, M. O., Lima, J. D., Einloft, S., & Ligabue, R. (2017). CO_2 conversion to propylene carbonate catalyzed by Ionic liquid containing organosilane groups supported on titanate nanotubes/nanowires. *Applied Catalysis A, General*, *544*, 46–54. doi:10.1016/j.apcata.2017.07.011

Motokura, K., Kashiwame, D., Miyaji, A., & Baba, T. (2012). Copper-Catalyzed Formic Acid Synthesis from CO_2 with Hydrosilanes and H_2O. *Organic Letters*, *14*(10), 2642–2645. doi:10.1021/ol301034j PMID:22540994

Nagashima, S., Miyagawa, T., Matsumoto, M., Suzuki, S., Komaki, H., Takagi, M., & Murai, S. (2011). *Life Cycle Assessment Performed on a CCS Model Case in Japan and Evaluation of Improvement Facilitated by Heat Integration*. Academic Press.

Naik, S. P., Vy Bui, T. R., Miller, J. D., Drinnan, N. B., & Zmierczak, W. (2011). Synthesis of DME from CO2/H2 gas mixture. *Chemical Engineering Journal*, *167*, 362–368.

Nichols Roth, G. (2015). *G7 Climate Change The New Economy*. World News -Climate Change The New Economy Ltd.

Nomura, R., Hasegawa, Y., Ishimoto, M., Toyosaki, T., & Matsuda, H. (1992). Carbonylation of Amines by Carbon Dioxide in the Presence of an Organoantimony Catalyst'. *The Journal of Organic Chemistry*, *57*(26), 7339–7342. doi:10.1021/jo00052a060

Oar-Arteta, L., Aguayo, A. T., Remiro, M., Arandia, A., Bilbao, J., & Gayubo, A. G. (2016). Kinetics of the steam reforming of dimethyl ether over $CuFe_2O_4/g\text{-}Al_2O_3$. *Chemical Engineering Journal*, *306*, 401–412. doi:10.1016/j.cej.2016.07.075

Ohmiya, H., Tanabe, M., & Sawamura, M. (2011). Copper-Catalyzed Carboxylation of Alkylboranes with Carbon Dioxide: Formal Reductive Carboxylation of Terminal Alkenes. *Organic Letters*, *13*(5), 1086–1088. doi:10.1021/ol103128x PMID:21275416

Otto, A., Grube, T., Schiebahn, S., & Stolten, D. (2015). Closing the loop: Captured CO_2 as a feedstock in the chemical industry. *Energy & Environmental Science*, *8*(11), 3283–3297. doi:10.1039/C5EE02591E

Peng, W., Zhao, N., Xiao, F., Wei, W., & Sun, Y. (2012). Recent progress in phosgene-free methods for synthesis of dimethyl carbonate. *Pure and Applied Chemistry*, *84*(3), 603–620. doi:10.1351/PAC-CON-11-06-02

Perez-Uriarte, P., Ateka, A., Gamero, M., Aguayo, A. T., & Bilbao, J. (2016). Effect of the operating conditions in the transformation of DME to olefins over a HZSM-5 zeolite catalyst. *Industrial & Engineering Chemistry Research*, *55*(23), 6569–6578. doi:10.1021/acs.iecr.6b00627

Peters, M., Kchler, B., Kuckshinrichs, W., Leitner, W., Markewitz, P., & Muller, T. E. (2011). Chemical technologies for exploiting and recycling carbon dioxide into the value chain. *ChemSusChem*, *4*(9), 1216–1240. doi:10.1002/cssc.201000447 PMID:21866580

Phosgene was used as a chemical weapon during World War I. (n.d.). Retrieved from https://www.opcw.org/chemical-weapons-convention/annexes/ annex-on-chemicals/schedule-3/

Prez-Uriarte, P., Ateka, A., Aguayo, A. T., Gayubo, A. G., & Bilbao, J. (2016). Kinetic model for the reaction of DME to olefins over a HZSM-5 zeolite catalyst. *Chemical Engineering Journal*, *302*, 801–810. doi:10.1016/j.cej.2016.05.096

Quaranta, E., & Aresta, M. (2010). The Chemistry of N–CO_2 Bonds: Synthesis of Carbamic Acids and Their Derivatives, Isocyanates and Ureas. In M. Aresta (Ed.), *Carbon Dioxide as Chemical Feedstock* (pp. 121–167). Weinheim: Wiley-VCH. doi:10.1002/9783527629916.ch6

Ren, Y., & Sophie, A. L. (2018). Metal-Free Synthesis of Unsymmetrical Ureas and Carbamates from CO_2 and Amines via Isocyanate Intermediates. *The Journal of Organic Chemistry*, *83*(2), 913–920. doi:10.1021/acs.joc.7b02905 PMID:29226676

Reutemann, W., & Kieczka, H. (2012). Formic acid. In *Ullmann's encyclopedia of industrial chemistry* (7th ed.). Wiley-VCH. Retrieved from https://en.wikipedia.org/wiki/Formic_acid

Ribeiro, A. P. C., Martins, L. M. D. R. S., & Pombeiro, A. J. L. (2013). Carbon dioxide-to-methanol single-pot conversion using a C-scorpionate iron(II) catalyst. *Green Chemistry*, *00*, 1–3.

Sakakura, T., Choi, J.-C., & Yasuda, H. (2007). Transformation of Carbon Dioxide. *Chemical Reviews*, *107*(6), 2365–2387. doi:10.1021/cr068357u PMID:17564481

Salvatore, R. N., Shin, S. I., Nagle, A. S., & Jung, K. W. (2001). Efficient carbamate synthesis via a three-component coupling of an amine, CO2, and alkyl halides in the presence of Cs2CO3 and tetrabutylammonium iodide. *The Journal of Organic Chemistry*, *66*(3), 1035–1037. doi:10.1021/jo001140u PMID:11430069

Santhoshkumar, R., Hong, Y.-C., Luo, C.-Z., Wu, Y.-C., Hung, C.-H., Hwang, K.-Y., ... Cheng, C.-H. (2016). Synthesis of Vinyl Carboxylic Acids using Carbon Dioxide as a Carbon Source by Iron-Catalyzed Hydromagnesiation. *ChemCatChem*, *8*(13), 1–5. doi:10.1002/cctc.201600279

Saylik, D., Horvath, M. J., Elmes, P. S., Jackson, W. R., Lovel, C. G., & Moody, K. (1999). Preparation of Isocyanates from Primary Amines and Carbon Dioxide Using Mitsunobu Chemistry. *The Journal of Organic Chemistry*, *64*(11), 3940–3946. doi:10.1021/jo982362j

Semelsberger, T. A., Borup, R. L., & Greene, H. L. (2006). Dimethyl ether (DME) as an alternative fuel. *Journal of Power Sources*, *156*(2), 497–511. doi:10.1016/j.jpowsour.2005.05.082

Shaikh, A. A. G., & Sivaram, S. (1996). Organic carbonates. *Chemical Reviews*, *96*(3), 951–976. doi:10.1021/cr950067i PMID:11848777

Shi, F., Deng, Y. Q., SiMa, T. L., Peng, J. J., Gu, Y. L., & Qiao, B. T. (2003). Alternatives to Phosgene and Carbon Monoxide: Synthesis of Symmetric Urea Derivatives with Carbon Dioxide in Ionic Liquids. Angewandte Chemie International Edition, 42, 3257–3260.

Six, C., & Richter, F. (2003). Isocyanates, Organic. In *Ullmann's Encyclopedia of Industrial Chemistry*. Weinheim: Wiley-VCH. doi:10.1002/14356007.a14_611

Sneeden, R. P. A. (1982). Article. In G. Wilkinson, F. A. G. Stone, & A. W. Abel (Eds.), Comprehensive Organometallic Chemistry (Vol. 8, p. 225). New York: Pergamon Press. doi:10.1016/B978-008046518-0.00109-4

Song, J., Feng, X., Yamamoto, Y., Almansour, A. I., Arumugam, N., Suresh Kumar, R., & Bao, M. (2016). Selective synthesis of δ-lactone via palladium nanoparticles-catalyzed telomerization of CO_2 with 1, 3-butadiene. *Tetrahedron Letters*, *57*(29), 3163–3166. doi:10.1016/j.tetlet.2016.06.024

Song, Q.-W., Zhao, Y.-N., He, L.-N., Gao, J., & Yang, Z.-Z. (2012). Synthesis of Oxazolidinones/Polyurethanes from Aziridines and CO_2. *Current Catalysis*, *1*(2), 107–124. doi:10.2174/2211544711201020107

Songa, H., Zhanga, N., Zhonga, C., Liua, Z., Xiaoa, M., & Gai, H. (2017). Hydrogenation of CO_2 into Formic Acid over Palladium Catalyst on Chitin. *New Journal of Chemistry*, *41*(17), 9170–9177. doi:10.1039/C7NJ00460E

Sopena, S., Fiorani, G., Martin, C., & Kleij, A. W. (2015). Highly Efficient Organocatalyzed Conversion of Oxiranes and CO_2 into Organic Carbonates. *ChemSusChem*, *8*(19), 3248–3254. doi:10.1002/cssc.201500710 PMID:26331287

Srivastava, R., Srinivas, D., & Ratnasamy, P. (2005). CO_2 activation and synthesis of cyclic carbonates and alkyl/aryl carbamates over adenine-modified Ti-SBA-15 solid catalysts. *Journal of Catalysis*, *233*(1), 1–15. doi:10.1016/j.jcat.2005.03.023

Srivastava, R., Srinivas, D., & Ratnasamy, P. (2006). Syntheses of polycarbonate and polyurethane precursors utilizing CO_2 over highly efficient, solid as-synthesized MCM-41 catalyst. *Tetrahedron Letters*, *47*(25), 4213–4217. doi:10.1016/j.tetlet.2006.04.057

Steinhauser, G. (2008). Cleaner production in the Solvay process: General strategies and recent developments. *Journal of Cleaner Production*, *16*(7), 833–841. doi:10.1016/j.jclepro.2007.04.005

Storace, L., Anzalone, L., Confalone, P. N., Davis, W. P., Fortunak, J. M., Giangiordano, M., ... Wood, C. C. (2002). For the use of phosgene in the multikilogram-scale preparation of a medicinally important compound, see. *Organic Process Research & Development*, *6*, 54–63. doi:10.1021/op015507o

Styring, P., Jansen, D., de Coninck, H., Reith, H., & Armstrong, K. (2011). *Carbon Capture and Utilisation in the Green Economy*. Centre for Low Carbon Futures. Available at: http://co2chem.co.uk/wpcontent/uploads/2012/06/CCU%20in%20the%20green%20economy%20report.pdf

Tai, C. C., Huck, M. J., McKoon, E. P., Woo, T., & Jessop, P. G. (2002). *Low-temperature synthesis of tetraalkylureas from secondary amines and carbon dioxide*. Academic Press.

Tao, J.-L., Jun, K.-W., & Lee, K.-W. (2001). Co-production of dimethyl ether and methanol from CO_2 hydrogenation: Development of a stable hybrid catalyst. *Applied Organometallic Chemistry*, *15*(2), 105–108. doi:10.1002/1099-0739(200102)15:2<105::AID-AOC100>3.0.CO;2-B

Tian, J.-S., Cai, F., Wang, J.-Q., Du, Y., & He, L.-N. (2008). Environmentally Benign Chemical Conversion of CO_2 into Organic Carbonates Catalyzed by Phosphonium Salts. *Phosphorus, Sulfur, and Silicon and the Related Elements*, *183*(2-3), 494–498. doi:10.1080/10426500701761540

Tsuda, T., Chujo, Y., & Saegusa, T. (1976). Copper(I) cyanoacetate as a carrier of activated carbon dioxide. *Journal of the Chemical Society. Chemical Communications*, (11): 415–416. doi:10.1039/c39760000415

Tsuda, T., Yamamoto, T., & Saegusa, T. (1992). Palladium-catalyzed cycloaddition of carbon dioxide with methoxyallene. *Journal of Organometallic Chemistry*, *429*(3), C46–C48. doi:10.1016/0022-328X(92)83194-M

Tsuda, T., Yasukawa, H., & Komori, K. (1995). Nickel(0)-Catalyzed Cycloaddition Copolymerization of Ether Diynes with Carbon Dioxide to Poly(2-pyrone)s. *Macromolecules*, *28*(5), 1356–1359. doi:10.1021/ma00109a004

Tuan Le, Q. A., Gon Kim, H., & Kim, Y. H. (2018). Electrochemical synthesis of formic acid from CO_2 catalyzed by Shewanella oneidensis MR-1 whole-cell biocatalyst. *Enzyme and Microbial Technology*, *116*, 1–5. doi:10.1016/j.enzmictec.2018.05.005 PMID:29887011

Valli, V. L. K., & Alper, H. (1995). A Simple, Convenient, and Efficient Method for the Synthesis of Isocyanates from Urethanes. *The Journal of Organic Chemistry*, *60*(1), 257–258. doi:10.1021/jo00106a044

Vincenzo, C., Nacci, A., Monopoli, A., & Fanizzi, A. (2002). Cyclic Carbonate Formation from Carbon Dioxide and Oxiranes in Tetrabutylammonium Halides as Solvents and Catalysts. *Organic Letters*, *4*(15), 2561–2563. doi:10.1021/ol026189w PMID:12123376

Waldman, T. E., & McGhee, W. D. (1994). Isocyanates from primary amines and carbon dioxide: 'dehydration' of carbamate anions. *Journal of the Chemical Society. Chemical Communications*, *0*(8), 957–958. doi:10.1039/C39940000957

Wang, J., & Zhang, Y. (2016). Boronic Acids as Hydrogen Bond Donor Catalysts for Efficient Conversion of CO_2 into Organic Carbonate in Water. *ACS Catalysis*, *6*(8), 4871–4876. doi:10.1021/acscatal.6b01422

Wei Chen, X., Zhao, C., Li, S., Song, Y., Ge, R., Wei, W., & Sun, Y. (2017). Exclusive formation of formic acid from CO_2 electro reduction by tunable Pd-Sn alloy. *Angewandte Chemie*, *129*(40), 12387–12391. doi:10.1002/ange.201707098

Weilhard, A., Qadir, M. I., Sans, V., & Dupont, J. (2018). Selective CO_2 Hydrogenation to Formic Acid with Multifunctional Ionic Liquids. *ACS Catalysis*, *8*(3), 1628–1634. doi:10.1021/acscatal.7b03931

Weisser, D. (2007). A guide to life-cycle greenhouse gas (GHG) emissions from electric supply technologies. *Energy*, *32*(9), 1543–1559. doi:10.1016/j.energy.2007.01.008

Whiteoak, C. J., Kielland, N., Laserna, V., Escudero-Adan, E. C., Martin, E., & Kleij, A. W. (2013). A Powerful Aluminum Catalyst for the Synthesis of Highly Functional Organic Carbonates. *Journal of the American Chemical Society*, *135*(4), 1228–1231. doi:10.1021/ja311053h PMID:23302007

Williams, C.M., Johnson, J.B., & Rovis, T. (2008), Nickel-catalyzed reductive carboxylation of styrenes using CO_2. *Journal of American Chemical Society, 130*, 14936 – 14937.

Wu, C., Cheng, H., Liu, R., Wang, Q., Hao, Y., Yu, Y., & Zhao, F. (2010). Synthesis of urea derivatives from amines and CO_2 in the absence of catalyst and solvent. *Green Chemistry*, *12*(10), 1811–1816. doi:10.1039/c0gc00059k

Wu, L., Liu, Q., Fleischer, I., Jackstell, R., & Beller, M. (2014). Ruthenium-catalysed alkoxycarbonylation of alkenes with carbon dioxide. *Nature Communications*, *5*, 1–6. PMID:24518431

Xiang, X., Guo, L., Wu, X., Ma, X., & Xia, Y. (2012). Urea formation from carbon dioxide and ammonia at atmospheric pressure. *Environmental Chemistry Letters*, *10*(3), 295–300. doi:10.100710311-012-0366-2

Xiaohe, J., Ningning, Z., Jian-Gong, M., & Peng, C. (2017). Conversion of CO_2 into cyclic carbonates by a Co (II) metal-organic framework and the improvement of catalytic activity via nanocrystallization. *Dalton Transactions (Cambridge, England)*, *00*, 1–3.

Yang, H., Xu, Z., Fan, M., Gupta, R., Slimane, R. B., Bland, A. E., & Wright, I. (2008). Progress in carbon dioxide separation and capture: A review. *Journal of Environmental Sciences (China)*, *20*(1), 14–27. doi:10.1016/S1001-0742(08)60002-9 PMID:18572517

Yeung, C. S., & Dong, V. M. (2008). Beyond Aresta's Complex: Ni- and Pd-Catalyzed Organozinc Coupling with CO_2. *Journal of the American Chemical Society*, *130*(25), 7826–7827. doi:10.1021/ja803435w PMID:18510323

Yonemoto-Kobayashi, M., Inamoto, K., Tanaka, Y., & Kondo, Y. (2013). Carboxylation of alkynylsilanes with carbon dioxide mediated by cesium fluoride in DMSO. *Organic & Biomolecular Chemistry*, *11*(23), 3773–3775. doi:10.1039/c3ob40760h PMID:23674078

Yoo, W.-J., Capdevila, M. G., Du, X., & Kobayashi, S. (2012). Base-Mediated Carboxylation of Unprotected Indole Derivatives with Carbon Dioxide. *Organic Letters*, *14*(20), 5326–5329. doi:10.1021/ol3025082 PMID:23035967

Yoshida, M., Hara, N., & Okuyama, S. (2000). Catalytic production of urethanes from amines and alkyl halides in supercritical carbon dioxide. *Chemical Communications (Cambridge)*, *0*(2), 151–152. doi:10.1039/a908819i

Yu, B., Diao, Z.-F., Guo C.-X., & He, L.-N. (2013). Carboxylation of olefins/alkynes with CO_2 to industrially relevant acrylic acid derivatives. *Journal of CO_2 Utilization, 1*, 60–68.

Yu, D.-Y., & Zhang, Y.-G. (2011). The direct carboxylation of terminal alkynes with carbon dioxide. *Green Chemistry*, *13*(5), 1275–1279. doi:10.1039/c0gc00819b

Yuan, G., Qi, C., Wu, W., & Jiang, H. (2017). Curr. Opin. *Green and Sustainable Chemistry*, *3*, 22–27. doi:10.1016/j.cogsc.2016.11.006

Yuan, Y., Chen, C., Zeng, C., Mousavi, B., Chaemchuen, S., & Verpoort, F. (2017). Carboxylation of Terminal Alkynes with Carbon Dioxide Catalyzed by an *in-situ* Ag_2O/ N-Heterocyclic Carbene Precursor System. *ChemCatChem*, *9*(5), 882–887. doi:10.1002/cctc.201601379

Zapp, P., Schreiber, J., Marx, A., Haines, M., Hake, J. F., & Gale, J. (2012). Overall environmental impacts of CCS technologies—A life cycle approach. *International Journal of Greenhouse Gas Control*, *8*, 12–21. doi:10.1016/j.ijggc.2012.01.014

Zhang, W., Wang, S., Zhao, Y., & Ma, X. (2018). Hydrogenation of CO_2 to formic acid catalyzed by heterogeneous Ru-PPh_3/Al_2O_3 catalysts. *Fuel Processing Technology*, *178*, 98–103. doi:10.1016/j.fuproc.2018.05.024

Zhang, Z., Fan, F., Xing, H., Yang, Q., Bao, Z., & Ren, Q. (2017). Efficient Synthesis of Cyclic Carbonates from Atmospheric CO_2 Using a Positive Charge Delocalized Ionic Liquid Catalyst. *ACS Sustainable Chemistry & Engineering*, *5*(4), 2841–2846. doi:10.1021/acssuschemeng.7b00513

Chapter 3
Review on Fisher-Tropsch Synthesis Method in Liquid Fuel Production:
FTS Method in Fuel Production

Rahman Shakibaei Lalehloo
Young Researchers and Elite Club IAU, UAE

Gomaa A. M. Ali
https://orcid.org/0000-0002-7152-531X
Al-Azhar University, Egypt

Hamidreza Sadegh
https://orcid.org/0000-0002-0163-9370
West Pomeranian University of Technology Szczecin, Poland

ABSTRACT

This chapter describes the Fisher-Tropsch Synthesis (FTS) method. Although it has been already applied at industrial scale for a century, the FTS has gained renewed interests as it is a key step for converting alternative feedstocks, including biomass to transportation fuels. It is the means by which synthesis gas containing hydrogen and carbon monoxide is converted to hydrocarbon products. The chapter explores that interest in FTS technology is increasing rapidly. In addition, the FTS process and products upgrading are discussed.

DOI: 10.4018/978-1-5225-8033-1.ch003

INTRODUCTION

During the past twenty years, tremendous investments from the major players in the energy market have been directed towards Gas-to-Liquid (GTL) technology. This has resulted from a significant increase in global demand for crude oil. The advantages of GTL technology in energy production are that it furnishes a broad range of environmentally clean fuels, additives and value-added chemicals that can be obtained from abundant natural resources such as natural gas, coal, or biomass (Hall, 2005; Keshav & Basu, 2007). The primary commercial technology of GTL for the production of liquid hydrocarbons and value-added chemicals involve the Fisher-Tropsch Synthesis (FTS) process (Bao *et al.*, 2010). Among the various options for the conversion of gas to liquids, FTS is a proven process for the production of linear hydrocarbons in the diesel range, from synthesis gas (produced via the reforming or partial oxidation of natural gas) (Demirbas, 2007).

The FTS looks back to a lively history of about seventy years (Fischer & Tropsch, 1923; Rofer-DePoorter, 1981). Fischer himself was well aware of the importance of the discovery, giving access to industrial organic chemistry on the basis of simple inorganic molecules. The FTS represents technology from the 1920s that has continuously been revived to provide synthetic hydrocarbon fuels and chemicals from initially coal, later natural gas, and nowadays also biomass (Fischer & Tropsch, 1923). FTS stands for the reactions of synthesis gas to predominantly straight-chain hydrocarbons, which can be kinds of paraffin from CH_4 to waxes (C_nH_{2n+2} with n from 1 to over 100), olefins from ethylene to much longer molecules (C_nH_{2n}, with n_2), and to a lesser extent oxygenated products such as alcohols. It produces as main byproducts water and/or carbon dioxide, that is, due to the water-gas shift reaction. Being a highly exothermic reaction, it generates large amounts of heat. The process is represented by the simplified reaction equations:

$$FTS: \quad CO + 2H_2 \rightarrow -CH_2 - + H_2O \quad \Delta H = -164\,kjmol^{-1}$$

$$WGS: \quad CO + H_2O \leftrightarrow H_2 + CO_2 \qquad \Delta H = -42\,kjmol^{-1}$$

Reaction represents a polymerization, implying that the product will be a mixture of hydrocarbons with distribution in molecular weights. Selectivity and control thereof are therefore of key importance in FTS technology (Rofer-DePoorter, 1981). FTS represents a subject of intensive research both in industry and in academia (Dry & Hoogendoorn, 1981).

As the FTS process is a catalyst process in which liquid hydrocarbons are produced from a mixture of gases including H_2 and CO. The catalysts applied in the method are heterogeneous gas-solid catalytic systems (Davis & Occelli, 2006). The metals are the most common mediators used as catalysts in FTS from which iron and cobalt are mostly used due to reasons such they are easily availability, and are cheaper (Ali *et al.*, 2013a; Fouad *et al.*, 2012; Fouad *et al.*, 2011; Iglesia, 1997; Steynberg, 2004). As iron catalyst is cheaper than cobalt ones, researchers have promoted this catalyst which makes iron catalyst superior to cobalt ones (Pour *et al.*, 2010; Sadegh *et al.*, 2014). Increasing effect of some of the catalytic enhancers such as manganese, copper, zeolite and zirconium have been investigated on selectivity and the range of conversion to products (Bae *et al.*, 2009; Ding *et al.*, 2011; Pour *et al.*, 2008).

Zamany et al., reported the effect of temperature on the distribution of the product in a catalyst is evaluated which reveals that shift reaction between water and gas increases while increasing temperature, while the rate of conversion of carbon monoxide (CO) reportedly is 87.6% (Zamani *et al.*, 2016). In another research by Feyzy et al. shows that the FTS on the iron catalyst is increased light alfins and decreases methane and CO (Feyzi & Jafari, 2012). Also, regarding the effect of strontium enhancers on iron nanostructure catalyst in FTS, Zamani et al., reported that iron nanocatalyst containing enhancers, improves FTS and water-gas shift reaction, decreases methane selectivity up to 7.44%, increase CO conversion up to 63.3% and the probability of Alfa (the probability of chain development) increases up to 0.66.

A study on iron oxide nanoparticles as a catalyst in FTS revealed that tableting has no considerable effect on the properties of iron oxide nanoparticles and that the applied catalyst may be used as a nanocatalyst. The results of improving process shows, decreasing temperature, increases H_2/CO proportion, the spatial velocity of gas synthesizing entering the reactor and also the target function of heavy hydrocarbons production ($>C_5$) in proportion to CO usage (Murzin, 2009; Pour *et al.*, 2014).

In research on the distribution of products in FTS in iron base catalysts, Mohajeri et al. showed that water/gas reaction decreases with temperature rise and the rate of CO conversion increases to 89.1% (Mohajeri *et al.*, 2014). Also, in a research performed on FTS in iron nanoparticles based catalyst, it is depicted that water/gas shift reaction will increase in contrast to alfin/paraffin (Ali *et al.*, 2013b). In addition, a study on zirconium enhancers in copper/iron catalyst on the basis of FTS process performed by Zamani et al. it is shown that increasing enhancers will improve Gas/ Water shift reaction and FTS (Zamani *et al.*, 2016).

Different types of catalysts were investigated for FTS including mono, bimetallic (Fe–Co, Co–Mn) and ternary (Fe–Co–Mn) catalysts. Co-precipitated Fe–Co–Mn catalysts were applied for light olefins production (Arsalanfar *et al.*, 2014). Addition of MgO (10%) has shown the better catalytic performance. Bimetallic (Fe-Co/$CaCO_3$) catalyst for carbon nanotubes growth using factorial experimental design and it has shown a formation of tubular networks of graphitic carbon materials with few traces of by-products (Bankole *et al.*, 2018).

BACKGROUND

Due to the large volume of existing literature on FTS, the diversity of the subject and the interest, it seemed indicated to write a historical sketch about the process, putting also an emphasis on present trends and future options. The twentieth-century coal-to-petroleum, or synthetic fuel, the industry evolved in following stages (Dry, 2002; Fischer & Tropsch, 1923; Schulz, 1999; Storch, 1951): (1) invention and early development of the Bergius coal liquefaction (hydrogenation) and FTS from 1910 to 1926; (2) Germany's industrialization of the Bergius and FTS processes from 1927 to 1945; and (3) global transfer of the German technology to Britain, France, Japan, Canada, the United States, South Africa, and other nations from the 1930s to the 1990s; (4) in the mid-1990s two developments triggered another synthetic fuel revival in the United States.

In fact, a combination of four factors, therefore, has led industrialized nations at various times during the twentieth century to conclude that synthetic fuel could contribute to their growing liquid fuel requirements: (1) the shift from solid to liquid fuel as a major energy source, (2) the invention of the Bergius and FTS coal-to-petroleum conversion or synthetic fuel processes, (3) recognition that global petroleum reserves were finite and much less than global coal reserves and that petroleum's days as a plentiful energy source were limited, and (4) the desire for energy independence.

Metal oxides are used as catalyst supports such as Al_2O_3, TiO_2, SiO_2, MgO, $CaCO_3$ and ZrO_2 modified SiO_2 or Al_2O_3 (Arsalanfar *et al.*, 2014; Jacobs *et al.*, 2002). These supports are used to increase the dispersion of the clusters, improve attrition resistance, enhance sulfur tolerance, or electronically modify the active metal site. In addition, catalyst promoters, including noble metals and metal cations, are used to improve its performance in a chemical reaction by gain access to the active sites (Jacobs *et al.*, 2002). Since the noble metals are reduced at lower temperature than the catalyst, it reduces the reaction temperature. Catalyst deactivation is the loss over time of catalytic activity and/or selectivity. It might be caused by poisoning, fouling, thermal degradation, vapor compound formation accompanied by transport, vapor-solid and/or solid-solid reactions, or attrition/crushing (Bartholomew, 2001).

FISHER-TROPSCH SYNTHESIS PROCESS

Performance of the FTS depends strongly on reaction parameters. The overall process from original carbon source for the syngas to the FTS product is named after the feedstock employed, hence the terminology coal-to-liquids (CTL), GTL and 'biomass-to-liquids' (BTL), collectively known as XTL ('anything'-to-liquids). In all instances, the carbon source is first converted to synthesis gas (or 'syngas' for short), which is a mixture of CO and H_2. Solid feedstocks such as coal or biomass are gasified, usually noncatalytically, by partial oxidation with oxygen (supplying the heat for the endothermic gasification reactions) and reaction with steam (which acts as a gasification agent, hydrogen source, and coolant) (Bahome *et al.*, 2005; Steynberg, 2004).

When the starting material is natural gas, it can also be adiabatically reformed in the presence of oxygen and steam. There are different embodiments of this approach, such as autothermal reforming, noncatalytic partial oxidation, and catalytic partial oxidation, while, the chemistry of all is the same and very similar to that of coal gasification. Alternatively, heat can be supplied externally; in which case the gas is reformed only with steam and/or CO_2, but no oxygen is added (Dry, 1990; Steynberg *et al.*, 2004).

Typical reforming catalysts are based on nickel as the active metal. The second step in the XTL process is to catalytically convert the syngas to a range of hydrocarbons via the FTS, which mainly yields linear alkanes and 1-alkenes. The third and last step is usually the workup of the hydrocarbons to final products, which are typically fuels, but optionally also chemicals. A popular application at present is to target the production of long-chain waxes in the FTS, followed by hydrocracking to middle distillate range components, such as diesel (C_9–C_{22}) and jet fuel (C_9–C_{15}). Hydrocracking catalysts are bifunctional in nature, with either a noble metal (e.g., Pt) or sulfided base metals (e.g., Ni/W or Co/Mo) as the hydrogenation function on a catalytically active acidic support, such as a silica-alumina (Bahome *et al.*, 2005; Dry, 1990).

FISHER-TROPSCH SYNTHESIS PRODUCTS

At the chemistry level, the FTS is both a CO hydrogenation reaction and a polymerization reaction. The former is reflected by the fact that the C–O bond must be broken and new C–H bonds formed. Additionally, C–C bonds must be formed in order to effect hydrocarbon chain growth. Since the product carbon number distribution approximately follows a statistical function called the Anderson–Schulz–Flory relationship, it is widely accepted that chain growth occurs one carbon atom at a time via a polymerization mechanism. Proposals for the monomer of chain growth, which is produced in situ, have included adsorbed CO, an enol species and a CH_x species and will be discussed further in the section on mechanism and kinetics (Steynberg, 2004; Tavakoli *et al.*, 2008; van Santen *et al.*, 2011).

The competition between chain growth (yielding a surface intermediate with one higher carbon number) and chain termination (yielding a desorbed final product) is determined by the probability for growth, called the a-value. A higher a-value will result in longer hydrocarbons and thus a heavier product spectrum (Figure 1). If a is independent of carbon number, the scheme presented in Figure 2 shows the total amount of carbon contained in products with n carbon atoms (namely C_n) can be formulated on a relative basis.

$$C_n = n\left(1-\alpha\right)\alpha^{n-1}$$

The total amount of carbon in the product spectrum then forms a convergent infinite sum with an analytical solution:

$$\sum_{1}^{\infty} C_n = \sum_{1}^{\infty} n\left(1-\alpha\right)\alpha^{n-1} = \frac{1}{1-\alpha}$$

Figure 1. Hydrocarbon products spectrum that is produced during FTS for varying chain growth probability (α). (HTFT: High-temperature Fischer–Tropsch technology corresponds approximately to 0.70<α<0.75, and LTFT: Low-temperature Fischer–Tropsch to about 0.85<α<0.95).

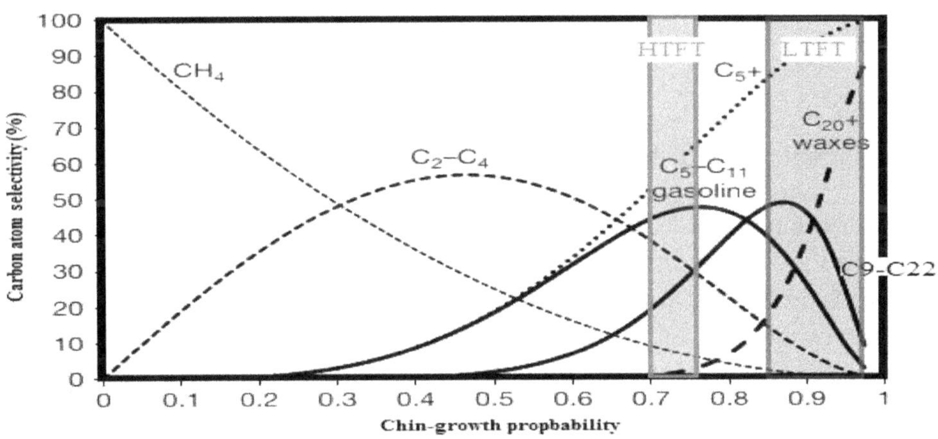

Figure 2. Carbon chain growth and termination scheme for the derivation of the Anderson–Schulz–Flory equation, with α chain growth probability factor, C_n the final products with n carbon atoms, and C_n the intermediates with n carbon atoms.*

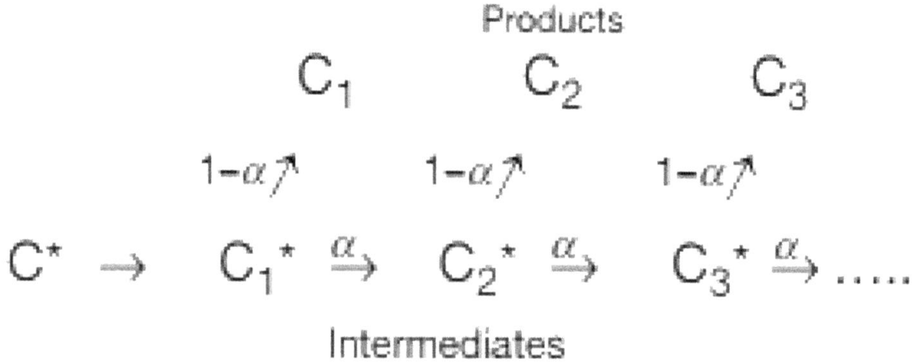

This means that the selectivity toward products with n carbon atoms on a carbon atom basis, namely S_n, can be expressed as follows:

$$S_n = \frac{C_n}{\sum_1^{\infty} C_n} = n\left(1-\alpha\right)^2 \alpha^{n-1}$$

After converting this equation to the logarithmic domain and rearranging, it is found that:

$$\ln\left(\frac{S_n}{n}\right) = n\ln\alpha + \ln\frac{\left(1-\alpha\right)^2}{\alpha}$$

As a result, a plot of $\ln\left(\frac{S_n}{n}\right)$ versus carbon number (n) gives rise to a straight line with a slope equal to ln(α) (Espinoza *et al.*, 1999). However, deviations in the actual FTS product spectrum from the ideal Anderson–Schulz–Flory distribution are usually observed. These include higher methane and a lower C_2 selectivity than predicted by the equation. There is also an increase in the chain growth probability factor and a concomitant decrease in the olefin/paraffin ratio with hydrocarbon chain length.

In addition to linear alkanes and 1-alkenes, a variety of other products are also formed, including branched aliphatic compounds, alcohols, aldehydes, ketones, and acids even aromatics. This alludes to the complexity of the reaction and many unresolved issues remain regarding the reaction mechanism. A further complicating factor is that it is not always clear which of these compounds are primary FTS products and which are formed subsequently by secondary reactions (Schulz *et al.*, 1994).

FISHER-TROPSCH SYNTHESIS CATALYSTS

Metals are known to catalyze the FTS reaction mainly include iron, cobalt, ruthenium, and nickel (Botes & Govender, 2007). Ruthenium is a scarce and expensive metal, whereas nickel only forms methane at reaction temperatures sufficiently high to suppress nickel carbonyl formation (note that methanation is the reverse reaction of methane reforming, for which nickel-based catalysts are commonly used) (Dry, 1990). As a result, only iron- and cobalt-based FTS catalysts have found commercial application (Deshmukh *et al.*, 2011; Steynberg *et al.*, 1999). Iron generally produces more olefins and oxygenates than cobalt (i.e., a less hydrogenated product spectrum), which may be related to the lower hydrogenating ability of iron. While cobalt is active in the metallic state, iron catalysts change under FTS conditions to a complex mixture of iron carbides and oxides (Espinoza *et al.*, 1999).

Iron-Based Catalysts

The iron-catalyzed FTS process is, along with ammonia synthesis, one of the most studied systems in the field of heterogeneous catalysis. The reason for this is possibly the fact that the application of the process is so versatile. Not only can iron FTS produce a light hydrocarbon product stream ideal for the fuel and chemical industry, it can also produce heavier hydrocarbons ($>C_{35}$) suited for the waxes market (Li *et al.*, 2002). Iron is also a cheap raw material when compared to its cobalt counterpart and it has been commercially applied (Aboelazm *et al.*, 2018; Sadegh *et al.*, 2016).

Iron is believed to be more tolerant of poisons, for example, sulfur in synthesis gas than cobalt. It is also known to be responsive to selectivity manipulation by the addition of promoters and a variation of typical process parameters, for example, temperature, pressure, and H_2/CO ratio. As already mentioned, the iron FTS process can be manipulated to produce a range of carbon number distributions with the final product stream depending mainly on the temperature applied during FTS. At lower temperatures, for example, 220–250 °C the chain growth probability (α) of the catalyst is approximately 0.94 indicating that the bulk of the products will consist of hydrocarbons longer than C_{21} (Dry, 2002; Steynberg, 2004).

In the case of higher temperatures, for example, 320–350 °C, the chain growth probability decreases to 0.7 and even lower with the main products being light hydrocarbons utilized for the production of transportation fuel and chemical feedstocks. Although there are many advantages regarding noncatalyze FTS, the transformations of the iron catalyst during activation and FTS are rather complex and still not fully understood. During catalyst preparation, iron oxides (e.g., hematite (Fe_2O_3) and magnetite (Fe_3O_4)) are produced and these are transformed to either a-Fe or iron carbides during activation depending on the conditions (Sadegh *et al.*, 2018; Sadegh *et al.*, 2016; Zare *et al.*, 2016).

Cobalt-Based Catalysts

Cobalt as an FTS catalyst was already claimed by Fischer and Tropsch in their original patent of 1925 (Fischer & Tropsch, 1923; Fischer *et al.*, 1925). The commercialization of the FTS by Germany and Japan in the period 1938–1945 relied fully on cobalt catalysts. Only after World War II did the focus shift to the use of iron catalysts for FTS applications. Since the oil crises of the 1970s, the interest in cobalt-based FTS catalysts reappeared, which has resulted in numerous scientific papers and patents. Almost all focused-on wax production, followed by hydrotreating to produce diesel. This is also the application that will receive the most attention in this section. Cobalt FTS catalysts are exclusively utilized in low-temperature synthesis or LTFT and are applied in fixed-bed, slurry-phase, and micro-channel FTS reactors (Iglesia *et al.*, 1993; Zamani *et al.*, 2016).

Catalyst design needs to be adjusted to the targeted reactor as well as the applied FTS conditions important for catalyst design are the composition, method of preparation, activity and selectivity behavior, deactivation and regeneration, and mechanical integrity. Exciting academic and industrial research in the last two decades has increased the fundamental knowledge of cobalt FTS catalysts substantially on topics like the nature of the active site, impact of crystallite size on activity and selectivity, and deactivation mechanisms, owing to the application of surface science techniques, model catalysts, in situ analyses at relevant industrial conditions, and molecular modeling (Tsakoumis *et al.*, 2010).

CONCLUSION AND FUTURE PERSPECTIVES

Fisher-Tropsch synthesis represents proven technology, which has secured its position in modern energy technology. Originally it is used to convert coal into liquid fuels, nowadays the emphasis is on monetizing natural gas, by converting it to diesel fuel, waxes, and naphtha. The technology has much potential for wider use, for example, in emerging economies, or at a smaller scale in the utilization of biomass. Interesting applications of Fisher-Tropsch synthesis have been proposed for conversion of remote natural gas at off-shore oil production locations. In addition, can be important tools in strategies aimed at the reduction of CO_2 emissions. Fisher-Tropsch synthesis continues to pose challenges from an industrial perspective. From more academic perspective, understanding the mechanism of the Fisher-Tropsch synthesis is a challenge. It is more and more realized that mechanisms may differ with conditions and catalysts. It is highly unlikely that one unique mechanism can account for all different forms of Fisher-Tropsch synthesis. Therefore, Fisher-Tropsch synthesis is expected to remain an inspiring source of industrial and academic research.

REFERENCES

Aboelazm, E. A., Ali, G. A. M., & Chong, K. F. (2018). Cobalt oxide supercapacitor electrode recovered from spent lithium-ion battery. *Chemistry of Advanced Materials*, *3*(4), 67–74.

Ali, G. A. M., Fouad, O. A., & Makhlouf, S. A. (2013a). Structural, optical and electrical properties of sol–gel prepared mesoporous Co3O4/SiO2 nanocomposites. *Journal of Alloys and Compounds*, *579*, 606–611. doi:10.1016/j.jallcom.2013.07.095

Ali, N. P., Riyahi, F., Housaindokht, M. R., Irani, M., Shahri, S. M. K., & Hatami, B. (2013b). Hydrocarbon production rates in Fischer-Tropsch synthesis over a Fe/Cu/La/Si catalyst. *Journal of Energy Chemistry*, *22*(1), 119–129. doi:10.1016/S2095-4956(13)60016-9

Arsalanfar, M., Mirzaei, A. A., Bozorgzadeh, H. R., Samimi, A., & Ghobadi, R. (2014). Effect of support and promoter on the catalytic performance and structural properties of the Fe–Co–Mn catalysts for Fischer–Tropsch synthesis. *Journal of Industrial and Engineering Chemistry*, *20*(4), 1313–1323. doi:10.1016/j.jiec.2013.07.011

Bae, J. W., Park, S.-J., Kang, S.-H., Lee, Y.-J., Jun, K.-W., & Rhee, Y.-W. (2009). Effect of Cu content on the bifunctional Fischer–Tropsch Fe–Cu–K/ZSM5 catalyst. *Journal of Industrial and Engineering Chemistry*, *15*(6), 798–802. doi:10.1016/j.jiec.2009.09.002

Bahome, M. C., Jewell, L. L., Hildebrandt, D., Glasser, D., & Coville, N. J. (2005). Fischer–Tropsch synthesis over iron catalysts supported on carbon nanotubes. *Applied Catalysis A, General*, *287*(1), 60–67. doi:10.1016/j.apcata.2005.03.029

Bankole, M. T., Mohammed, I. A., Abdulkareem, A. S., Tijani, J. O., Ochigbo, S. S., Abubakre, O. K., & Afolabi, A. S. (2018). Optimization of supported bimetallic (Fe-Co/CaCO3) catalyst synthesis parameters for carbon nanotubes growth using factorial experimental design. *Journal of Alloys and Compounds*, *749*, 85–102. doi:10.1016/j.jallcom.2018.03.150

Bao, B., El-Halwagi, M. M., & Elbashir, N. O. (2010). Simulation, integration, and economic analysis of gas-to-liquid processes. *Fuel Processing Technology*, *91*(7), 703–713. doi:10.1016/j.fuproc.2010.02.001

Bartholomew, C. H. (2001). Mechanisms of catalyst deactivation. *Applied Catalysis A, General*, *212*(1), 17–60. doi:10.1016/S0926-860X(00)00843-7

Botes, F. G., & Govender, N. S. (2007). Secondary reactions of ethylene as studied in a laboratory-scale recycle slurry reactor. *Energy & Fuels*, *21*(6), 3095–3101. doi:10.1021/ef7003464

Davis, B. H., & Occelli, M. L. (2006). *Fischer-Tropsch synthesis, catalysts and catalysis* (Vol. 163). Elsevier.

Demirbas, A. (2007). Converting biomass derived synthetic gas to fuels via Fisher-Tropsch synthesis. *Energy Sources. Part A, Recovery, Utilization, and Environmental Effects*, *29*(16), 1507–1512. doi:10.1080/15567030601003676

Deshmukh, S. R., Tonkovich, A. L. Y., McDaniel, J. S., Schrader, L. D., Burton, C. D., Jarosch, K. T., ... LeViness, S. (2011). Enabling cellulosic diesel with microchannel technology. *Biofuels*, *2*(3), 315–324. doi:10.4155/bfs.11.17

Ding, M., Yang, Y., Wu, B., Wang, T., Xiang, H., & Li, Y. (2011). Effect of reducing agents on microstructure and catalytic performance of precipitated iron-manganese catalyst for Fischer–Tropsch synthesis. *Fuel Processing Technology*, *92*(12), 2353–2359. doi:10.1016/j.fuproc.2011.08.011

Dry, M., & Hoogendoorn, J. (1981). Technology of the Fischer-Tropsch process. *Catalysis Reviews. Science and Engineering*, *23*(1-2), 265–278. doi:10.1080/03602458108068078

Dry, M. E. (1990). Fischer-Tropsch synthesis over iron catalysts. *Catalysis Letters*, *7*(1-4), 241–251. doi:10.1007/BF00764506

Dry, M. E. (2002). The Fischer–Tropsch process: 1950–2000. *Catalysis Today*, *71*(3-4), 227–241. doi:10.1016/S0920-5861(01)00453-9

Espinoza, R., Steynberg, A., Jager, B., & Vosloo, A. (1999). Low temperature Fischer–Tropsch synthesis from a Sasol perspective. *Applied Catalysis A, General*, *186*(1-2), 13–26. doi:10.1016/S0926-860X(99)00161-1

Feyzi, M., & Jafari, F. (2012). Study on iron-manganese catalysts for Fischer-Tropsch synthesis. *Journal of Fuel Chemistry and Technology*, *40*(5), 550–557. doi:10.1016/S1872-5813(12)60021-8

Fischer, F., & Tropsch, H. (1923). The preparation of synthetic oil mixtures (synthol) from carbon monoxide and hydrogen. *Brennstoff-Chem*, *4*, 276–285.

Fischer, F., Tropsch, H., & Dilthey, P. (1925). Reduction of carbon monoxide to methane in the presence of various metals. *Brennst. Chem*, *6*, 265–271.

Fouad, O. A., Ali, G. A. M., El-Erian, M. A. I., & Makhlouf, S. A. (2012). Humidity sensing properties of cobalt oxide/silica nanocomposites prepared via sol–gel and related routes. *Nano*, *07*(05), 1250038. doi:10.1142/S1793292012500385

Fouad, O. A., Makhlouf, S. A., Ali, G. A. M., & El-Sayed, A. Y. (2011). Cobalt/silica nanocomposite via thermal calcination-reduction of gel precursors. *Materials Chemistry and Physics*, *128*(1), 70–76. doi:10.1016/j.matchemphys.2011.02.072

Hall, K. R. (2005). A new gas to liquids (GTL) or gas to ethylene (GTE) technology. *Catalysis Today*, *106*(1-4), 243–246. doi:10.1016/j.cattod.2005.07.176

Iglesia, E. (1997). Design, synthesis, and use of cobalt-based Fischer-Tropsch synthesis catalysts. *Applied Catalysis A, General*, *161*(1-2), 59–78. doi:10.1016/S0926-860X(97)00186-5

Iglesia, E., Reyes, S. C., Madon, R. J., & Soled, S. L. (1993). Selectivity control and catalyst design in the Fischer-Tropsch synthesis: Sites, pellets, and reactors. *Advances in Catalysis*, *39*, 221–302. doi:10.1016/S0360-0564(08)60579-9

Jacobs, G., Das, T. K., Zhang, Y., Li, J., Racoillet, G., & Davis, B. H. (2002). Fischer–Tropsch synthesis: Support, loading, and promoter effects on the reducibility of cobalt catalysts. *Applied Catalysis A, General*, *233*(1), 263–281. doi:10.1016/S0926-860X(02)00195-3

Keshav, T. R., & Basu, S. (2007). Gas-to-liquid technologies: India's perspective. *Fuel Processing Technology*, *88*(5), 493–500. doi:10.1016/j.fuproc.2006.12.006

Li, S., Krishnamoorthy, S., Li, A., Meitzner, G. D., & Iglesia, E. (2002). Promoted iron-based catalysts for the Fischer–Tropsch synthesis: Design, synthesis, site densities, and catalytic properties. *Journal of Catalysis*, *206*(2), 202–217. doi:10.1006/jcat.2001.3506

Mohajeri, A., Zamani, Y., & Shirazi, L. (2014). Investigation of products distribution in Fischer-Tropsch synthesis by nano-sized iron-based catalyst. *Journal of Nanoanalysis*, *1*(2), 58–64.

Murzin, D. Y. (2009). Thermodynamic analysis of nanoparticle size effect on catalytic kinetics. *Chemical Engineering Science*, *64*(5), 1046–1052. doi:10.1016/j.ces.2008.10.066

Pour, A. N., Housaindokht, M. R., Irani, M., & Shahri, S. M. K. (2014). Size-dependent studies of Fischer–Tropsch synthesis on iron based catalyst: New kinetic model. *Fuel*, *116*, 787–793. doi:10.1016/j.fuel.2013.08.080

Pour, A. N., Housaindokht, M. R., Tayyari, S. F., Zarkesh, J., & Alaei, M. R. (2010). Deactivation studies of Fischer–Tropsch synthesis on nano-structured iron catalyst. *Journal of Molecular Catalysis A Chemical*, *330*(1-2), 112–120. doi:10.1016/j.molcata.2010.07.010

Pour, A. N., Zamani, Y., Tavasoli, A., Shahri, S. M. K., & Taheri, S. A. (2008). Study on products distribution of iron and iron–zeolite catalysts in Fischer–Tropsch synthesis. *Fuel*, *87*(10-11), 2004–2012. doi:10.1016/j.fuel.2007.10.014

Rofer-DePoorter, C. K. (1981). A comprehensive mechanism for the Fischer-Tropsch synthesis. *Chemical Reviews*, *81*(5), 447–474. doi:10.1021/cr00045a002

Sadegh, H., Ali, G. A. M., Makhlouf, A. S. H., Chong, K. F., Alharbi, N. S., Agarwal, S., & Gupta, V. K. (2018). MWCNTs-Fe_3O_4 nanocomposite for Hg (II) high adsorption efficiency. *Journal of Molecular Liquids*, *258*, 345–353. doi:10.1016/j.molliq.2018.03.012

Sadegh, H., Helmi, H., Hamdy, A. S., Masjedi, A., Dastjerdi, M. J. H.& Shahryari-ghoshekandi, R. (2016). A developed simple, facile, economic and fast technique for preparation of high quality Fe_3O_4 magnetic nanoparticles. *Chemistry of Advanced Materials*, *1*(1), 27-32.

Sadegh, H., Shahryari-ghoshekandi, R., & Kazemi, M. (2014). Study in synthesis and characterization of carbon nanotubes decorated by magnetic iron oxide nanoparticles. *International Nano Letters*, *4*(4), 129–135. doi:10.100740089-014-0128-1

Schulz, H., vein Steen, E., & Claeys, M. (1994). Selectivity and mechanism of Fischer-Tropsch synthesis with iron and cobalt catalysts. In Studies in Surface Science and Catalysis (Vol. 81, pp. 455-460). Elsevier.

Schulz, H. (1999). Short history and present trends of Fischer–Tropsch synthesis. *Applied Catalysis A, General, 186*(1-2), 3–12. doi:10.1016/S0926-860X(99)00160-X

Steynberg, A., Dry, M., Davis, B., & Breman, B. (2004). Fischer-tropsch reactors. *Studies in Surface Science and Catalysis, 152*, 64–195. doi:10.1016/S0167-2991(04)80459-2

Steynberg, A., Espinoza, R., Jager, B., & Vosloo, A. (1999). High temperature Fischer–Tropsch synthesis in commercial practice. *Applied Catalysis A, General, 186*(1-2), 41–54. doi:10.1016/S0926-860X(99)00163-5

Steynberg, A. P. (2004). Introduction to Fischer-Tropsch Technology. In A. Steynberg & M. Dry (Eds.), *Studies in Surface Science and Catalysis* (Vol. 152, pp. 1–63). Elsevier.

Storch, H. H. (1951). *The Fischer-Tropsch and related syntheses: including a summary of theoretical and applied contact catalysis*. Wiley.

Tavakoli, A., Sohrabi, M., & Kargari, A. (2008). Application of Anderson–Schulz–Flory (ASF) equation in the product distribution of slurry phase FT synthesis with nanosized iron catalysts. *Chemical Engineering Journal, 136*(2-3), 358–363. doi:10.1016/j.cej.2007.04.017

Tsakoumis, N. E., Rønning, M., Borg, Ø., Rytter, E., & Holmen, A. (2010). Deactivation of cobalt based Fischer–Tropsch catalysts: A review. *Catalysis Today, 154*(3-4), 162–182. doi:10.1016/j.cattod.2010.02.077

van Santen, R. A., Ghouri, M. M., Shetty, S., & Hensen, E. M. (2011). Structure sensitivity of the Fischer–Tropsch reaction; molecular kinetics simulations. *Catalysis Science & Technology, 1*(6), 891–911. doi:10.1039/c1cy00118c

Zamani, Y., Rahimizadeh, M., & Seyedi, S. M. (2016). The Effect of Temperature on Product Distribution over Fe-Cu-K Catalyst in Fischer-Tropsch Synthesis. *Journal of Petroleum Science and Technology, 6*(1), 46–52.

Zare, K., Sadegh, H., Shahryari-ghoshekandi, R., Asif, M., Tyagi, I., Agarwal, S., & Gupta, V. K. (2016). Equilibrium and kinetic study of ammonium ion adsorption by Fe_3O_4 nanoparticles from aqueous solutions. *Journal of Molecular Liquids, 213*, 345–350. doi:10.1016/j.molliq.2015.08.045

Chapter 4
Advanced Catalysis and Processes to Convert Heavy Residues Into Fuels and High Value Chemicals

Feras Ahmed Alshehri
King Abdulaziz City for Science and Technology, Saudi Arabia

Abdulaziz F. Al-Ghashem
King Abdulaziz City for Science and Technology, Saudi Arabia

Saeed M. Al-Shihri
King Abdulaziz City for Science and Technology, Saudi Arabia

Ali A. Algarni
King Abdulaziz City for Science and Technology, Saudi Arabia

Mohammed C. Al-Kinany
King Abdulaziz City for Science and Technology, Saudi Arabia

Sami D. Alzahrani
King Abdulaziz City for Science and Technology, Saudi Arabia

Bandar M. Al-Hudaib
King Abdulaziz City for Science and Technology, Saudi Arabia

Peter P. Edwards
University of Oxford, UK

Tiancun Xiao
University of Oxford, UK

ABSTRACT

The petroleum refining process begins with distillation, first at atmospheric pressure and after at reduced pressure. The volatile fractions, in both cases, have greater economic value, and the distillation residue-produced atmospheric residue and vacuum residue represent a significant portion of a barrel of crude. The need to convert bottom of the barrel into cleaner and more valuable olefins and liquid products is continuously increasing. Thus, residue must be converted into more valuable products, and further processes can be employed for upgrading residue. Examples are delayed coking, visco-reduction, and fluidized catalytic cracking.

DOI: 10.4018/978-1-5225-8033-1.ch004

On the other hand, the optimization of refining facilities to deal with such feeds brings economic competitiveness since these oils have low prices in the international market. Studies on processes and catalytic cracking are quite important under this aspect. The conversion of heavy petroleum fraction into valuable liquid products and high value chemicals has been important objectives for upgrading heavy petroleum oils.

INTRODUCTION

The petroleum refining process begins with distillation, first at atmospheric pressure and after at reduced pressure. The volatile fractions, in both cases, have greater economical value and the distillation residues produced – atmospheric residue and vacuum residue represent a significant portion of a barrel of crude. The need to convert bottom of the barrel into cleaner and more valuable olefins and liquid products is continuously increasing. Thus, residue must be converted into more valuable products and further processes can be employed for upgrading residue. Examples are delayed coking, visco-reduction and fluidized catalytic cracking. On the other hand, the optimization of refining facilities to deal with such feeds brings economical competitiveness, since these oils have low prices in the international market. Studies on processes and catalytic cracking are quite important under this aspect. The conversion of heavy petroleum fraction into valuable liquid products has been one of the important objectives for upgrading heavy petroleum oils (Corma, 2017). At present this conversion is mainly achieved by thermal cracking, catalytic cracking and hydro-cracking.

ATMOSPHERIC DISTILLATION RESIDUE

Crude petroleum as it is produced from the field is a relatively low-value material since, in its native state, it is rarely usable directly. Due to the presence of contaminants such as sulfur, heavy metals and asphaltenes (Sahu, 2015) .However, it can be refined and further processed into any number of products whose value is many times that of the original oil.

The first step in any petroleum refinery is the separation of the crude into various fractions by the process of distillation. These fractions may be products in their own right or may be feed-stocks for other refining or processing units.

In most refineries, this process is carried out in two stages; the oil is first heated to the maximum temperature allowable for the crud being processed and for the operation being practiced and then fed to a fractionating tower which operates at slightly above atmospheric pressure. This tower is usually called the atmospheric tower or atmospheric distillation unit. It yields several distillation products and bottoms product or atmospheric residue which is the residual liquid material which could not be vaporized under the conditions of temperature and pressure existing in the atmospheric tower.

Most crude oils are distilled in the atmospheric crude oil tower to cut the atmospheric residue at a 650°F (343.33°C) up to a 680°F (360°C) cut point. The atmospheric residue has a boiling point higher than 375 °C, consists of large molecules with more than 30 carbon atoms and makes up about 50% of the crude.

This bottoms liquid is then reheated to the maximum allowable temperature-usually higher than the maximum allowed for the feed to the atmospheric tower-and fed to a fractionating tower that operates at sub-atmospheric pressure. This tower is usually called the vacuum tower or vacuum distillation unit.

VACUUM DISTILLATION RESIDUE

Vacuum distillates and vacuum residue can have several uses depending upon the type of crude oil feed, the type of refinery and its downstream capabilities and, most important of all, the anticipated sales of products. Thus, the general function if vacuum tower is to remove the maximum possible amount of distillate from the charge stock consistent with meeting products specifications on the residuum as well as the distillates.

Vacuum residue of Ural crude contains mainly hydrocarbons with boiling points above 520 °C. Moreover, it contains about 3% sulphur and 0.5% nitrogen in the form of hetero-organic compounds, and minor amounts of organo-metallic compounds, predominantly nickel and vanadium compounds.

The methods of vacuum tower operation are pitch and asphalt operations. One other method of vacuum tower operation is the production of large volumes of bunker fuels. The resulting residual fuels are slightly lighter than in normal operations but are still quite adequate from the end use standpoint.

Bunker Oil

Bunker fuel is a type of liquid fuel which is fractionally distilled from crude oil. It is also known as fuel oil, and a number of different classifications around the world are used to describe fuel oil; these classifications break bunker fuel into different categories based on its chemical composition, intended purpose, and boiling temperature. In comparison with other petroleum products, bunker fuel is extremely crude and highly polluting. Bunker fuel is technically any type of fuel oil used aboard ships. It gets its name from the containers on ships and in ports that it is stored in, called bunker.

Fuel oil is classified into six classes, numbered 1 through 6, according to its boiling point, composition and purpose. The boiling point, ranging from 175 to 600 °C, and carbon chain length, 9 to 70 atoms, of the fuel increases with fuel oil number. Viscosity also increases with number, and the heaviest oil has to be heated to get it to flow.

Bunker A is No. 2 fuel oil, bunker B is No. 4 or No. 5, and bunker C is No. 6. Bunker fuel is often used as a synonym for No.6. No.5 fuel oil is also called navy special fuel oil or just navy special, No. 6 or 5 may also be known as furnace fuel oil (FFO), and its high viscosity requires it to be heated, usually by a reticulated low-pressure steam system, before it can be pumped from a bunker tank. The chain length of No. 6 is 20-70 and the chain length of No. 5 is 12-70.

CATALYTIC PYROLYSIS

Catalytic pyrolysis, a promising technology for the production of light olefins, is usually conducted at high temperature over special catalysts. Compared with conventional steam pyrolysis, catalytic pyrolysis cannot only reduce reaction temperature and energy cost, but also allow one to flexibly adjust product distribution. Catalytic pyrolysis can also produce light olefins from a wide range of lower quality feedstocks, such as heavy oils. Studies of catalytic pyrolysis on heavy oil have been carried out since the 1960s, but these studies did not attract great interest until the 1980s. Up to now, many technologies of heavy oil catalytic pyrolysis have been studied and developed, together with pyrolyzing catalysts .It is reported that the reaction mechanisms of catalytic pyrolysis vary with catalysts and technologies. Experimental research shows that both product yields and the effects of operating conditions on product distribution vary greatly with both catalyst properties and feed properties, indicating that the processes of hydrocarbon catalytic pyrolysis are very complicated.

Most of the cracking catalysts of commercial interest are dual functional in nature, consisting of an acidic support. The reactions catalyzed by the individual components are quite different. In specific catalysts, the relative strengths of the two components can be varied. The reactions occurring and the products formed depend critically upon the balance between these two components.

The acidic function of the catalyst is supplied by the support. The acidic support consists of (a) amorphous oxides (e.g. silica alumina) (b) a crystalline zeolites (mostly modified Y zeolites) plus binder (e.g. alumina) or (c) a mixture of crystalline zeolites and amorphous oxides. Moreover, among the supports are silica-alumina, silica-zirconia, silica-magnesia, alumina-boria, silica-titania, acid-treated clays, acidic metal phosphates, alumina, and other such solid acids.

Historically, refinery heavy oils have been disposed of in one of three ways: immediate use, without further processing, as fuel for power production, industrial processes, and marine bunkers; conversion to higher demand products through various technologies including deep conversion by two main routes, carbon rejection (e.g. deasphalting, pyrolysis, resid cat cracking, visbreaking) and hydrogen addition (i.e. hydro treating); and gasification to produce power and/or hydrogen. Several new processes that will upgrade the heavy feedstocks to olefinic products have been recently developed. Heavy oils are defined as atmospheric gas oil (AGO), atmospheric residue, vacuum gas oil (VGO), and vacuum residue. In the past the gas oil has been a valuable source of feedstock for fluid catalytic cracking (FCC) or hydrocracking units to produce olefins. This research project will concentrates on developing catalysts for bunker oil, atmospheric, and vacuum residues catalytic cracking to produce olefins.

FLUID CATALYTIC CRACKING

General

Many important processes exist within the oil refining industry, none more so perhaps, than the FCC process. FCC is considered by many to be the central technology in the oil refineries arsenal of processes. It is used primarily to upgrade lower quality (heavy) gas oils and resid fuels to gasoline, diesel and it also produces a degree of light gases such as (C_3 - C_4) (Chen, 2006).

Over the last 95 years FCC has been revolutionised from the process developed in 1915 which employed an aluminium chloride based catalyst. Highlights of these developments are shown in Figure 1 (Harding, *et al*, 2001). Arguably some of the most significant discoveries have come with the advent and subsequent maturity of zeolite science. For example, in the 1960's implementation of faujasite Y was shown to increase the selectivity of gasoline to such an extent that approximately 2 US dollars could be added to the value of one barrel of processed crude oil (Gambicki, 2000). In the 1980's a relatively new zeolitic system discovered by Mobil in 1965 and known as ZSM-5 would prove useful in increasing olefin yields (C_3 - C_4) from the FCC process, but perhaps most important was the notable increase in gasoline

Figure 1. Time line detailing important dates in FCC development (adapted from Harding et al, 2001)

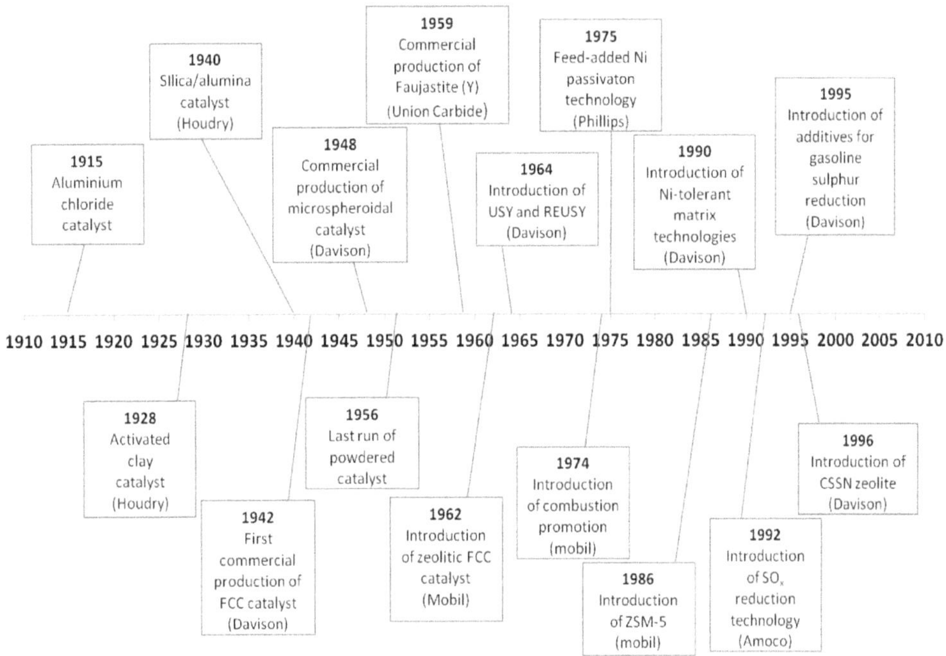

quality (octane number) (Degnan, *et al*, 2000). In 1990 Davison developed new matrix based technologies (alumina), which allowed through a higher tolerance to transition metal contaminants (Nickel and Vanadium) more flexibility in the FCC process and a subsequent ability to process heavy feed-stocks (Harding *et al*, 2001).

Even though the FCC process is relatively mature, improvements and optimisation of the catalysts used continues. There are two key driving points for this development; complex environmental legislations and the need to process heavier feed-stocks containing a larger degree of contaminants (Ni, V, N, S, O etc). Three areas specific to these driving points will be examined (environmental, residual oil processing and product and selectivity control) although the author acknowledges that many other additional areas exist, these however will not be covered within this review.

Environmental Issues Associated with Existing Technology

Arguably the major contributor to this need for new catalytic systems arises from the complex structure of government borne environmental legislations, which continually evolve to lower emissions from various aspects of not only the process but also the product. These regulations apply to air and water quality emission standards and cleaner fuel initiatives. Air quality emission standards require a significant decrease in airborne particulate emissions as well as carbon monoxide, NO_x and SO_x. The implemented cleaner fuel initiatives are designed to reduce the amount of aromatics, olefins and sulphur in transportation fuels (Cheng *et al*, 1998).

Particulates

During the normal operational lifetime of an average FCC unit, approximately 1% of its catalyst load may be replaced, due to the breakdown of the formed catalyst particles under process conditions (attrition). The physical properties of the catalyst particle are primarily designed for optimized fluidisation and low attrition. However, a compromise is reached between the mechanical strength of the particle and the need to limit the erosion of the FCC unit (Boerefijn *et al*, 1998).

Catalyst manufacturers have developed methods designed to limit lose due to attrition. A Trade-off however is reached between the hardness of the particle, which is a function of the ratio of binder to catalyst and the catalyst selectivity. As a general rule of thumb, increasing the binder ratio increases the attrition resistance, but may invariably decrease the selectivity through the formation of coke and gaseous products (Harding, *et al*, 2001, Scheffer, *et al.* 1998)

Sulphur Oxides SO_x

The oxides of sulphur are a major contributor to atmosphere borne pollution and are well documented as a precursor for acid rain. In the FCC process the major contributor to SO_x is the regeneration stage where the sulphur containing coke is removed hydrothermally. In the 1990's additives were developed to absorb the SO_x and carry it over the riser side of the FCC (Bhattacharyya, 1988)unit where it could be released as H_2S. Further to this the H_2S would be treated via the Claus process whereby Hydrogen and elemental sulphur could be produced (Harding, *et al*, 2001).

For a SO_x transfer catalyst to perform three primary objectives must be met:

1. The oxidation of SO_2. Under the conditions found in a FCC regenerator SO_2 is favoured over SO_3. Therefore, the first component in the catalyst is designed to promote the oxidation mentioned above (Bhattacharyya, *et al* 1988).
2. The chemisorption of SO_3, which in the regenerator must be strong. The second component used is a metal oxide component (magnesium alumina spinel) (Bhattacharyya, *et al* 1988).
3. The catalysed process whereby sulphuris released as H_2S in the riser. This final component is a metal with multiple valence states (for example vanadium) (Davey, 1999).

CO and NO_x

The reduction of NO_x is becoming significantly more important and from this view point a relationship is observed between CO and NO_x levels produced in the regenerator. It is known that carbon monoxide can be readily converted to carbon dioxide with the introduction of a combustion promoter based on noble metals such as platinum or palladium etc De (Carvalho*et al*, 2004). Experimentally it was found that the use of such a promoter resulted in an increase in NO_x due to the decrease in CO. This suggests that without the presence of the promoter, a direct mechanism exists whereby a redox reaction occurs where NO_x is reduced to N_2 and CO is oxidised to CO_2(Peters *et al*, 2002).

Although this mechanism does indeed exist this is not the dominant one under FCC regenerator conditions. Harding *et al.* (2001) suggest that nitrogen containing species found in coke are one of the last species burned off the catalyst due to the strong acid-base interaction between nitrogen and the acidic catalyst. It is proposed that through the burning a mixture of species are formed (N_2, NO_x and hydrocarbon based species such as HCN).

It has been observed that platinum based promoters catalyse the reaction of HCN to NO (Harding, *et al*, 2001; Mehlberg, *et al.*, 2011) whereas, non-platinum containing combustion results in HCN producing nitrogen (Harding, *et al*, 2001). Hence for the joint objective of reducing both CO and NO from the flue gas a non-platinum based promoter is required.

Controlling Sulphur Content in Gasoline

In most conventional metal supported catalystssulphur poisons the catalytic site causing almost irreversible deactivation. This is the same in the automotive industry, where sulphur in fuel and eventually exhaust gases will deactivate a cars catalytic converter, which results in increased amounts of other airborne pollutants. Hence, the world's governments are offering tax incentives to petroleum industries in an attempt to reduce the sulphur contents of their fuels (Marsh*et al*, 2000). For refiners to achieve the levels required there are numerous options available to them in the form of pre/post-treatment methods. These include gas oil Hydrotreating (Sambi, *et al.* 1982); FCC feed Hydrotreating (Ancheyta, *et al.* 2004), gasoline adsorption processes (Honeywell UOP,Gasoline Desulfurization Process) and gasoline hydrofinishing (Zhao, *et al.* 2007). However, each of these techniques has very specific requirements which may degrade the quality of the fuel (decrease in octane number), or require additional feed gases (such as hydrogen) which would increase the cost of manufacturing the fuel.

Strategies currently exist where sulphur can reduced with minimal expenditure, the basics such as zeolite modifications (unit cell size, matrix activity and aluminium content) result in very small decreases in sulphur content approximately 10% and will result in shifts in selectivity which may not be beneficial (Harding, *et al*, 2001). Whereas more concerted efforts such as, designing catalysts and additives specifically to reduce sulphur (Can, *et al*, 2007, Liu, *et al*, 2004, Sun, *et al*, 2006) have shown reductions up to 35% (Harding, *et al*, 2001). It would be expected that to meet legislation, that reactor technology will be designed to work in tandem with sulphur reduction catalysts.

Gasoline Range Olefins

Currently technology is pushing towards limiting the formation of higher order olefins (C_5>) and also, by increasing their rate of destruction (due to the formation of volatile organic compounds (VOC's) during combustion). One of the most effective methods is to use a high-hydrogen transfer catalyst (Cormier, *et al*, 1994). Which, when combined with new matrix technologies, is designed to limit the formation of these troublesome olefins and also allows for metallic species (Ni + V) to become trapped (Ishutenko, *et al*, 2017).

One of the most complicated challenges faced in this area is the selective retention of propylene, butylenes and octane number, in conjunction with maximised removal of gasoline range olefins. Some companies such as Grace Davison RFG catalyst technology have shown promise in this field with purpose designed catalysts showing a reasonable reduction in gasoline range olefins, while still retaining the yields of lighter olefins and octane number (Zhao *et al*, 2003).

Residual Oil Processing: Obstacles in Materials and Catalyst Development

As feed-stocks have become heavier, containing higher levels of polycyclic aromatics and contaminants such as metals and heteroatom species, new technologies including catalyst and process design is expected to improve. In this section the focus will be on a selection of the problems associated with metallic species, larger, more diffusion limiting molecules in the feed-stock.

Contaminant Metals

As indicated in Figure 1 technology was developed in the 1990's to improve the resistance of FCC catalysts to contaminated metals in the feed (Ni, Fe, Cu and V) (Amoli, *et al*, 2006). Ni acts as dearomatisation catalyst, which can result in coke formation and subsequent fouling issues. It is observed that as the Ni content increases the amount of coke also increases (Poels *et al*, 1995). However Nickel functions as more than just a dearomatisation catalyst, it also has a high degree of dehydrogenation activity (Al-Ayed &Kunzru, 1988). This results in an increase in molecular hydrogen in the system. Due to the difficulty associated with compression of hydrogen, an increase can limit the wet gas compressor in the FCC unit.

A catalytic phase's tolerance to Ni can be increased by employing very specific binders and additives to the catalyst formulation. An example of this is the incorporation of a series of aluminas based on selective active matrix aluminas into the existing FCC catalyst. Experimental evidence (Oxygen TPD) supports the idea that nickel diffuses in the solid state into the alumina and away from the surface and subsequently the reactants (Harding, *et al*, 2001, Nee *et al*, 2004).

Vanadium forms vanadic acid in the regenerator (hydrothermal process of coke removal/ highly exothermic) which may destroy the zeolite based components of the catalyst (Huh & Yan, 1990). Although the mobility of vanadium is dependent upon several factors including the excess oxygen in the regenerator (Rudesill & Peters, 1998), one thing remains clear that vanadium must be kept away from the zeolite components. Two ways of achieving this would be to trap the vanadium out prior to interaction with the zeolitic phases of the catalyst and to develop increasingly stable catalysts.

Catalyst Connectivity and Mass Transport

In the 1990's the introduction of materials with good bottom end cracking performance became common place. These materials were designed at constant conversion to convert a greater percentage of less volatile compounds (> 370 °C boiling point). The ability of the feed molecules to reach the active sites of the zeolite is one of the five important phases in a catalytic process. One can argue that this is the fundamental prerequisite of all catalysis. In the case of very large molecules however, there are diffusion limitations through the system (Andersson &Myrstad, 1997) and also in the case of a zeolites internal surface, reagent selectivity which inhibits access through size discrimination (Luckily not all catalysis occurs on a zeolite internal surface). These challenges have been addressed and as such the use of high surface area aluminas is common practice in FCC catalyst formulations (Al-Khattaf, 200). These aluminas have small crystal sizes and large pore volumes for pores in the 50 - 600 Å range. Small pores (50 - 100 Å) increase the cracking of bottom of the barrel oil, but at the expense of coke formation, whereas, intermediate pores (100 - 300 Å) have a lower matrix surface area for the same pore volume, but tend to convert resid molecules into more useful materials (Falco *et al*, 2006). It is apparent that the catalyst needs to be designed in such a way as to ease mass transport limitations to and from the zeolites surface.

Product Selectivity and Control

The main objective for most if not all refiners is the drive to produce fuel based product, however if the refiner in question has close connections to chemical plants then chemical feed-stocks become a viable source of product. This is particularly true for producing olefinic (gaseous) species such as C_2-C_4 molecules. Selectivities to these species are generally enhanced with the addition of ZSM-5 into the FCC catalyst formulation. Other materials such as syngas and xylenes are also particularly useful as feed gases and intermediates respectively.

Production of Chemical Intermediates

Although the primary interest is in the production of olefins, other desirable and intermediate species can also be produced selectively; these include the xylene isomers as well as other alkylated benzene species (Zhao, *et al*, 1996). However to completely utilise this technology to its fullest the separation processes must be refined, for example UOP's para-xylene separation technology (Honeywell UOP, Aromatics).

Production of Light Olefinic Species

The requirement for light olefinic material is circuitous; as such it is highly dependent on the production and subsequent use of polymers. There are several key factors in maximising the olefinic yield from oil, increasing reactor temperature (Golden,*et al*, 2004) using ZSM-5 (size exclusion), other methods include using catalysts with lower proton transfer affinity (which would decrease the conversion of olefin to paraffin) and moderate matrix activity (Haas, *et al*, 1997).

The use of ZSM-5 is arguably the single most important method to increase the yields of light olefins, however, it should be noted that this is at the expense of gasoline yield. Efforts have been made to stabilise ZSM-5 against hydrothermal deactivation at higher concentrations in the additive (Blasco *et al*, 2006). Technologies indicate maximum loadings of 10 to 25% can be employed in the catalyst particle (Harding, *et al*, 2001). It was found that when attempting to alter the C_4:C_3 olefin ratio by increasing the Si/Al ratio (decreasing the number of acid sites, but effectively increasing the acidity per site) in ZSM-5 that the overall activity decreased quite markedly. Although valuable olefins are often defined as propylene and butylenes, ethylene is also rather valuable in its own right and as such can be selectively produced from ever improving FCC catalyst design. As of 2001 significant increases of up to 6 wt% were observed (Harding, *et al*, 2001).

RESIDUAL OIL PROCESSING TECHNOLOGIES

General

Residual oil makes up a significant proportion of a barrel of crude oil (up to 85 volume %) (Rana, *et al*, 2007). A whole host of technologies exist for residual oil processing which can not only act as a precursor method to FCC, but in some instances produce product molecules in their own right. They can also upgrade the quality of feed-stocks to help reduce some of the aforementioned problems in previous sections. The economies of processes that convert residual oil to products are determined by two main factors, these are; the amount of low value by-products that remain and the required input of hydrogen (Bridjanian& KhademSamimi, 2011). There are two main types of residual oil (source dependent); atmospheric residue (AR, > 343°C) from the atmospheric distillation tower, the second type is vacuum residue (VR, > 565°C) (Jones & Pujado, 2006) obtained from the vacuum (25-100 mm Hg) distillation tower. The processing of these residues requires the use of various technologies, which may be assigned to either of two different routes (Table 1).

Table 1. Residual oil processing technologies (reproduced from Rana, et al, 2007).

Non-catalytic processes	Catalytic processes
Solvent deasphalting	Residue fluid catalytic cracking
Thermal processing	Hydroprocessing
Gasification	Fixed bed hydrotreating
Delayed coking	Fluid bed hydrotreating
Fluid coking	Slurry hydrocracking
Flexicoking	Ebullated bed hydrotreating
Visbreaking	Ebullated bed hydrocracking

Non-Catalytic Processes

Solvent Deasphalting

Solvent deasphalting (SDA) is a unique separation process, by which the residual oil is separated into fractions by molecular weight as opposed to the more traditional boiling point methods. This process produces low contaminant deasphalted oil (DAO), which is particularly rich in paraffinic hydrocarbons (Sattarin, *et al*, 2006). SDA is relatively low cost and is highly flexible in producing a wide range of DAO qualities. During SDA, the feed is mixed with a light paraffinic solvent (C_4-C_7), the oil solubilises in the solvent and the insoluble fraction precipitates out as asphaltene and various other fractions (Jacquin, *et al*, 1984). The separation of phases occurs in the extractor, where contamination of the DAO phase is minimised. The quality of the DAO can be controlled in the extractor, by altering solvent composition, pressure and temperature. DAO is commonly used as a FCC and hydro-conversion feed-stock due to its relatively low metal content (Ni and V) (Jacquin, *et al*, 1984; Rana, *et al*, 2007; Sattarin, *et al*, 2006).

Thermal Processing (Carbon Rejection)

Thermal conversion is one of the most important processes for residue treatment. Thermal cracking of the residue is carried out at moderate pressures and is referred to as coking processes (Dickenson, *et al*, 1997). Coking processes occur via hydrogen transfer from larger/heavier molecules to smaller/lighter molecules, which results in formation of coke. The residue acts a hydrogen donor molecule at high temperatures, which further decreases the H:C ratio further to 1 - 0.5 and produces yet more coke. A number of processes are designed in this manner and will be detailed below.

Gasification

Gasification is a process whereby the residue is converted via cracking processes to gaseous products (Sanchez &Toral, 2011). Due to its uncontrolled selective coal conversion it has received significantly less attention than other methods mentioned herein. Gasification of residue is performed at high temperatures (> 1000°C) which results in a multitude of products such as ash, carbon black and syn gas as major products (Sanchez &Toral; 2011Tsujino, 1996). An alternate approach is the integrated gasification combined cycle (IGCC), which is a technology designed for efficient power generation (from coal and a whole host of other carbonaceous materials) which has minimised consequences for the environment (carbon capture and lower NO_x and SO_x) (Mitsubishi Hitachi Power Systems, Integrated Coal Gasification Combined Cycle (IGCC) Power Plants).

Delayed Coking

The choice of many refiners for bottom of the barrel conversion, delayed coking has a large amount of inherent flexibility towards feed-stock choice (Houk, 1996). The process essentially rejects metals and carbon while still providing a degree of conversion to liquid products (primarily diesel and naphtha). The selectivity is dictated by operating conditions (temperature and pressure) of the process (Zhou, *et al*, 2010). Due to the relatively large energy requirement the process is more expensive than SDA, but, is significantly less expensive to operate than some other thermal processes (Reinoso, *et al*, 1998). The major drawback with this process is the low yield of liquid products and the large yield of coke (Rana, *et al*, 2007). Never the less, delayed coking remains the process of choice for refining residue.

Fluid Coking and Flexicoking

Both are fluidised bed processes that were developed from FCC technology. Both technologies exhibit real advantages over other methods of residue processing. It is therefore quite interesting that refiners still favour delayed coking (Furimsky, 2000). In these processes coke is circulated from the burner back to the reactor, where from the absorption of heat within the burner, coke acts as reaction sites for the cracking of residue to lighter products. If conditions are set up accordingly fluid coking can produce larger yields of liquid products and less coke, but the products are of lower quality (Mcketta, 1992). Flexicoking is an extended form of fluid coking that uses a gasifier to convert coke to syngas, but the temperature (900°C >) (Furimsky, 2000) is not sufficient to burn all of the coke. Fluid coking is a slightly better process than delayed coking due to its slight increase in liquid yields. Utility costs and fuel consumption are also higher for the latter process (Mcketta, 1992; Rana, *et al*, 2007).

Visbreaking

This mature process can be applied to AR and VR. Commonly operated in conjunction with a thermal cracker designed to minimise fuel oils, while producing additional light distillates. Vis breaking is performed by heating the residue stream in a furnace to 440 – 480°C and subsequently cracking during a low specified residence time. This helps to avoid coking within the soaking zone under mild process specific temperature and pressure conditions (Vezirov, 2011). Visbreaking generally used to increase a refineries distillate production.

Although liquid products are produced from thermal process in most instances they will contain an appreciable amount of impurities (S, N and metals), hence require further purification through a wide variety of hydrotreating processes (Rana, *et al*, 2007). Thermal processes are an excellent option for relatively low investment costs (Biasca, *et al*, 2003), but in real terms cannot compete with the larger invested catalysed processes.

CATALYTIC PROCESSES

Residue Fluid Catalytic Cracking

Residue fluid catalytic cracking (RFCC) is essentially an extension of the already well-established FCC process. This method offers improved selectivity to high gasoline range hydrocarbons and lower yields of gases than those observed for thermal and Hydroprocessing technologies. One drawback with RFCC is the need for better quality feed-stocks (high H:C ratio and low metal content) than thermal and hydroprocessing methods. The RFCC process is designed with mix temperature control, two regeneration stages and a catalyst cooler. This ensures that some heat can be recovered for generating steam and allows excellent heat balance control. The catalyst used is a robust mixture of ultra-stable Y (USY) or a rare earth exchanged USY (REUSY) in conjunction with an inorganic matrix (alumina etc). This mixture allows the catalyst to be very resilient to the adverse conditions found within this process (metal contaminants, coking, hydrothermal etc). The biggest limiting factor as mentioned previously is the use of good quality feed-stocks, which should avoid high yields of coke and minimise catalyst consumption (attrition, deactivation etc). This means that the process is limited to using AR as feed (Rana, *et al*, 2007). These feeds are highly limited in refinery operation, and the crude from which they are refined tend to be of a high value.

Residuehydroprocessing

Given the sheer number of hydroprocessing reactions (and literature volume), a general overview will be given in this section. Table 2 details a number of different residue hydroconversion processes, including reactor type employed and licenser.

Table 2. Residue hydroconversion processes (reproduced from Rana, et al, 2007).

Reactor type	Process	Licenser
Fixed bed	Continuous catalyst replacement (OCR)	Chevron Lumus Global (CLG)
	Up-flow rector (UFR)	Shell (Bunker flow)
		Axen (Swing reactor)
Ebullated bed	Bunker type reactor (Hycon)	Shell
	Swing reactor concept (Hyvahl)	IFP (Axen)
	H-Oil	Axen (HRI/IFP)
	T-Star	Chevron
	LC-Fining	ABB Lummus
		Amoco oil (BP)
Slurry system	Microcat - RC	Exxon Mobil
	Vebacombi-cracking	VebaOel
	Hydrocracking distillation hydrotreating (HDH)	Intevep
	Chevron activated slurry hydroprocessing (Cash)	Chevron
	Eni slurry technology (ENI)	Eni Technologies
		Snamprogetti
	CanMet	Energy Research Laboratories (Canada)

Hydroprocessing consumes a significant quantity of hydrogen and requires relatively large investment and operating costs. It however offers a drastic improvement in the selectivity and quality of liquid products (Marafi, *et al*, 2003). Hydrotreatment of the residue improves the quality of liquid products (which may also be used in RFCC) (Marques, *et al*, 2011), whereas, hydrocracking residue is much more rigorous (Morel, *et al*, 1997). A number of different processes have been developed over the years to allow the refining industry to not only improve their selectivity but also obtain cleaner fuel specifications. These include, for example, hydrodesulphurisation, hydrodenitrogenation, hydrodematallisation and hydrocracking (Bridge, *et al*, 1975; Chianelli & Daage, 1989; Rana, *et al*, 2007). The diversity that hydroprocessing offers not only comes from the development of catalysts (for specific features) but also process technology, which can be selected based upon the product distribution required. These processes have a large scope for modification to their parameters and by extension the product selectivities, by doing this the commercial importance of the technology can be shifted (Fernandez-Baujin*et al*, 1993).

For hydroprocessing catalysts tend to be high surface area supported mixed transition metal sulphides. Most common are mixtures of the oxides of metals found in Group VI (3-15 wt%) and Group VIII (2-8 wt%), however the catalyst must be converted from the oxide to the sulphide form (Kim, *et al*, 2017). These processes are generally catalysed via the use of multi-component catalysts that are used to generate a number of different processes in a single reactor. They may also be used individually in different reactor zones, whereby process steps must be undertaken consecutively in different reactor heating zones. The use of hydrogen generally inhibits coke formation on the catalyst (Sahu, *et al*, 2016) via hydrogenation of coke precursor molecules and also helps in the presence of a suitable catalyst the removal of heteroatoms (Laccino, 1999).

Simply put residue hydroprocessing is a combination of catalyst development, process operations and conditions.

DEEP CATALYTIC CRACKING

General

In recent years the increase in the demand for propylene, has led to the requirement of new technologies for the selective production of this elusive olefinic species. Generally, there are considered three main technologies that precede all others for the production of propylene (Liu,*et al*, 2011) these are; steam cracking (KBR, Petrochemical technologies, Olefins) which accounts for up to 70% of propylene production (Liu, *et al*, 2011). However with this technology the C_2:C_3 ratio is very low and the required feedstock (light oils) (KBR, Petrochemical technologies, Olefins) is becoming increasingly difficult to acquire, presumably due to the shift in refiners product slate. The remaining propylene is produced from the fluid catalytic cracking process (Nawaz, *et al*, 2009) as well as other processes (for example propane dehydrogenation (Chang, *et al*, 2002))). But still the production capacities are limited. In response to this challenge SINOPEC has been developing a series of novel technologies designed around FCC technology to selectively increase the yields of specific target molecules (Sinopec Corp., Oil Refining). A summary of some of these technologies along with target molecules are given in Table 3.

DCC is the only commercial process in the world using heavy oils as feed-stocks for the production of propylene. It has been found that with specific feed and suitable operating and process conditions that the propylene yield can be as high as 20% (Liu, *et al*, 2011).

Table 3. A summary of SINOPEC's FCC based technologies for gas/liquid production (Sinopec Corp., Oil Refining).

Process	Target molecules
Deep Catalytic Cracking (DCC)	Propylene
Maximum liquid Gas and high octane Gasoline (MGG)	LPG and naphtha
Maximising Iso-Olefins (MIO)	Iso-olefins and naphtha
Catalytic Pyrolysis Process (CPP)	Ethylene and propylene
Maximising Iso-Paraffins – Clean Gasoline and more Propylene (MIP-CGP)	Clean gasoline and propylene

In 1990 the first demonstration unit for DCC came online and in 1994 was commercialized successfully to a full scale process (Li, *et al*, 1994). To date eight units are in operation around the world, recently a new unit was set up in the Middle East (Petro-Rabigh) and been commissioned in just over a year (Liu, *et al*, 2011). The DCC process has shown to be not only a highly profitable venture, but also a true innovation in the field of heavy oil processing. As a result DCC technology is making progress in many countries and catalyst development companies (Hamler & Upson, 1999). As such SINOPEC have not rested on their laurels, but have realised that continued development is necessary to stay on top (Liu, *et al*, 2010).

Comparison of Steam Cracking and DCC Technologies

As discussed previously steam cracking and DCC are two of the three technologies used primarily for the production of propylene. However, these two technologies couldn't be more different.

Steam cracking uses no catalytic phase for the formation of olefins, but it was shown by Bach *et al.* (1995) that process equipment (dependent upon materials used) offers up defect sites which can act as catalytic centres for coke formation through the adsorption and subsequent polymerisation of olefinic species. Temperatures of 750 – 950°C (Siemens, 2007) are used to heat the chromium/nickel alloy tubes containing the feedstock (naphtha, fuel oil etc) in the presence of steam. The reaction proceeds through the cracking of feedstock to smaller molecules via a non-catalytic free radical mechanism. To avoid possible further degradation of the product stream the reaction is quenched through a series of transfer line exchangers (Ren, *et al*, 2006).

DCC is a catalysed process that can be modified by process control to fine tune the product selectivity. The most recent catalyst developed by SINOPEC for DCC is the DMMC-1 catalyst (Long, *et al*, 2007) which consists of (Long, *et al*, 2006):

1. A zeolite Y component (may be USY of rare/alkaline earth derivatives) contributing 20-80% of the total pore volume to the catalyst. (1-50 wt%)
2. A second zeolite component with a SiO_2/Al_2O_3 of 20 or greater (Si/Al = 10 or greater) (not specified but suspect based upon the pore volume ratios and work done in FCC that this could be a ZSM-5 based component). (1-40 wt%)
3. Metal dopant(s) from groups IIIa, IVa, Va, Ib, IIb, Vb, VIb, VIIb and VIII (none noble metals). (0.5-20 wt%)
4. Heat resistant inorganic oxides (once again not specified, but suspect that they will alumina mixtures to accommodate diffusion of larger oil molecules and may act as a primary zeolite binding agent as well as potential support for the metal component). (2-80 wt%)
5. Clay presumably used as a secondary binding agent. (0-80 wt%)

The catalyst mixtures were observed to work under a relatively wide range of conditions. Temperatures from 400-700°C were usable although 450-600°C were optimum. Weight hourly space velocities (WHSV) were best between 1-80 hr^{-1} but were tested up to 120 hr^{-1}. The catalyst to oil weight ratio worked particularly well within the 3 to 15 range. This reflects well with the flexible conditions under which DCC are run (Long, *et al*, 2006; Long, *et al*, 2007)

Advances in the DCC Process

DCC process technology is similar to that found in FCC (Figure 2), but the operating conditions in DCC are regulated (higher temperature, generally more severe conditions) to achieve deeper molecular cracking.

Figure 2. An overview of FCC (Yokogawa, Top Gas Density Measurement in a Fluid Catalytic Cracker (FCC)) and DCC (Pennwell, DCC reactor and generator) (simplified) reactor technology.

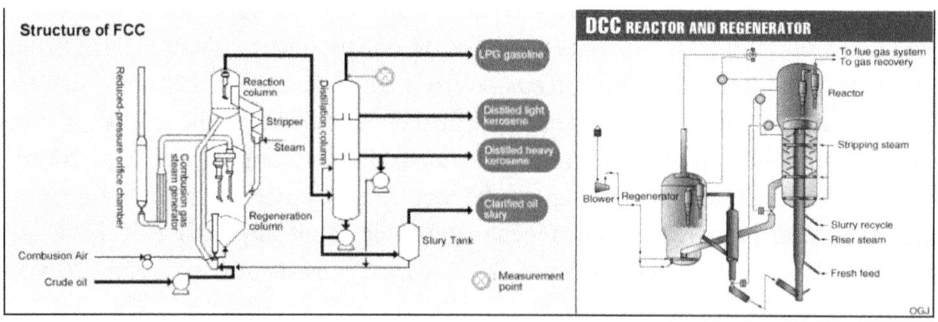

DCC has two very different operating modes and presumably catalytic phases, these are referred to DCC I and DCC II. DCC I is designed to maximise the yield of propylene production, whereas DCC II still maximises the yield of propylene, but also maximisesiso-olefins and naphtha yields. DCC feedstockshave been extended to include lower quality heavy oils similar to FCC feedstock. Unlike to the FCC process however, DCC employs significantly higher temperature, increased catalyst-to-oil ratio, longer residence time, lower hydrocarbon pressure and an increase in steam injection.

The DCC reactor unit consists of a riser (or a combination of riser and bed), the DCC's fractionators needs higher vapour to liquid ratios, which results in the need of a larger compressor capacity. The propylene yield is found to be between 3 to 5 times that of FCC, with increased isobutylene yield and decreased naphtha and LCO yields. Given the need of the refiner the DCC units can be switched to either process DCC I or DCC II as the market demand changes (Liu, *et al*, 2011).

Advances in the DCC Catalyst Formulation

DCC catalyst development has been notable with an improvement upon every preceding generation. The first gen DCC catalyst was the CHP series, which were designed with high propylene selectivity in mind (Shi, 1993). The second gen catalysts CRP and CIP series, exhibited improved hydrothermal stability and increased higher molecular weight (bottom barrel) cracking affinity (Liu, *et al*, 2011; Long, *et al*, 2007). The most recent catalysts (3rd gen) MMC have improved selectivity to propylene and increased metal tolerance (Long, *et al*, 2006; Long, *et al*, 2007). To date most DCC units are currently employing these catalysts. The most recent permutation of the MMC catalyst DMMC-1 is already adopted in commercial application (Liu, *et al*, 2011; Long, *et al*, 2007).

The first and second gen catalysts have high matrix activity for bottom of the barrel primary conversion, a mesoporous zeolite was used for secondary cracking reactions. Isomerisation performance of the catalyst is needed in conjunction with controlled hydrogen transfer (Shi, 199; Shi, 1996). Catalysts from the same type and generation were evaluated in different refineries for their performance with different feed-stock oils (displayed in Table 4). Refinery A showed the largest yield of propylene at 23 m% from the typical paraffinic oil. Yields at refineries B, C and D were found to be lower from feeds based upon intermediate oil. It was noted that as the aromatic content of the feed oil increased that the propylene yield decreased.

To further enhance propylene selectivity and metal tolerances the MMC series of catalysts the zeolite ZSP as the main active ingredient. Compared to the reference catalyst the MMC series has an increased deeper cracking activity and also an improved yield of propylene. It was found that for both MMC-1 and MMC-2 that

Table 4. Comparison of feedstock on light olefin yields (Liu, et al, 2011).

Refinery	A	B	C	D
Feedstock	VGO + ATB	VGO	VGO + DAO	VGO + DAO
Feedstock type	Paraffinic	Intermediate	Intermediate	Intermediate
Ethylene m%	3.7	3.5	5.3	5.3
Propylene m%	23.0	18.6	18.5	19.2
Butylenes m%	17.3	13.8	13.3	13.2

propylene content in both cases improved (Table 5). The MMC-1 catalyst was developed for refineries where both an increase in propylene and LPG are required, whereas MMC-2 increases the yield of propylene at the expense of LPG yield (Li, *et al*, 2007; Liu, *et al*, 2011).

Innovations Used in the Creation of the 3rd Generation of DCC Catalysts

There are two major innovations for DMMC-1; the new preparation technique and the use of novel catalytic phases. The new preparation techniques have been applied with the aim of improving bottom of the barrel cracking capabilities in conjunction with improving the accessibility and hence the availability of the active sites (Liu, *et al*, 2011; Long, *et al*, 2007). The DCC process is regulated by a series of cracking reactions in a specific sequence, the larger molecules can be cracked to smaller ones in a sequence when accessing a series of pores and active sites. In this particular mechanism, the catalyst with macroporous structure is required to crack these larger molecules, a catalyst exhibiting a larger availability of active sites can further boost the deep cracking reactions.

This type of catalyst design is undertaken to improve and optimise catalysis kinetics. By undertaking N_2 adsorption measurements of the fresh catalysts, the

Table 5. Pilot plant results from the MMC range of catalysts.

Products	CIP	MMC-1	MMC-2
Dry gas yield (m%)	Base	+ 0.5	- 0.8
LPG yield (m%)	Base	+ 2.2	- 0.3
Naphtha yield (m%)	Base	- 2.7	+ 2.1
Propylene in total gas (m%)	Base	+ 0.7	+ 3.3
Propylene yield (m%)	Base	+ 1.3	+ 1.0

cumulative pore volume is found to be larger for DMMC-1. Upon steaming the DMMC-1 retains a larger degree of pore volume than the equivalent MMC catalyst. Scanning electron microscopy showed an increased permeability of the DMMC-1 catalyst over the MMC-2 catalyst. Pilot plant tests further verified that this enhancement was beneficial for the reaction, where enhancements were observed for olefinic species (selectivity and yield) with a subsequent decrease in gasoline range hydrocarbons. Commercial application mirrored the enhancements observed in pilot plant tests (Long, *et al*, 2007).

POTENTIAL FUTURE DEVELOPMENTS BY SINOPEC OF DCC TECHNOLOGY

Given the already vast improvements made in this technology, more should be expected through various potential routes; unit operation and R + D on the DCC process.

Unit operations will attempt to process heavier feedstocks, which would reduce valuable carbon rich waste from refineries. Milder operations would desirable, which would be expected to decrease dry gas yield while still maintaining the same or even enhancing propylene yield.

Research and development activities related to the DCC process should look to establish petrochemicals based operations in pursuit of maximising yields of industrially important chemicals (olefins and light by-product aromatics etc). In comparison to FCC technology, the bench march set for DCC will be 6-8 times the propylene yield observed. It is hoped that the total ethylene and propylene yield would be equal to that of steam cracking processes. To achieve this it is anticipated that new reactor designs and catalyst formulations would be required. Investigating the worth in recycling lower value products (C_4 fractionate, lighter gasoline and even LCO) to produce propylene and ethylene obtained as a by-product. Heavy gasoline fractions (gas oil etc) can be used as feedstock for aromatic extraction to obtain xylene rich cuts (Liu, *et al*, 2011).

REFERENCES

Al-Ayed, O., & Kunzru, D. (1988). Cyclohexane dehydrogenation on a nickel catalyst: Kinetics and catalyst fouling. *Journal of Chemical Technology and Biotechnology (Oxford, Oxfordshire)*, *43*(1), 23–38. doi:10.1002/jctb.280430104

Al-Khattaf, S. (2003). The Influence of Alumina on the Performance of FCC Catalysts during Hydrotreated VGO Catalytic Cracking. *Energy & Fuels*, *17*(1), 62–68. doi:10.1021/ef020066a

Ancheyta, J., Morales, P., Betancourt, G., Centeno, G., Marroquin, G., & Munoz, J. (2004). Individual Hydrotreating of FCC Feed Components. *Energy and Fuels, 181*, 1-4.

Andersson, S., & Myrstad, T. (1997). Optimum performance of residue FCC catalysts. *Oil Gas European*, *23*, 19–22.

Bach, G., Zimmermann, G., Kopinke, F., Barendregt, D., Oosterkamp, P., & Woerde, H. (1995). Transfer-Line Heat Exchanger Fouling during Pyrolysis of Hydrocarbons. 1. *Deposits from Dry Cracked Gases Indstrial and Engineering Chemistry Research*, *34*, 1132–1139.

Bhattacharyya, A., Woltermann, G., Yoo, J., Karch, J., & Cormier, W. (1988). Catalytic SOx abatement: The role of magnesium aluminate spinel in the removal of SOx from fluid catalytic cracking (FCC) flue gas. *Industrial & Engineering Chemistry Research*, *27*(8), 1356–1360. doi:10.1021/ie00080a004

Biasca, F., Dickenson, R., Chang, E., Johnson, H., Bailey, R., & Simbeck, D. (2003). *Upgrading Heavy Crude Oils and Residue to Transportation Fuel: Technology Economics and Outlook, Phase 7*. SAF Pacific Inc. Engineering and Economic Consultant.

Blasco, T., Corma, A., & Matinez-Triguero, J. (2006). Hydrothermal stabilization of ZSM-5 catalytic-cracking additives by phosphorus addition. *Journal of Catalysis*, *237*(2), 267–277. doi:10.1016/j.jcat.2005.11.011

Boerefijn, R., Gudde, N., & Ghadiri, M. (2000). A review of attrition of fluid cracking catalyst particles. *Advanced Powder Technology*, *11*(2), 145–174. doi:10.1163/156855200750172286

Bridge, A., Scott, J., & Reed, E. (1975). *Hydrocarbon Processing*. Paper presented at the API Midyear Meeting.

Bridjanian, H., & Samimi, K. (2011). Bottom of The Barrel, An Important Challenge of the Petroleum Refining industry A. *Petroleum and Coal*, *53*, 13.

Can, F., Travert, A., Ruaux, V., Gilson, J., Mauge, F., Hu, R., & Wormsbecher, R. (2007). FCC gasoline sulfur reduction additives: Mechanism and active sites. *Journal of Catalysis*, *249*(1), 79–92. doi:10.1016/j.jcat.2007.04.001

Chang, J., Roh, H., Park, M., & Park, S. (2002). Propane Dehydrogenation over a Hydrogen Permselective Membrane Reactor. *Bulletin of the Korean Chemical Society*, *23*(5), 674–678. doi:10.5012/bkcs.2002.23.5.674

Chen, Y. (2006). Recent advances in FCC technology. *Powder Technology*, *163*(1-2), 2–8. doi:10.1016/j.powtec.2006.01.001

Cheng, W., Kim, G., Peters, A., Zhao, X., Rajagopalan, K., Ziebarth, M. S., & Pereira, C. J. (1998). Environmental Fluid Catalytic Cracking Technology. *Catalysis Reviews. Science and Engineering*, *40*(1-2), 39–79. doi:10.1080/01614949808007105

Chianelli, R., & Daage, M. (1989). Structure/Function Relations in Transition Metal Sulfide. *Studies in Surface Science and Catalysis*, *50*, 1–19. doi:10.1016/S0167-2991(08)60486-3

Corma, A., Corresa, E., Mathieu, Y., Sauvanaud, L., Al-Bogami, S., Al-Ghrami, M. S., & Bourane, A. (2017). Crude oil to chemicals: Light olefins from crude oil. *Catalysis Science & Technology*, *7*(1), 12–46. doi:10.1039/C6CY01886F

Cormier, W., Woltermann, G., Magee, J., Baars, F., & Upson, L. (1994). *Increasing metal-tolerance of FCC catalyst by sulfur oxide removal.* US Patent: 5,371,055.

Davey, S. (1999). *Proceedings of the European Refining Technology Conference.* Paris, France: Academic Press.

De Carvalho, M., Morgado, E., Cerqueira, H., De Resende, N., & Schmal, M. (2004). Behavior of Fresh and Deactivated Combustion Promoter Additives. *Industrial & Engineering Chemistry Research*, *43*(12), 3133–3136. doi:10.1021/ie034292z

Degnan, T., Chitnis, G., & Schipper, P. (2000). History of ZSM-5 fluid catalytic cracking additive development at Mobil. *Microporous and Mesoporous Materials*, *35-36*, 245–252. doi:10.1016/S1387-1811(99)00225-5

Dickenson, R., Biasca, F., Schulman, B., & Johnson, H. (1997). Refiner options for converting and utilizing heavy fuel oil. *Hydrocarbon Processing*, *76*, 57–62.

Falco, M., Morgado, E., Amadeo, N., & Sedran, U. (2006). Accessibility in alumina matrices of FCC catalysts. *Applied Catalysis A, General*, *315*, 29–34. doi:10.1016/j.apcata.2006.08.028

Fernandez-Baujin, J., Sundaram, K., & Chien, J. (1993). *Inhibition of coke formation during vaporization of heavy hydrocarbons.* US Patent: 5,190,634.

Furimsky, E. (2000). Characterization of cokes from fluid/flexi-coking of heavy feeds. *Fuel Processing Technology*, *67*(3), 205–230. doi:10.1016/S0378-3820(00)00103-X

Gembicki, S. (2000). New solid acid based breakthrough technologies. *Studies in Surface Science and Catalysis*, *130*, 147–156. doi:10.1016/S0167-2991(00)80951-9

Golden, S., Pulley, R., & Dean, C. (2004). Catalyst Changes, Downstream Improvements Increase FCC Propylene yields. *Oil & Gas Journal*, *102*, 44–46.

Haas, A., Nee, J., & Kanz, B. (1997). *Proceedings of the DGMK Conference*. Aachen, Germany: Academic Press.

Harding, R., Peters, A., & Nee, J. (2001). New developments in FCC catalyst technology. *Applied Catalysis A, General*, *221*(1-2), 389–396. doi:10.1016/S0926-860X(01)00814-6

Hemler, C., & Upson, L. (1999). Maximising FCC propylene production. *Petroleum Technology Quarterly*, *4*, 31–35.

Honeywell, U. O. P. Aromatics. (n.d.). Retrieved from www.uop.com/processing-solutions/petrochemicals/benzene-para-xylene-production

Honeywell, U. O. P. Gasoline Desulfurization Process. (n.d.). Retrieved from www.uop.com/fcc-gasoline-hds-selectfining

Houk, B. (1996). Model predictive control of a delayed coking unit with [DMC]: Development issues. *Fuel Energy, 37,* 176.

Huh, B., & Yan, T. (1990). *Catalytic cracking process with partial CO combustion.* US Patent: 4,980,050.

Ishutenko, D., Minaev, P., Anashkin, Yu., Nikulshina, M., Mozhaev, A., Maslakov, K., & Nikulshin, P. (2017). Potassium effect in K-Ni(Co)PW/Al2O3 catalysts for selective hydrotreating of model FCC gasoline. *Applied Catalysis B: Environmental*, *203*, 237–246. doi:10.1016/j.apcatb.2016.10.043

Jacquin, Y., Gimenez-Coronado, M., & Dai-Nghia, H. (1984). *Process for solvent deasphalting heavy hydrocarbon fractions.* US Patent: 4,440,633.

Jones, D., & Pujado, P. (2006). *Handbook of Petroleum Processing*. Berlin: Springer. doi:10.1007/1-4020-2820-2

KBR. (n.d.). *Petrochemical technologies, Olefins*. Retrieved from https://www.kbr.com/en/solutions/technologies/process-technologies/petrochemical-technologies

Kim, S. H., Kim, K. D., & Lee, Y. K. (2017). Effects of dispersed MoS2 catalysts and reaction conditions on slurry phase hydrocracking of vacuum residue. *Journal of Catalysis*, *347*, 127–137. doi:10.1016/j.jcat.2016.11.015

Laccino, L. (1999). *Heteroatom removal through countercurrent sorption.* US Patent: 5,928,497.

Li, Z., Shi, W., Wang, X., & Jiang, F. (1994). Deep Catalytic Cracking III. In Process for Light-olefins Production (pp. 33-42). American Chemical Society.

Li, Z., Xie, C., Luo, Y., Zhao, L., & Shu, X. (2007). MMC-High Propylene Selectivity DCC Catalyst. *China Petroleum Processing and Petrochemical Technology*, *1*, 13–18.

Liu, C., Gao, X., Zhang, Z., Zhang, H., Sun, S., & Deng, Y. (2004). Surface modification of zeolite Y and mechanism for reducing naphtha olefin formation in catalytic cracking reaction. *Applied Catalysis A, General*, *264*(2), 225–228. doi:10.1016/j.apcata.2003.12.048

Liu, Y., Long, J., & Tian, H. (2010). *Innovations in DCC catalyst for propylene production from heavy oils.* Beijing: Technical Progress in Refining and Petrochemical Industry China Petrochemical Press.

Liu, Y., Long, J., Tian, H., Xu, Y., & Zhao, L. (2011). Advances in DCC Process and Catalyst for Propylene Production from Heavy Oils. *China Petroleum Processing and Petrochemical Technology*, *13*, 1–5.

Long, J., Tian, H., Liu, Y., Xie, C., & Li, J. (2007). Study on the Performance and Commercial Application of New Generation DMMC-1 Type Catalyst for Deep Catalytic Cracking. *China Petroleum Processing and Petrochemical Technology*, *4*, 1–6.

Long, J., Tian, H., Zhang, W., Zhu, Y., Liu, Y., Chen, Z., . . . He, M. (2006). *Cracking catalyst for hydrocarbon containing molecular sieve and preparation process thereof.* Chinese Patent Application: 03147987.1.

Marafi, A., Al-Bazzaz, H., Al-Marri, M., Maruyama, F., Absi-Halabi, M., & Stanislaus, A. (2003). Residual-Oil Hydrotreating Kinetics for Graded Catalyst Systems: Effect of Original and Treated Feedstocks. *Energy & Fuels*, *17*(5), 1191–1197. doi:10.1021/ef020282j

Marques, J., Guillaume, D., Merrignac, I., Espinat, D., & Brunet, S. (2011). Effect of catalysts acidity on residues hydrotreatment. *Applied Catalysis B: Environmental*, *101*(3-4), 727–737. doi:10.1016/j.apcatb.2010.11.015

Marsh, G., Hill, N., & Sully, J. (2000). *Consultation on the Need to Reduce the Sulphur Content of Petrol and Diesel Fuels Below 50 PPM: - A Policy Makers Summary.* Restricted – Commercial AEAT/ENV/R/*0372* issue 5.

McKetta. (1992). *Petroleum Processing Handbook.* New York: Marcel Dekkar.

Mehlberg, R., Rosser, F., Fei, Z., & Stevens, C. (2011). *Low NOx FCC catalyst regeneration process.* US Patent: 7,914,666.

Mitsubishi Hitachi Power Systems. (n.d.). *Integrated Coal Gasification Combined Cycle (IGCC) Power Plants.* Retrieved from https://www.mhps.com/products/igcc/

Morel, F., Kressmann, S., Harle, V., & Kasztelan, S. (1997). Processes and catalysts for hydrocracking of heavy oil and residues. *Studies in Surface Science and Catalysis, 106*, 1–16. doi:10.1016/S0167-2991(97)80003-1

Nawaz, Z., Zhu, J., & Wei, F. (2009). Drastic enhancement of propene yield from 1-hexene catalytic cracking using a shape intensified MESO-SAPO-34 catalyst. *Journal of Engineering Science and Technology, 4*, 409–418.

Nee, J., Harding, R., Yaluris, G., Cheng, W., Zhao, X., Dougan, T., & Riley, J. (2004). *Kirk-Othmer Encyclopedia of Chemical Technology. Fluid Catalytic Cracking (FCC)* Catalysts and Additives.

Occelli, M., & Anthony, R. (1989). *Advances in Hydrotreating Catalysts.* Amsterdam: Elsevier.

Occelli, M., & O'Connor, P. (1994). *III, Materials and Processes.* American Chemical Society.

Occelli, M., & O'Connor, P. (1998). *Fluid Cracking Catalysts.* New York: Marcel Dekker.

Pennwell. (n.d.). *DCC reactor and generator.* Retrieved from http://images.pennwellnet.com/ogj/images/ogj2/9602jfufu02.gif

Peters, A., Rakiewicz, E., Weatherbee, G., & Zhao, X. (2002). *Reduced NOx combustion promoter for use in FCC processes.* US Patent: 6,358,881 B.

Poels, E., van Beek, W., den Hoed, W., & Visser, C. (1995). Deactivation of fixed-bed nickel hydrogenation catalysts by sulfur. *Fuel, 74*(12), 1800–1805. doi:10.1016/0016-2361(95)80011-6

Rana, M., Samano, V., Ancheyta, J., & Diaz, J. (2007). A review of recent advances on process technologies for upgrading of heavy oils and residua. *Fuel, 86*(9), 1216–1231. doi:10.1016/j.fuel.2006.08.004

Reinoso, F., Santana, P., Palazon, E., Diez, M., & Marsh, H. (1998). Delayed coking: Industrial and laboratory aspects. *Carbon, 36*(1-2), 105–116. doi:10.1016/S0008-6223(97)00154-1

Ren, T., Patel, M., & Blok, K. (2006). Olefins from conventional and heavy feedstocks: Energy use in steam cracking and alternative processes. *Energy*, *31*(4), 425–451. doi:10.1016/j.energy.2005.04.001

Rudesill, J., & Peters, A. (1996). The Effect of Oxidation State of Vanadium on the Selectivity of Fluid Catalytic Cracking Catalysts. *ACS Division of Petroleum Chemistry, Inc. Preprints*, *41*, 407–409.

Sahu, R., Song, B. J., Im, J. S., Jeon, Y.-P., & Lee, C. W. (2015). A review of recent advances in catalytic hydrocracking of heavy residues. *Journal of Industrial and Engineering Chemistry*, *27*, 12–24. doi:10.1016/j.jiec.2015.01.011

Sahu, R., Song, B.-J., Jeon, Y. P., & Lee, C. W. (2016). Upgrading of vacuum residue in batch type reactor using Ni-Mo supported on goethite catalyst. *Journal of Industrial and Engineering Chemistry*, *35*, 115–122. doi:10.1016/j.jiec.2015.12.017

Salar, H. (2006). Analysis of metal ions in crude oil by reversed-phase high performance liquid chromatography using short column. *Journal of Chromatography. A*, *1118*(1), 82–84. doi:10.1016/j.chroma.2006.04.068 PMID:16723133

Sambi, I., Khulbe, K., & Mann, R. (1982). Catalytic hydrotreatment of heavy gas oil. *Industrial & Engineering Chemistry Product Research and Development*, *21*(4), 575–580. doi:10.1021/i300008a012

Sanchez, J. M., & Toral, M. R. (2011). Model for Gasification of Residual Fuels from Petroleum Refineries Using the Equation Oriented (EO) Approach. *Industrial & Engineering Chemistry Research*, *50*(5), 2628–2640. doi:10.1021/ie100663w

Sattarin, M., Modarresi, H., Talachi, H., & Teymori, M. (2006). Solvent deasphalting of vacuum residue in a bench-scale unit. *Petroleum and Coal*, *48*, 14–19.

Sinopec Corp. (n.d.). *Oil Refining*. Retrieved from http://www.sinopec.com/listco/en/products_service/License/Oil_Refining/

Tsujino, T. (1996). Recent development of Texaco gasification technology and its applications. *Fuel and Energy Abstracts*, *37*(3), 183. doi:10.1016/0140-6701(96)88538-9

Use of Process Analytics for Cracking Processes in Refineries. (2007). *Case Study*. Siemens.

Vezirov, R. (2011). Visbreaking – technologies tested by practice and time. *Chemistry and Technology of Fuels and Oils*, *46*(6), 367–374. doi:10.100710553-011-0236-9

Yokogawa. (n.d.). *Top Gas Density Measurement in a Fluid Catalytic Cracker (FCC)*. Retrieved from https://www.yokogawa.com/library/resources/application-notes/top-gas-density-measurement-in-a-fluid-catalytic-cracker-fcc/

Zhao, L., Pang, H., You, B., Fang, X., Wang, J., Duan, W., & Liu, J. (2007). *Gasoline hydrofinishing startup method and gasoline hydrofinishing operation method.* Chinese Patent: 101376835.

Zhao, X., Roberie, T., & Rajagopalan, K. (1996). p-Xylene yield in fluid catalytic cracking products. *Applied Catalysis A, General*, *145*(1-2), 407–418. doi:10.1016/0926-860X(96)00180-9

Zhao, Y., Luo, Q., Gao, W., Gao, S., & Song, Y. (2003). Industrial application test of a FCC gasoline olefin reducing catalyst RFG-FS3. *Petroleum and Petrochemicals*, *1*, 33–37.

Zhou, X., Di, X., Yu, G., Lu, R., & Li, C. (2010). Simulation of Delayed Coking Reaction in Coke Drum. *Petroleum Science and Technology*, *28*(3), 277–285. doi:10.1080/10916460802706505

Chapter 5
"Catalyst in Biorefineries" Solution to Promote Environment Sustainability in India

Vikas Gupta
HPCL, India

ABSTRACT

India is one of the fastest growing economies in the world. India has taken various initiatives to promote environmental sustainability. One of the important initiatives is increasing production and utilization of renewable energy by adopting ethanol biorefineries. Lignocellulosic biomass, which has a complex structure that can be broken down by enzymes to produce products like biodiesel, bioethanol, and various high-value products, is the major feedstock for biorefineries. The chapter presents a review of the various type of catalysts used in biorefineries and their positive impact on the environment.

INTRODUCTION

India is one of the fastest growing economy of the world. Energy is not only required to sustain the lives of citizens of a country but it also acts as an impetus to it's socio-economic development. Gross domestic Product (GDP) of a country by sector and energy consumption by type are cointegrated. Fossil fuels have been powering economies around the globe to meet the energy requirement. There is empirical evidence to suggest that developing countries should increase their renewable

DOI: 10.4018/978-1-5225-8033-1.ch005

energy consumption to achieve growth in their GDP (Al-mulali & Mohammed, 2015). However, given the focus on environmental issues and depleting fossil fuel reserves, there is a major transformation in the global energy system that will take place over the next few decades in which Renewable energy (RE) and natural gas is going to play a major role in meeting global energy requirements.

India's energy demand has almost doubled and responsible for around 10% of the increase in Global energy demand since the year 2000 (Energy Agency, 2015). Primary energy consumption for India will grow by 4.2% (from 701 Mtoe to 1603 Mtoe) in the next two decades till 2035. RE will play a major contributory factor, that may contribute around 8% of primary energy consumption by the year 2035, with a net change of (+) 712% from the last two decades (BP, 2017).

With the ever-increasing energy requirements of the country, India has established goals to expand its use of RE and more efficient technologies. The Biofuel apart from photovoltaics, wind and solar are the area that the RE technologies are showing maturity for the positive changes in energy needs. India has undertaken ambitious target to set up 12 second generation bio refineries in the country which shall produce around 360 million litres of bioethanol to support ethanol blending program and utilizing surplus lignocellulosic agricultural residue.

India has 140MMT surplus lignocellulosic feedstock agricultural residue such as rice straw, cotton stalk, wheat straw, which has the potential to yield around 27000 Million litres of ethanol. This can help in reducing the dependency on fossil fuels and since Biofuels are generally considered green fuels, with a zero-net carbon dioxide emission, the proposed bio refineries will help in promoting environment sustainability as well.

LIGNOCELLULOSIC BIOMASS

Lignocellulosic materials have been the most promising feedstock because of its renewable nature. Lignocellulosic materials includes biomass from agriculture (corn stover, corn cobs, rice straw, sugarcane bagasse), forestry (paper and saw mill discards), naturally occurring terrestrial trees/bushes/grass and energy crops such as elephant grass, Miscanthus which are grown as low-cost and low-maintenance harvest with high yield of lignocellulosic biomass. Energy crops are preferably used for production of second generation biofuels as feedstock.

Physico-Chemical Characteristics of Lignocellulosic Biomass

The composition of Lignocellulosic biomass includes carbohydrate polymers (cellulose, hemicellulose) and an aromatic polymer (lignin).On a broader scale, most of the lignocellulosic biomass derived from agriculture is composed of about 40-50% cellulose, 20-30% hemicellulose and 10-25% lignin.

Cellulose

Cellulose is the major component of the plant cell walls, which is responsible for the mechanical strength. It is the principal structural compound in plant material which is unbranched but highly organized polymer of several thousand units of β-D-glucopyranose moieties which are linked via β-(1,4) glycosidic (cellobiases)bonds. Hydrogen bonds hold the cellulose molecules. The chemical formula of cellulose is $(C_6H_{10}O_5)n$ (Watling & Thiel, 2013). Figure 1 depicts the typical structure of cellulose.

Hemicellulose

Hemicellulose is the second most abundant polymer which is heterogeneous in nature and mainly consists of glucuronoxylan, glucomannan and traces of other polysaccharides (Anwar, Gulfraz, & Irshad, 2014). It has a random and amorphous structure with high solubility and more reactive nature unlike cellulose. Xylan, Mannan and Xyloglucan are the types of different hemicellulose. The bonding between cellulose and hemicellulose is through non-covalent attractions. Figure 2 depicts typical structure of hemicellulose.

Figure 1. Typical structure of cellulose

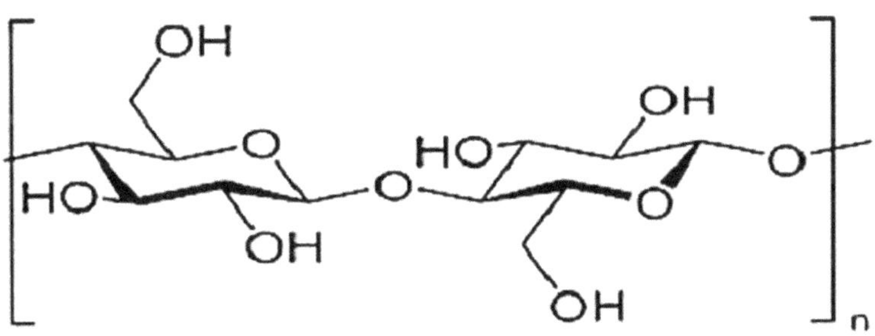

Figure 2. Typical structure of hemicellulose

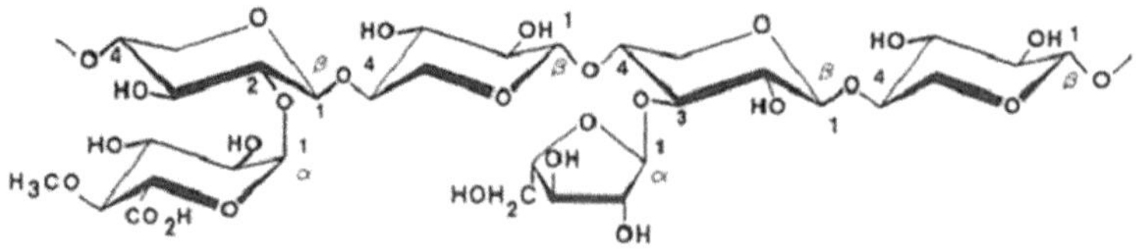

Lignin

Lignin is the most complex component of the biomass which has a long-chain. It is a heterogeneous polymer, comprising of phenyl- propane units, which are linked with ether bonds. The presence of methoxy groups, phenyl-propane and non-carbohydrate poly phenolic substance enables the bonding of cell walls together which makes it resistant to degradation. (Anwar et al., 2014). Figure 3 depicts complex structure of lignin.

Figure 3. Lignin complex structure

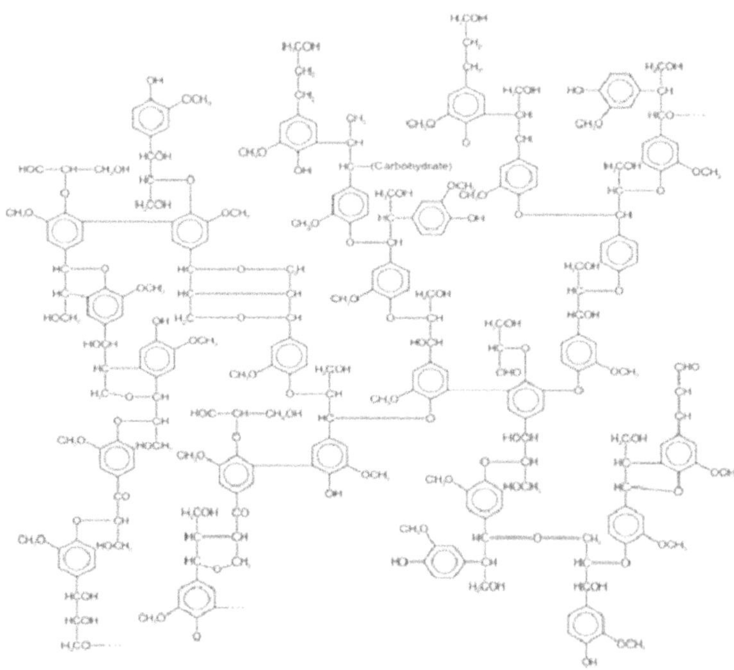

Biomass Conversion Technologies and Need for Catalysts

The plant dry matter, i.e. biomass having lignocellulose properties can be converted into different high value products such as valuable chemicals, bio-fuels including bio ethanol, sustainable polymers etc. In the biofuel production process, the carbohydrates (cellulose and hemi cellulose) need to be first converted to simple sugars, which are further fermented to produce biofuel. Cellulose is not soluble in water or any conventional organic solvents, due to its high crystalline nature resulting from the hydrogen bonds. There are four types of conversion technologies currently available, depending upon the specific biomass types yielding to specific energy products:(Demirbaş, 2001). *Thermal conversion* is the use of heat, with or without the presence of oxygen, to convert biomass materials or feedstock into other forms of energy. Thermal conversion technologies include direct combustion, pyrolysis, and torrefaction. *Thermochemical conversion* is the application of heat and chemical processes in the production of energy products from biomass. A key thermochemical conversion process is gasification. *Biochemical conversion* involves use of enzymes, bacteria or other microorganisms to break down biomass into liquid fuels, and includes anaerobic digestion, and fermentation. *Chemical conversion* involves use of chemical agents to convert biomass into liquid fuels.

The catalyst, plays a vital role in Biochemical and chemical conversion process. The catalyst enhances the rate of reaction to result into the desired products. If the hydrolysis was to be done without use of catalyst, the energy required to break the bond would be very high and also the reaction time would be very large. Catalyst when used during the hydrolysis phase (under required temperature, pressure and medium), reacts with the cellulose to produce water soluble sugars without altering itself. Similarly, catalysts can also be used to hydrolyze the hemicellulose present in the lignocellulosic biomass to produce xylose and other soluble sugars (Figure 4). Hence, catalysts play a major role in conversion of lignocellulosic biomass to biofuels.

Figure 4. Typical process of enzymatic hydrolysis

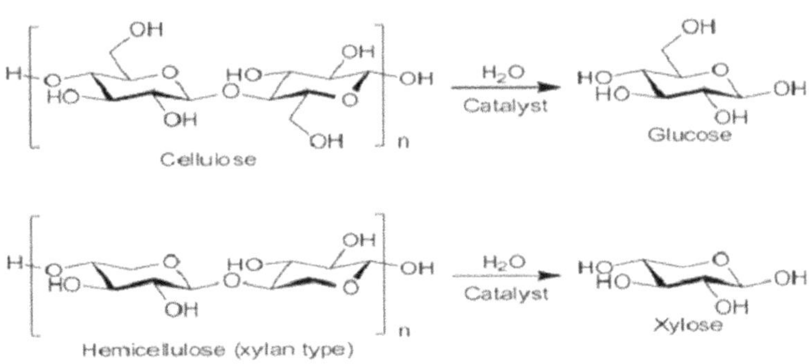

Figure 5. Catalyst providing low activation energy
(Source:https://www.chemicool.com/definition/catalyst.html)

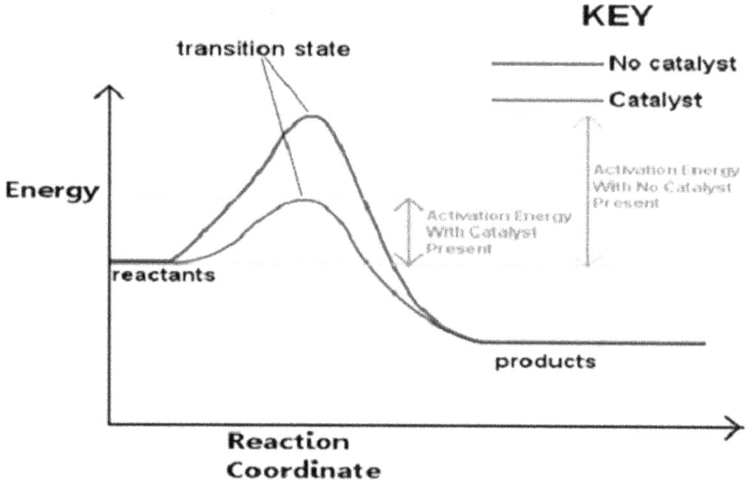

CATALYST-BASIC PERSPECTIVE

Catalysts are defined as any substance that accelerates the rate of any chemical reaction without being affected in the whole process. Hence, it can be used to act repeatedly. The cause of any reaction occurring fast is because the activation energy required is very less in case of usage of catalysts. Catalyst reacts to form a temporary intermediate and then the original catalyst is regenerated in a cyclic process. A catalyst provides an alternative route for the reaction with a lower activation energy. It does not lower the activation energy of the reaction

Catalysts are broadly classified as homogeneous, heterogeneous and biocatalysts. Homogenous catalyst is the one whose dispersion of molecules occur in the same phase as the molecules of the reactant (usually gaseous or liquid) whereas in case of heterogeneous catalyst, the dispersion of molecules is not in the same phase as the reactants molecules. In this case, the reactant molecules which is typically in gaseous or liquid phase gets adsorbed to the solid catalyst. Some of the key differences has been enumerated in the Table 1. In bio catalysis process, natural catalysts such as protein enzymes are used to perform the chemical transformations on organic compounds.

Table 1. Key differences between homogenous and heterogeneous catalyst

Characteristics	Homogeneous	Heterogeneous
Phase	Liquid/gas	Gas/solid
Activity	Moderate	High
Selectivity	High	Low
Diffusion	Facile	Important
Catalyst Recycle	Expensive	Simple
Catalyst modification	Easy	Difficult

HISTORY OF USAGE OF CATALYST IN PETROLEUM SECTOR

Economies around the globe are being run with the help of various chemicals and the role played by catalysts is substantial. *Catalysis* has been defined as the process by which chemical reaction rates are altered by the addition of a substance (the catalyst) which remains unaltered during the chemical reaction. Today more than 90% of all industrial chemicals are produced with the help of catalyst (*Technology Vision 2020: The U.S. Chemical Industry*, 1996). Catalysis is also a vital component of a number of national critical technologies and an important factor in energy security (JACKSON N.B., 1999; Phillips W, 1991).

The history dates back to the time of art of producing alcohol from sugar by fermentation process. For centuries before 1750, catalysts were used to make food and beverages.

Catalysis Era

The history of catalysis can be majorly classified in six era (Figure 4). The first Era was disordered & was followed by the second era, which was marked by the systematic research. The third era which began at the end of the nineteenth century was noticeable with the contribution from academics which helped in giving birth to Industrial catalysis and contributed a lot during World War I, when the demand on explosives based on nitric acid were booming (Lindström & Pettersson, 2003).

Era 3 was marked by the discovery and implementation of new catalytic process and ammonia process was one of the most significant process of this Era. The first important advancement in the development of ammonia process came in 1903 by Haber and Nernst. Haber received the Nobel prize in 1918, for the synthesis of ammonia.

Figure 6. History of catalysts

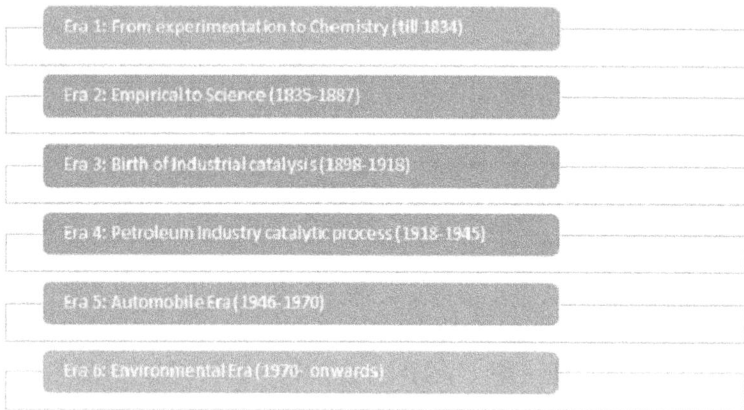

Era 4 is the most significant, which began after World War I,as the focus shifted towards manufacturing of synthetic fuels. This Era saw major innovations happening in the field of Petroleum industry which included the notable Fisher-Tropsch process. The first significant development happened in the year 1920 when the American oil company, Standard Oil Company started large scale industrial production of isopropanol from petroleum. Year 1921 witnessed a change for both automobile and petroleum sector by the discovery of tetraethyl lead by Thomas Midgley, renowned chemist at Dayton research laboratory, a subsidiary of General Motors Chemical Company, USA. TEL (Tetra ethyl lead) was branded under the name "Ethyl" and was proposed as anti-knocking agent which can be added to gasoline. In the year 1922, Franz Fischer and Hans Tropsch developed process to synthesize hydrocarbons from CO & H_2, which industry utilizes as Fischer-Tropsch technology (F-T).This process was significant during World War II when the demand for hydrocarbon had peaked. In the year 1926, Dupont began the production of synthetic methanol which had variety of applications in chemical industries. In the year 1927, Hinshelwood presented his kinetic theory based upon earlier findings of Irving Langmuir. This formed the basis of catalysis modelling and Irving Langmuir was awarded with Nobel prize in the year 1932 for his work in the field of surface chemistry.

The exemplary discovery of fixed bed catalytic cracking of petroleum in 1936,by Eugene Houndry, an American French mechanical engineer was originally developed to find an alternative to TEL which was being used as additive for anti-knocking in gasoline. There were several modifications carried out in the cracking units and notably was when in the year 1941, fluid bed catalytic cracking (FCC) technology was developed by Warren K Lewis and Edwin Gilliland of MIT for Standard Oil Company who were unable to afford the license fees for Houndry process. This resulted in a refinery design that remains the primary method of high-quality gasoline production. The more immediate result was the introduction of 100-octane aviation fuel that gave a crucial technological advantage to the United States and its allies during World War II. The fourth era of catalysis ended with World War II in 1945.

Era 5 of catalysis saw the world economies moving from World war II to rapid demand in automobile market and thereby increase in the demand of petroleum products. In the year 1949, first commercial Naptha reforming plant became operational which was constructed by UOP (Universal oil products), thereby starting an era of catalytic reforming. Catalytic reforming is a chemical process used to convert petroleum refinery naphthas, typically having low octane ratings, into high-octane liquid products called reformates which are components of high-octane gasoline (also known as petrol). In 1953, Karl Ziegler and in 1954 Giulio Natta independently developed catalysts that enabled polymerization at room temperature and normal atmospheric pressure and exactly controlled the positioning of atoms attached to the polymer chains. Ziegler discovered that organic compounds of metals, such as aluminum alkyl, could catalyze the production of polyethylene without using high pressure and temperature. Natta's group experimented with catalyst systems to make a wide variety of polymers, resulting eventually in the long-sought goal of a synthetic rubber molecularly identical with natural rubber.

Significant advances were made in 1954 by introduction of chromium based catalysts for the production of HDPE (polyethylene) by Phillips Petroleum and then Gulf Oil Company to use Ziegler-Natta Catalyst in the production of synthetic rubber. In 1963, Ziegler and Natta were awarded the Nobel Prize in Chemistry for their discoveries in the science and technology of polymers.

In the year1960, there were three important technologies which were launched on commercial scale:

- Benzene to Cyclohexene (Hydrar process) by UOP. The Hydrar-process (UOP) uses a series of three reactors with a Ni support and a stepwise increasing temperature (400°to 600° C) at 30 bar.
- Ethylene to Acetaldehyde by Wacker process. The Wacker process or the Hoechst-Wacker process refers to the oxidation of ethylene to acetaldehyde in the presence of palladium (II) chloride as the catalyst. This chemical reaction was one of the first homogeneous catalysis with organopalladium chemistry applied on an industrial scale.
- First commercial Acrylonitrile plant by SOHIO (now BP International) used a catalyst based on Bi_2O_3 .MoO_3. Acrylonitrile (AN) is one of the leading chemicals & important applications are acrylic fibers, thermoplastics (SAN, ABS), technical rubbers, adiponitrile, as well as speciality polymers.

Year 1960-70, witnessed astonishing development of new processes and catalysts. Mobil Oil discovery of rare earth stabilization of X-zeolite for catalytic cracking in 1964,which revolutionized the petroleum industry apart from discovery of new structures as BETA and MFI by Robert L Wadlinger etal.

In the year 1965, Professor Geoffrey Wilkinson, Imperial College, London, discovery of homogeneous catalyst for hydrogenation started a new branch of catalysis and Wilkinson was awarded with Nobel prize in 1973 for his pioneering work performed independently on the chemistry of the organometallic, so called sandwich compounds.

Era 6, the current period of catalysis has been characterized and driven primarily by environmental forces. Environmental catalysis has continuously grown in importance over the last 4 decades not only in terms of the worldwide catalyst market, but also as a driver of advances in the whole area of catalysis. The development of innovative environmental catalysts is also the crucial factor towards the objective of developing a new sustainable industrial chemistry. The first commercial exhaust after treatment system was developed by Engelhard in mid-1960's. Mobil oil's new process of converting methanol to gasoline with ZSM-5 zeolite catalyst & oxidation of benzene to phenol over a Fe-ZSM-5 zeolite in the year 1990 paved further way of role of zeolites in the industry. This era also witnessed the introduction of selective catalytic reduction (SCR) for controlling the NOx emissions apart from considerable expansion of traditional process, venturing new areas as reduction of the environmental impact in the disposal of catalyst, reduced crude cracking (RCC), production of alpha-olefins, catalyst for green house control, catalyst process for sustainable eco-compatible refineries among others.

APPLICATION OF CATALYSTS IN BIOREFINERIES

Biorefinery by definition means a plant processing bio based feedstock to generate biofuel and various biochemical. Based on feedstock bio-refineries can be divided in mainly in four types: First Generation, Second Generation, Third Generation and Forth Generation.

First generation biorefinery utilizes starch rich and sugar rich feedstock like sugarcane juice, sugar beet, molasses, oil seeds etc to produce biofuels while Second generation biorefinery uses residue biomass or agro-waste which are non-food feedstock to generate biofuels and other bio-based chemicals.

The Third Generation of biofuels is based on specially engineered energy crops such as algae while fourth generation biofuel refer to genetically modified algae for rapid consumption of CO_2 and high production of lipids (Dutta, Daverey, & Lin, 2014).

Figure 7,8,9 and 10 provides overview of 1st, 2nd, 3rd and 4th generation biofuels

The upcoming sections will be focusing about applications of catalysts in first and second generation biorefineries.

Application of catalysts in biorefineries can be divided into following ways:

Figure 7. First Generation Biofuel process flowchart

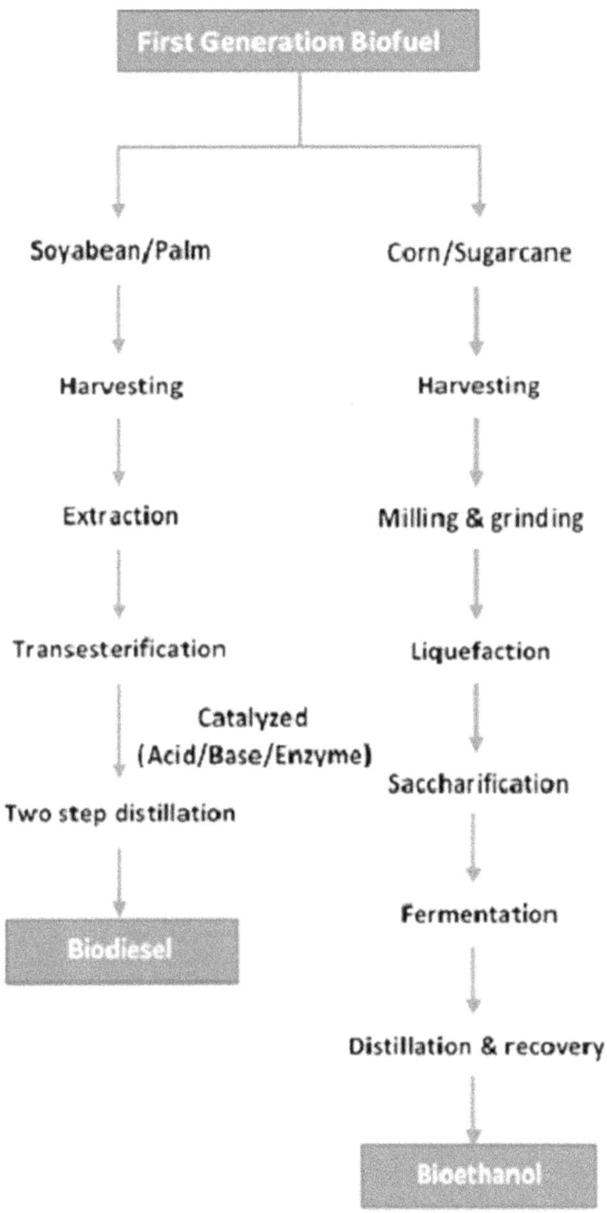

Figure 8. Second Generation Biofuel process flowchart

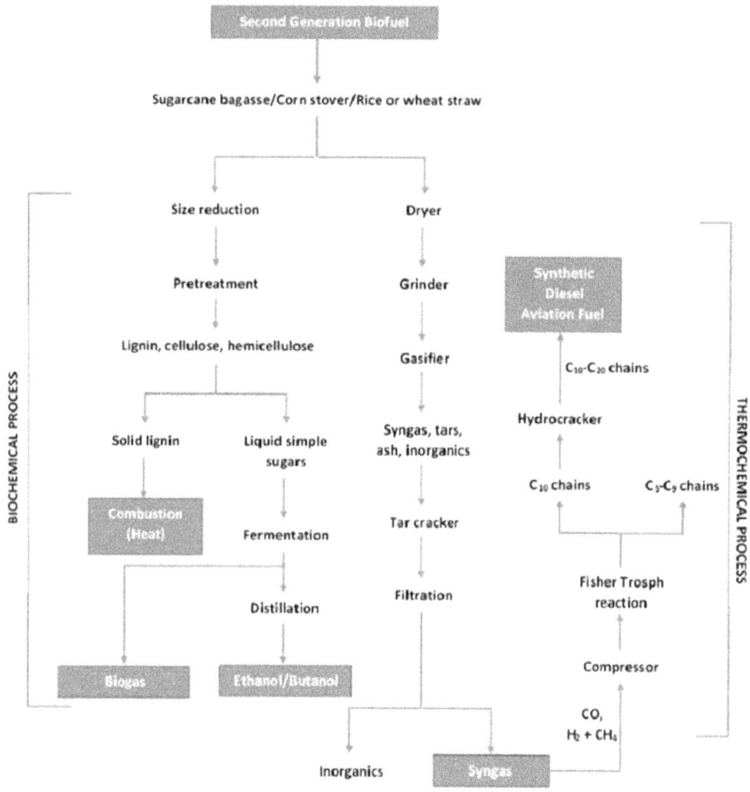

1. For Biodiesel production.
2. For Bioethanol production through hydrolysis
3. For Bio-oil production through pyrolysis

Catalyst Application in Production of Biodiesel

Biodiesel is a biodegradable fuel consisting of long-chain alkyl (methyl, ethyl, or propyl) esters. Presently, used cooking oil, non-edible oil or fat is used in the production of biodiesel. The main problem utilizing vegetable oil/ non-food oil as fuel is the high viscosity of oil. Hence, it is required to lower the viscosity of oil by methods such as preheating the oils, blending or dilution with other fuels, transesterification and thermal cracking/pyrolysis. The most efficient and viable process of reducing viscosity of oil is transesterification. In this process the oil or fat

Figure 9. Third Generation Biofuel process flowchart

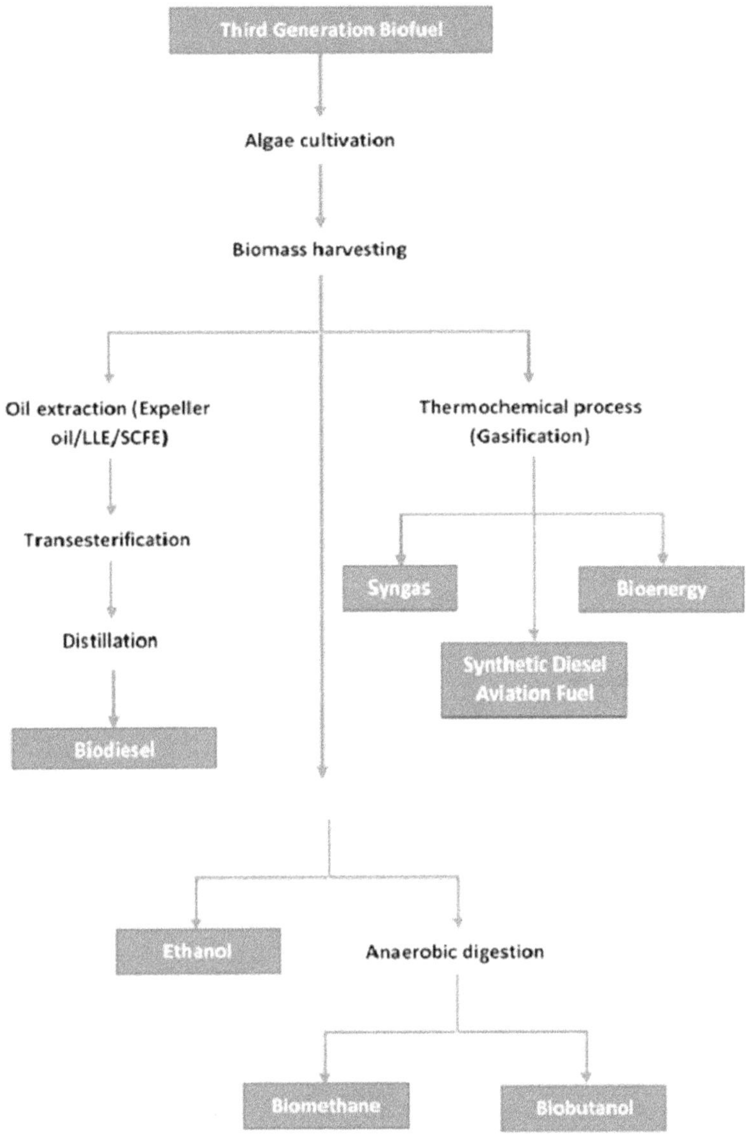

is reacted with alcohol, such as methanol or ethanol, in the presence of a catalyst to form alkyl ester and glycerol. The glycerol is byproduct which also has commercial value. The viability of the process largely depends on the catalyst being used for transesterification.

Figure 10. Fourth Generation Biofuel process flowchart

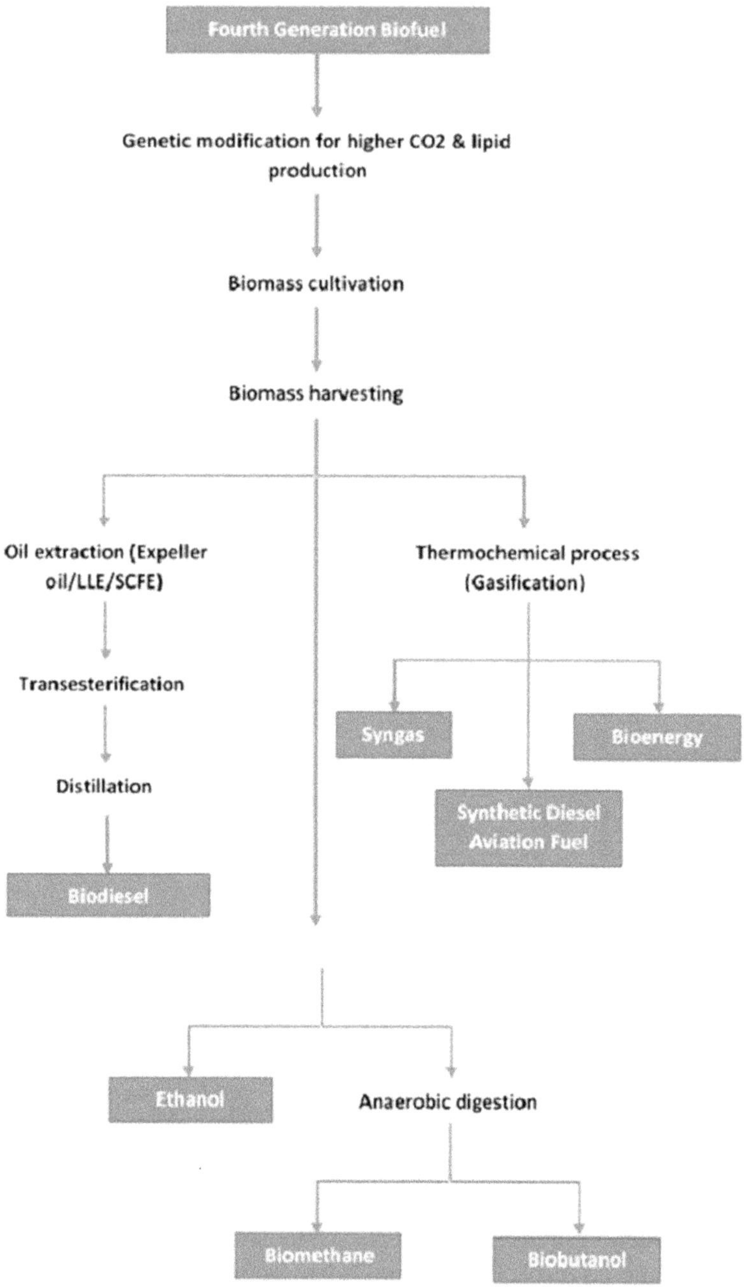

Table 2. Type of catalyst with example

Sr. No.	Type of Catalyst	Example
1	Homogeneous base-catalysts	Potassium hydroxide (KOH), Potassium methoxide (KOCH3), sodium hydroxide (NaOH), sodium methoxide (NaOCH3), sodium ethoxide (NaOCH2CH3)
2	Homogeneous acid-catalyst	Sulfuric acid (H2SO4), Sulfonic acid (C6H6O3S), hydrochloric acid (HCL), and ferric sulphate (Fe2(SO4)3)
3	Heterogeneous acid-catalyst	Ion-exchange resins, tungstated and sulphated zirconia, metal complexes and zeolite
4	Homogeneous acid- catalyst	Single component metal oxides, Zeolites (e.g. Alkali ion-exchanged zeolite), Supported alkali metal, clay minerals, non-oxides (Alkaline alkoxide Alkaline carbonate Guanidine-containing catalysts)
5	Enzymes	Lipaze, Immobilized lipase on SiO2, Immobilized lipase on crystalline PVA

The catalysts used in production of biodiesel can be divided into three categories "Alkalis, Acids, Enzymes". These catalysts can further be classified as heterogeneous and homogeneous. The most preferred type of catalyst is enzymes as it avoids soap formation and purification of the end product is relatively easy. Table 2 depicts different type of catalysts with their chemical names.

Based on the catalyst used the process is identified in following manner:

1. Homogeneous base-catalyzed transesterification
2. Homogeneous acid-catalyzed transesterification
3. Heterogeneous base-catalyzed transesterification
4. Heterogeneous acid-catalyzed transesterification
5. Enzyme catalyzed transesterification

Table 3 highlights the salient features of each transesterification process for producing Biodiesel.

Catalyst Application in Production of Bioethanol

As mentioned in the starting of this topic the production of bioethanol can be categorized in two categories based on feedstock utilized during production in First Generation Bioethanol & Second Generation Bioethanol

Table 3. Differences between various esterification processes

Homogeneous base-catalyzed transesterification	Homogeneous acid-catalyzed transesterification	Heterogeneous base-catalyzed transesterification	Heterogeneous acid-catalyzed transesterification	Enzyme catalyzed transesterification
These catalysts are commonly used because of several benefits such as able to catalyze reaction at low reaction temperature and atmospheric pressure, high conversion in shorter time, and economically available (Lam, Lee, & Mohamed, 2010)	Waste oils contain free fatty acids which cannot be converted to biodiesel using an alkaline catalyst (M. Canakci & J. Van Gerpen, 1999). Hence, liquid acid-catalyzed transesterification is proposed in order to overcome lots of conundrum caused by liquid base catalysts	Most of the heterogeneous catalysts developed for production of biodiesel are either alkaline oxide or alkaline earth metal oxide supported over large surface area (Helwani, Othman, Aziz, Fernando, & Kim, 2009)	Despite of the effectiveness of homogeneous acid catalyst, it can lead to absolute contamination problems ("BIODIESEL DEFINITION AND BENEFITS," n.d.). This will be resulted in higher production cost. It is believed that heterogeneous acid catalysts have the potential as alternative to homogeneous acid catalysts.	enzyme catalysis occurs without the generation of by-products, easy recovery product, mild reaction condition, insensitive to high FFA oil and catalyst can be reuse (Kulkarni & Dalai, 2006).
Alkaline catalyst is more commonly used in commercial biodiesel production because it do not form water during transesterification reaction (Sharma & Singh, 2008)	In the production of biodiesel, hydrochloric acid and sulfuric acid are favoured as catalyst (Lam et al., 2010). Despite of its insensitivity to FFA in the feedstock and can catalyzes esterification and transesterification simultaneously, acid catalyst has been less popular in transesterification reaction because it has relatively slower reaction rate (Arzamendi et al., 2007; Helwani et al., 2009; Kouzu et al., 2008)	Heterogeneous basic catalysts are more active than heterogeneous acid catalyst, similar to their homogeneous counterparts [10].	solid acid catalysts are preferred over liquid acid catalysts due the fact that they contain a multiple sites with different strength of Bronsted or Lewis acidity (Atadashi, Aroua, Abdul Aziz, & Sulaiman, 2013),(Helwani et al., 2009).	enzyme catalyzed biodiesel production has proven to have high potential to be an eco-friendly process and a promising alternative to the chemical process.
NaOH and KOH are the most common homogeneous base catalyst in biodiesel production.	Sulfuric acid, sulfonic acid, hydrochloric acid, organic sulfonic acid, and ferric sulphate are most commonly acids used as catalysts in transesterification (Atadashi et al., 2013)	calcium oxide (CaO) is most commonly used catalyst and provide many advantages such as higher activity, long catalyst life times, and could run in moderate reaction condition(Kouzu et al., 2008)	Types of acid catalysts that were commonly used in esterification and transesterification reaction works include tungsten oxides, sulphonated zirconia (SZ), sulphonated saccharides, Nafionl resins, and organosulphoric functionalized mesoporous silica (Chopade S.G.;Kulkarni K S;Topare N S, 2012)	Lipase is often used in enzymatic catalyzed transesterification.

continued on the following page

Table 3. Continued

Homogeneous base-catalyzed transesterification	Homogeneous acid-catalyzed transesterification	Heterogeneous base-catalyzed transesterification	Heterogeneous acid-catalyzed transesterification	Enzyme catalyzed transesterification
Advantage: - Do not form water during transesterification reaction (Atadashi et al., 2013) - 4000 times faster reaction rate than acid catalyzed transesterification (Atadashi et al., 2013; Lam et al., 2010) - Two-step alkaline-catalyzed transesterification from used vegetable oil is an economic method for biodiesel production (Narasimharao, K.; Lee, Adam; Wilson, 2007) - Reaction can occur at mild reaction condition and thus less energy required(Atadashi et al., 2013; Lam et al., 2010) - NaOH and KOH are economically feasible and widely available (Lam et al., 2010)	Advantage:- Insensitive to FFA and water content in the oil (Edgar Lotero et al., 2005) - Preferred-method if low-grade oil is used - Esterification and transesterification occur simultaneously (Lam et al., 2010) - Saponification can be avoided (Leung, Wu, & Leung, 2010) - Produce high yield of biodiesel (Atadashi et al., 2013)	Advantage:- - Reusable - Easy to separate from product (AYTEN SAGIROGLU, ŞEBNEM SELEN ISBILIR, HAKKI MEVLUT OZCAN, HATICE PALUZAR, & NESLIHAN M. TOPRAKKIRAN, 2011) - Relatively faster reaction rate than acid catalyzed transesterification (Lam et al., 2010) - Reaction can occur at mild reaction condition and relatively lower energy (Lam et al., 2010) - Long catalyst life times (Atadashi et al., 2013)	Advantage:- - Insensitive to FFA and water content in the oil (Edgar Lotero et al., 2005) - Preferred-method if low-grade oil is used (Lam et al., 2010) - Esterification and transesterification occur simultaneously - Easy separation of catalyst from product (Lam et al., 2010) - High possibility to reuse and regenerate the catalyst (Lam et al., 2010; Narasimharao, K.; Lee, Adam; Wilson, 2007) - Recyclable (Leung et al., 2010)	Advantage:- - Prevent saponification (Leung et al., 2010) - Only simple purification step is required (Kulkarni & Dalai, 2006; Leung et al., 2010) - Environmental friendly and nonpolluting (Atadashi et al., 2013)
Disadvantage: - Sensitive to FFA content in the oil (Lam et al., 2010) - Saponification can occur if the FFA content in the oil is more than 2 wt.% (Lam et al., 2010) - Saponification will decrease the biodiesel yield and cause problem during product purification (Atadashi et al., 2013) - Produce more wastewater from purification (Leung et al., 2010)	Disadvantage: - Low reaction rate (Atadashi et al., 2013; Lam et al., 2010; Leung et al., 2010) - Can lead to equipment corrosion (Lam et al., 2010; Leung et al., 2010) - Harder separation of catalyst from product	Disadvantage: - Poisoning of the catalyst when exposed to ambient air - Sensitive to FFA content in the oil due to its basicity property (Lam et al., 2010) - Saponification can occur if the FFA content in the oil is more than 2 wt.% - Saponification will decrease the biodiesel yield and cause problem during product purification (Atadashi et al., 2013; Lam et al., 2010) - Higher molar ratio of alcohol to oil - Produce more wastewater from purification -Diffusion limitation (Leung et al., 2010)	Disadvantage: - Low reaction rates (Atadashi et al., 2013) - Unfavourable side reaction - Higher cost (Leung et al., 2010) - High reaction conditions and longer reaction times (Atadashi et al., 2013) - More energy requirement - Leaching of catalyst active sites may result to product contamination (Lam et al., 2010)	Disadvantage: - Very slow reaction rate (Lam et al., 2010) - High cost (Lam et al., 2010; Leung et al., 2010) - Sensitive to alcohol, typically methanol that can deactivate the enzyme (Bajaj, Lohan, Jha, & Mehrotra, 2010) - Inactivation and denaturation of enzyme can lead to decreasing yield of biodiesel (Helwani et al., 2009)

Catalyst Application in Production of First Generation of Bioethanol

Majority of commercial production of bioethanol is being produced through 1st Generation route due to the ease of obtaining reducing sugar units which can be separated in water (hydrolysed) and subsequently fermented. Sucrose, a disaccharide of α-D-glucopyranose and β-D-fructofuranose, freely hydrolyses in water (Bungay, 1981) .

Starch is a combination of two polysaccharides:

1. Amylose.
2. Amylopectin.

Starches can be hydrolysed efficiently by the use of inexpensive enzymes, mainly alpha-amylase and gluco-amylase, and employing moderate reaction conditions (Mcaloon, Taylor, Yee, Ibsen, & Wooley, 2000). Thus obtained sugar is further fermented to produce 1st Generation bioethanol.

Catalyst Application in Production of Second Generation of Bioethanol

Second generation bioethanol is produced from biomass residue or agro-waste which is popularly known as ligno-cellulosic biomass. As name suggests it has mainly three components Hemi-cellulose, cellulose and lignin. These are basically complex structure which are required to be broken down in monomers of sugars to finally converted into bioethanol through fermentation.

Second generation bioethanol production process mainly involves following broad steps:

1. Size reduction
2. Pretreatment
3. Hydrolysis
4. Fermentation
5. Distillation

Catalysts play a major role in conversion of complex structure of hemi cellulose and cellulose during pretreatment and hydrolysis stages.

Pretreatment

There are various types of pretreatment which can be adopted for preparing biomass for hydrolysis to convert cellulose into hexose. During this process depending upon type of pretreatment hemicellulose also get converted in pentose. Different procedures adopted for pretreatment is enumerated below:

Steam Explosion

This pretreatment has been incorporated in many technologies across the globe depending upon the type of hydrolysis. Some of the major technologies utilizing steam explosion as pretreatment includes Iogen, Abengoa, China Alcohol Resources Corp., Celunol, and GreenField Ethanol. In this process, partially-comminuted biomass is subjected to high pressure steam (at 210-290oC) for required time before this steam is rapidly vented, resulting in an explosive decompression and instant cooling of the biomass. The result of this process is the biomass which is much more amenable to enzymatic hydrolysis. This possible due to the fact that at high temperature and pressure acetic and other acids are liberated from the biomass which acts as catalyst also at high temperature steam has lower pH which also helps in catalyzing the media and hydrolyze the hemicellulose (Weil, Brewer, Hendrickson, Sarikaya, & Ladisch, 1998).

It is seen that there is increase in yield of hemicellulose if the process is catalyzed with dilute sulfuric acid (Moniruzzaman, Dale, Hespell, & Bothast, 1997).

Chemical Pretreatments

- **Acid-Catalysed Pretreatment:** Acids in diluted form is utilized as catalysts such as sulfuric acid, nitric acid and hydrochloric acid during pretreatment mainly to separate hemicellulose from biomass under controlled condition. If the pressure and temperature is maintained at required level xylose yields of 75-90% are possible (Sun & Cheng, 2002). The cellulose does not go un affected by acid pretreatment. Some amorphous regions of the polymer can be hydrolysed (Zerbe, John I & Baker, 1988) and this can result in a strong decrease in the degree of polymerisation (DP) of cellulose (Dinus & R.J., 2000). The term 'Level-Off Degree of Polymerisation' (LODP) is often used for the DP of the remaining crystallites of cellulose after this preliminary acid hydrolysis (Krässig, 1993).
- **Alkaline Catalyzed Pretreatment:** Alkaline treatments are being used by paper-pulping processes for the years. Unlike acid pre-treatments which are focused on hydrolyzing hemicellulose, these treatments are more focused on the solubilisation and removal of lignin from the biomass (Hamelinck,

Hooijdonk, & Faaij, 2005). If alkaline pretreatment performed prior to acidic pretreatment, the basic nature of alkaline media helps in removing acetyl and uronic acid groups from hemicellulose, further improving the digestibility of the polysaccharides to enzymatic treatments (Chang & Holtzapple, 2000). The Conditions for Alkaline Catalysed Pretreatment are more mild than acid pre-treatments, hence the reaction time is more than that of acid pretreatment (Wyman et al., 2005). The most common alkaline catalysts are sodium, calcium, and potassium hydroxide but they are hard to recover and often costly. The Lime can be used, which is cheap and easily recoverable from water as insoluble calcium carbonate by reaction with carbon dioxide (Chang, Nagwani, & Holtzapple, 1998). If the biomass is treated with added oxygen under alkaline conditions, with superheated water having more than 185^{0}C which is kept in liquid state by maintaining elevated pressure, the hydrolysis of most of the lignin and hemicellulose happens leaving pure and less crystalline cellulose which provides ease in enzymatic hydrolysis (Schmidt & Thomsen, 1998).

- **Other Chemical Treatments:** There are other solvents available which acts as catalysts and help in the de-crystallisation and depolymerisation of cellulose. The examples are dioxane and phenol, glycerol, Hydrogen peroxide, ozone etc.

Hydrolysis Methods

Two stage hydrolysis with dilute acid: In this method dilute acid (e.g. sulfuric acid, nitric acid) are used as catalyst at 140 to 160 ^{0}C temperature to first hydrolyzed hemicellulose. After hydrolyzing hemicellulose, the temperature of the process is raised 200-250 ^{0}C. At this temperature the cellulose gets hydrolyzed and get converted into hexose.

Two stage hydrolysis with Concentrated Acid: In this method first stage with dilute or concentrated acid is performed to hydrolyze the hemicellulose and convert it into xylose. While in second stage the remaining biomass is completely dried and concentrated acid with upto 90% concentration is added to biomass (Dipardo, n.d.). This results into complete amorphization of cellulose. This amorphous cellulose forms homogeneous gelatin with acid and can easily be hydrolyzed into glucose using water at slightly high temperature. Of course this happens with slight degradation (Dinus & R.J., 2000) of sugar but experiments show negligible effect. A number of feedstock when processed at bench-scale shows 95% conversion of xylose and 87% conversion of glucose (Broder & Barrier, 1988).

Enzymatic Hydrolysis: Majority of technologies available across the globe utilizes enzymatic hydrolysis to convert cellulose into glucose. Biomass is first pretreated to make it suitable for hydrolysis with enzyme using steam, acidic or alkaline catalyst. In this process most of the hemicellulose is converted into xylose and cellulose becomes more susceptible for hydrolysis. The cellulase enzymes are used to hydrolyze the cellulose to glucose.

The cellulase can be obtained from various bacteria and fungi (micro-organisms). A cellulase is mixture of many other enzymes like endo-1,4-β-D-glucanases and cellobiohydrolases. The process is typically carried out at temperature of 30-70 ^{0}C hence the degradation of glucose is minimum compared to other types of hydrolysis. Naturally, this results into highest conversion of biomass to bioethanol.

Catalysts Application in Thermo-Chemical Technologies

This technology is basically used to produce bio-oil through heating the biomass. There are two paths available to produce bio-oil

1. **Pyrolysis:** Direct liquefaction
2. **Gasification:** Indirect liquefaction

Pyrolysis: Direct Liquefaction

Pyrolysis is process of heating organic matter at elevated temperature in the absence of oxygen such that decomposition of the organic matter happens with change in the chemical and physical properties. This process is irreversible. The gases produced during pyrolysis are rapidly quenched to prevent them from getting depolymerized and convert them in bio-oil. Catalysts have roles in numerous areas of pyrolysis. For example, catalyst like zinc chloride lets the process to occur at lower temperatures (Bungay, 1981) and using Acidic-catalyst in pyrolysis has made possible produce levoglucosene from wood which has applications in the pharmaceutical industry (Dobele et al., 2003).

Gasification: Indirect Liquefaction

In gasification process the biomass is heated at elevated temperature (>1000 ^{o}C) with higher amount of oxygen than that of used in pyrolysis. As the gasification process is endothermic additional oxygen is required in the quantum of 25% of the biomass quantity. In order to have synthesis of biomass chemicals require much cleaner gas than air. Hence pure oxygen or steam is used. Thus the result gas is known as synthesis gas or syngas.

The general formula for the gasification of glucose utilizing oxygen as fuel is shown below:

$$C_6H_{12}0_6 + 3/2\ O_2 \rightarrow 6\ CO + 3\ H_2 + 3\ H_2O$$

Several possible products which can be synthesized from these gases are shown in the following Figure, along with the catalysts used. The synthesis of this gases depends on the ratio of carbon monoxide to hydrogen in the syngas.

Figure 11 and 12 indicates the potential chemicals from syngas and some of the catalysts involved (Overend, 2004).

Catalysts application for Value added products

1st generation and 2nd generation feedstock as described above can be converted into pentose, hexose, lignin. When this first derived products are processed with help of specific catalysts, they produces many high value biochemical which have numerous applications in various industries. Authors (Werpy & Petersen, 2004) has identified 12 building blocks which have numerous applications in industries which are enumerated in Table 4.

Figure11. Chemicals derived from Syngas with help of catalysts

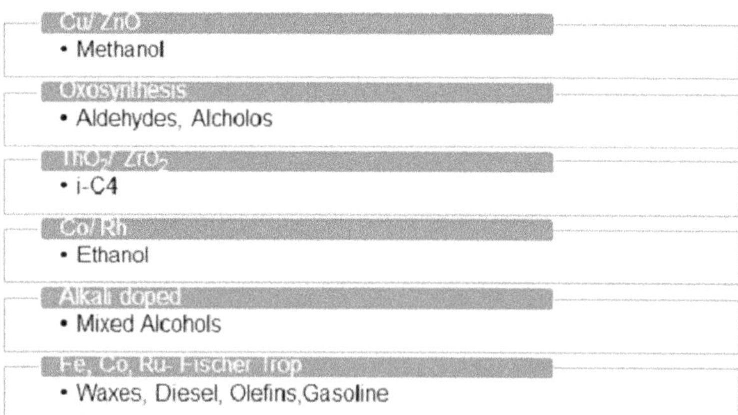

Figure 12. Chemicals derived from Methanol with help of catalysts

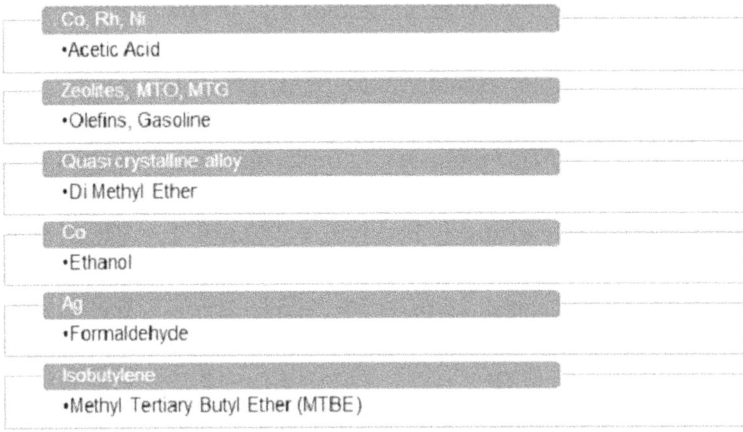

Table 4. Twelve building blocks

Building Blocks
1,4 succinic, fumaric and malic acids
2,5 furan dicarboxylic acid
3 hydroxy propionic acid
aspartic acid
glucaric acid
glutamic acid
itaconic acid
levulinic acid
3-hydroxybutyrolactone
Glycerol
Sorbitol
xylitol/arabinitol

PERSPECTIVES ON CLIMATE CHANGE AFTER BIOFUEL IMPLEMENTATION IN INDIA

Most of the biofuels are environmental friendly and help in reducing Green House Gases in atmosphere as compared to other conventional fossil fuels. Indian Government has implemented several policies, keeping in view the welfare of the country's environmental conditions and energy security viz. Ethanol Blended Petrol (EBP) Programme, National Mission on Biodiesel and National Policy on Biofuels. These policies served as a major milestone in country's progress towards sustainable energy and motivated further research in the field of biofuels.

The main focus has been on biodiesel and bioethanol. Bioethanol is currently being produced from sugarcane molasses. Jatropha plantation was promoted among farmers for biodiesel production on large scale. The main reasons for choosing Jatropha for biodiesel production is because it could be cultivated on barren or fallow land, its low demand for water and fertilizers and the resistance of crop to pests and drought. Advanced technologies for production of biofuel like conversion of lignocellulosic materials (i.e., agricultural residues, forest residues, algae etc.) to ethanol, are also being explored.

Land Use and Biofuels

Land is required for growing fuel crops for production of biofuels. One of the salient features of the biofuel policy is that it aims to use barren or wasteland or marginal land for plantation of crops for biofuels, thus avoiding any fuel vs food conflict. However, even if plants like Jatropha are promoted on marginal land, existing activities such as livestock grazing and gathering of forest products by local communities will be displaced. Land use change for biofuel production may contribute more to emissions which can significantly impact climate change. However, Jatropha cultivation will also help in providing benefits in the form of employment and rural development.

To reduce the land use change to minimum alternate biofuel technologies and feedstock are also being explored. Microalgae have reported highest potential producing 15-300 times more biodiesel than oilseed crops on the area basis (Chandel & Sukumaran, n.d.). Biofuels like second generation ethanol have little impact on land use as they can be produced from agricultural residues like rice straw, wheat straw etc. However, existing fodder resources can meet only 51.92% of the total requirements to sustain the livestock population of India as per Ministry of Agriculture and Farmers Welfare (Kothari & Mishra, 2017). So it is necessary to optimize combination of different advanced biofuel technologies keeping in mind the impact on land availability and its use.

Water Use and Biofuels

Large scale biofuel production affects water quality in many ways for instance by the direct use in growing crops and its processing into fuels. This also pollutes the surface and ground water by the release of chemical fertilizers, pesticides and other by-products of the process. Although Jatropha can be grown in water scarce area, increased yields can be obtained through irrigation, thus competing with resources for agriculture.

Ethanol can be produced from sugarcane, corn, sugar beet, corn, cassava, forest residues, agriculture residues etc. Sugarcane crop, which is a major feedstock for production of ethanol, puts stress on the already depleting water resources of our country, as it is having a water requirement of 20,000 – 30,000 m^3 ha^{-1} $crop^{-1}$ [1]. Also, for processing of molasses to ethanol in distillery, fresh water is required in large quantity which is typically in the range of 9 to 21 litres per litre of alcohol production [5]. On the other hand, feedstock for the second generation ethanol, i.e. rice straw, corn stover, cotton stalk etc., are already present as a waste in the field after harvesting of crops. So second generation ethanol does not require any additional water for production of feedstock. The water required in second generation biorefineries is in the range of 10 to 30 litre per litre of ethanol production. The quantity of water required for the conversion of agricultural residues to ethanol varies depending on the technology used and the feedstock used.

Impact of Biofuels on Ecosystem

For biofuel production, large-scale plantation of monospecific species like Jatropha may create problems in the biodiversity rich states of Chhattisgarh, Mizoram, Uttarakhand etc. Plantations of such crops has the potential to replace forests and grazing lands. These will affect the common property resources such as firewood, marginal grazing land for animals and watersheds which helps in providing support to the landless people in the area.

Jatropha plantations, at the same time, can provide environmental benefits like protection of crops or pasture land, and act as a fence for erosion control and also a source of organic manure. The environmental sustainability of Jatropha production and processing is still not clear as different LCA studies have given huge variations in the results. Several researchers have shown that the answer to this problem highly depends on previous use of land, cultivation intensity, processing efficiency and use of products & by-products produced by Jatropha.

Some of the agricultural residues which are not collected in the field are ploughed into the soil to serve as a fertilizer for the next crop. If the same agricultural residues are used for the production of second generation ethanol, then the need for fertilizer will increase for the next crop. However, in most of the places in India the uncollected agricultural residue is burnt in the field to prepare the field for the next crop because of it is much more convenient and cheaper than ploughing the field. It has been observed that open burning of agricultural residues contributes

Figure 13. Comparison of Well-to-Wheels (WTW) emissions from gasoline and first generation ethanol with second generation ethanol produced from different technologies
(Bioenergy, 2015; Detailed California-Modified GREET Pathway for Cellulosic Ethanol from Forest Waste, 2009, Lifecycle GHG Emissions from Poet-DSM Project Liberty Cellulosic Ethanol Plant, 2015; Gao, 2011; Groode, Tiffany Amber, 2008; Pont & Unnasch, 2014; Sabina Yunis & Renata Oliveira, 2015)

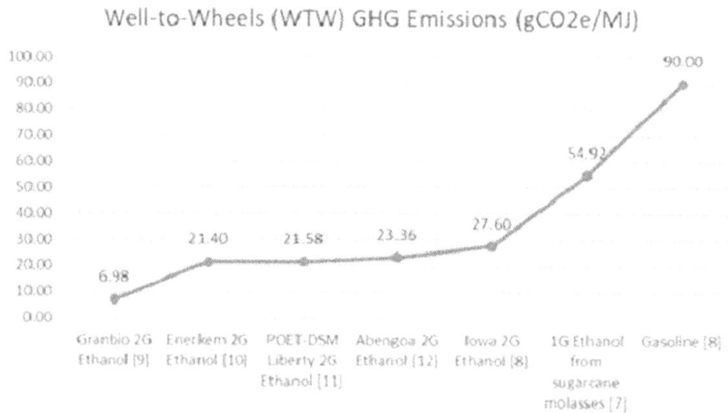

to emissions of harmful air pollutants, which can cause severe impacts on human health and environment. In Asia, the annual contribution form agriculture residues burning is calculated to be 0.10 Tg of SO_2, 0.96 Tg of NO_x, 379 Tg of CO_2, 23 Tg of CO and 0.68 Tg of CH_4 (Gadde, Bonnet, Menke, & Garivait, 2009; Streets, Yarber, Woo, & Carmichael, 2003). The surplus agricultural residues which are burnt in the field can be used for manufacturing of ethanol, thereby reducing emissions and contributing to the country's requirement for ethanol for blending in MS.

Carbon Accumulation Potential through Biofuels

Main aim for using biofuel is that they are environmental friendly and help in reducing GHG emissions. The biofuel feedstock uses CO_2 to grow. Thus, most of the CO_2 which is released when biomass is converted into biofuel and burned in vehicle engines is recaptured when new biomass is grown to produce more biofuels. Carbon sequestration is an important distinction between biofuels and conventional fossil fuels. A study by Francis et al. has projected a CO_2 sequestration potential of Jatropha plantation of 4.6 and 22.9 Mt $year^{-1}$ if 2 mha and 10 mha of wasteland is cultivated with Jatropha. Such plantations, when managed with regular pruning, could store 8-10 tons C ha^{-1} in above ground biomass and litter in India. The use of microalgae could incorporate an average of 0.26 x 10^6 tons C into the harvested biomass annually (Chandel & Sukumaran, n.d.) . However, for different types of algae carbon sequestration potential will be different.

Environmental Sustainability and LCA of biorefineries

Life Cycle Analysis (LCA) is an important tool to assess the capability of biofuel and its impact on the environment. The concept is based on the impact of biofuel on environment in all of its life-cycle stages viz: raw material production, transportation, processing, power consumption, waste management, biofuel transport and its final use in vehicles. Basically it covers all of its life-cycle phases (pre-operation, operation and post-operation) which have environmental and energy impacts.

Theoretically, greenhouse gas emissions from Biofuels may approach zero as the carbon dioxide is absorbed by biofuel plants. It has been found out in various studies that emission reduction from first generation biofuels is in the range of 20-60% of CO_2eq relative to fossil fuels while the expected reductions in emissions from future commercialized Second Generation biofuels is in the range of 70-90% of CO_2eq relative to fossil fuels (Chandel & Sukumaran, n.d.)

Well-to-Wheels Approach

A Well-to-Wheels (WTW) analysis is a systematic approach for assessing the energy consumption and GHG emissions related to a fuel. It consists of two stages: 1) a well-to-pump stage, 2) a pump-to-wheels stage. Well-to-pump stage analysis includes the production of feedstock for biofuel, conversion of the feedstock to fuel and its storage and the subsequent transportation, storage and distribution of the fuel. Pump-to-wheels stage analysis refers to the vehicle operation activities through its lifetime. Figure 1 gives WTW emissions from gasoline, first generation ethanol and second generation ethanol.

Second generation biofuels have high potential to reduce carbon dioxide emissions and does not raise the food vs fuel debate as compared to first generation biofuels. Biofuels can have a positive or negative impact on the environment depending on how the fuel is grown, processed and then used. So clear standards, guidelines and models are needed to distinguish advantages of different biofuels and their feedstock to fuel pathways.

REFERENCES

Al-mulali, U., & Mohammed, A. H. (2015). The relationship between energy consumption and GDP in emerging countries. *International Journal of Energy Sector Management*, *9*(1), 77–93. doi:10.1108/IJESM-04-2013-0006

Anwar, Z., Gulfraz, M., & Irshad, M. (2014). Agro-industrial lignocellulosic biomass a key to unlock the future bio-energy: A brief review. *Journal of Radiation Research and Applied Sciences*, *7*(2), 163–173. doi:10.1016/j.jrras.2014.02.003

Arzamendi, G., Campo, I., Arguiñarena, E., Sánchez, M., Montes, M., & Gandía, L. M. (2007). Synthesis of biodiesel with heterogeneous NaOH/alumina catalysts: Comparison with homogeneous NaOH. *Chemical Engineering Journal*, *134*(1–3), 123–130. doi:10.1016/j.cej.2007.03.049

Atadashi, I. M., Aroua, M. K., Abdul Aziz, A. R., & Sulaiman, N. M. N. (2013). The effects of catalysts in biodiesel production: A review. *Journal of Industrial and Engineering Chemistry*, *19*(1), 14–26. doi:10.1016/j.jiec.2012.07.009

Bajaj, A., Lohan, P., Jha, P. N., & Mehrotra, R. (2010). Biodiesel production through lipase catalyzed transesterification: An overview. *Journal of Molecular Catalysis. B, Enzymatic*, *62*(1), 9–14. doi:10.1016/j.molcatb.2009.09.018

Biodiesel Definition And Benefits. (n.d.). Retrieved from http://www.tristatebiodiesel.com/download/biodieselbenefits.pdf

Bioenergy, A. (2015). *Lifecycle GHG Emissions from Abengoa Bioenergy Biomass of Kansas, LLC Hugoton Cellulosic Ethanol Plant*. Retrieved from https://www.arb.ca.gov/fuels/lcfs/2a2b/apps/abbk-rpt-012915.pdf

BP. (2017). *BP Energy Outlook Country and regional insights – India*. Retrieved from https://www.bp.com/content/dam/bp/pdf/energy-economics/energy-outlook-2017/bp-energy-outlook-2017-country-insight-india.pdf

Broder, J. D., & Barrier, J. W. (1988). Producing ethanol and coproducts from multiple feedstocks. *American Society of Agricultural Engineers (Microfiche Collection) (USA)*. Retrieved from http://agris.fao.org/agris-search/search.do?recordID=US9023079

Bungay, H. R. (1981). *Energy, the biomass options. Energy, the biomass options*. New York: John Wiley & Sons. Retrieved from https://www.cabdirect.org/cabdirect/abstract/19840693586

Canakci, M., & Van Gerpen, J. (1999). Biodiesel production viaacid catalysis. *Transactions of the ASAE. American Society of Agricultural Engineers, 42*(5), 1203–1210. doi:10.13031/2013.13285

Chandel, A. K., & Sukumaran, R. K. (n.d.). *Sustainable biofuels development in India*. Academic Press.

Chang, V. S., & Holtzapple, M. T. (2000). Fundamental Factors Affecting Biomass Enzymatic Reactivity. *Applied Biochemistry and Biotechnology, 84–86*(1–9), 5–38. doi:10.1385/ABAB:84-86:1-9:5

Chang, V. S., Nagwani, M., & Holtzapple, M. T. (1998). Lime pretreatment of crop residues bagasse and wheat straw. *Applied Biochemistry and Biotechnology, 74*(3), 135–159. doi:10.1007/BF02825962 PMID:9429297

Chopade, S. G., Kulkarni, K. S., & Topare, N. S. (2012). Solid Heterogeneous Catalysts For Production Of Biodiesel From Trans-Esterification Of Triglycerides With Methanol: A Review [Abstract]. *Acta Chimica and Pharmaceutica Indica*, 8–14. Retrieved from http://www.tsijournals.com/abstract/solid-heterogeneous-catalysts-for-production-of-biodiesel-from-transesterification-of-triglycerides-with-methanol-a-revi-11071.html

Demirbaş, A. (2001). Biomass resource facilities and biomass conversion processing for fuels and chemicals. *Energy Conversion and Management, 42*(11), 1357–1378. doi:10.1016/S0196-8904(00)00137-0

Detailed California-Modified GREET Pathway for Cellulosic Ethanol from Forest Waste. (2009). Retrieved from https://www.arb.ca.gov/fuels/lcfs/022709lcfs_forestw.pdf

Dinus, R. J. (2000). *Genetic Modification of Short Rotation Poplar Biomass Feedstock for Efficient Conversion to Ethanol*. doi:10.2172/769297

Dipardo, J. (n.d.). *Outlook for Biomass Ethanol Production and Demand*. Retrieved from https://www.agmrc.org/media/cms/biomass_E6EE9065FD69D.pdf

Dobele, G., Dizhbite, T., Rossinskaja, G., Telysheva, G., Meier, D., Radtke, S., & Faix, O. (2003). Pre-treatment of biomass with phosphoric acid prior to fast pyrolysis: A promising method for obtaining 1,6-anhydrosaccharides in high yields. *Journal of Analytical and Applied Pyrolysis, 68–69*, 197–211. doi:10.1016/S0165-2370(03)00063-9

Dutta, K., Daverey, A., & Lin, J.-G. (2014). Evolution retrospective for alternative fuels: First to fourth generation. *Renewable Energy*, *69*, 114–122. doi:10.1016/j.renene.2014.02.044

Gadde, B., Bonnet, S., Menke, C., & Garivait, S. (2009). Air pollutant emissions from rice straw open field burning in India, Thailand and the Philippines. *Environmental Pollution*, *157*(5), 1554–1558. doi:10.1016/j.envpol.2009.01.004 PMID:19201513

Gao, L. (2011). Well-to-Wheels Analysis of Energy Use and Greenhouse Gas Emissions for Alternative Fuels. *International Journal of Applied Science and Technology*, *1*(6). Retrieved from http://www.ijastnet.com/journals/Vol_1_No_6_November_2011/1.pdf

Groode, T. A. (2008). *Biomass to ethanol: potential production and environmental impacts*. Massachusetts Institute of Technology. Retrieved from http://dspace.mit.edu/handle/1721.1/43144

Hamelinck, C. N., van Hooijdonk, G., & Faaij, A. P. (2005). Ethanol from lignocellulosic biomass: Techno-economic performance in short-, middle- and long-term. *Biomass and Bioenergy*, *28*(4), 384–410. doi:10.1016/j.biombioe.2004.09.002

Helwani, Z., Othman, M. R., Aziz, N., Fernando, W. J. N., & Kim, J. (2009). Technologies for production of biodiesel focusing on green catalytic techniques: A review. *Fuel Processing Technology*, *90*(12), 1502–1514. doi:10.1016/j.fuproc.2009.07.016

IEA. (2015). *India Energy Outlook - World Energy Outlook Special Report 2015*. Retrieved from https://www.iea.org/publications/freepublications/publication/IndiaEnergyOutlook_WEO2015.pdf

Jackson, N.B. (1999). *Research Teams and Partnerships: Trends in the Chemical Sciences*. Washington, DC: National Academies Press. doi:10.17226/9759

Kothari, B. L., & Mishra, N. (2017). *CHAPTER VI | Department of Animal Husbandry, Dairying & Fisheries*. Retrieved January 10, 2018, from http://dahd.nic.in/related-links/chapter-vi

Kouzu, M., Kasuno, T., Tajika, M., Sugimoto, Y., Yamanaka, S., & Hidaka, J. (2008). Calcium oxide as a solid base catalyst for transesterification of soybean oil and its application to biodiesel production. *Fuel*, *87*(12), 2798–2806. doi:10.1016/j.fuel.2007.10.019

Krässig, H. A. (1993). *Cellulose: structure, accessibility, and reactivity*. Yverdon: Gordon and Breach Science. Retrieved from http://www.worldcat.org/title/cellulose-structure-accessibility-and-reactivity/oclc/876452167?referer=di&ht=edition

Kulkarni, M. G., & Dalai, A. K. (2006). Waste Cooking OilAn Economical Source for Biodiesel: A Review. *Industrial & Engineering Chemistry Research*, *45*(9), 2901–2913. doi:10.1021/ie0510526

Lam, M. K., Lee, K. T., & Mohamed, A. R. (2010). Homogeneous, heterogeneous and enzymatic catalysis for transesterification of high free fatty acid oil (waste cooking oil) to biodiesel: A review. *Biotechnology Advances*, *28*(4), 500–518. doi:10.1016/j.biotechadv.2010.03.002 PMID:20362044

Leung, D. Y. C., Wu, X., & Leung, M. K. H. (2010). A review on biodiesel production using catalyzed transesterification. *Applied Energy*, *87*(4), 1083–1095. doi:10.1016/j.apenergy.2009.10.006

Lifecycle GHG Emissions from Poet-DSM Project Liberty Cellulosic Ethanol Plant. (2015). Retrieved from https://www.arb.ca.gov/fuels/lcfs/2a2b/apps/poet-lib-rpt-121715.pdf

Lindström, B., & Pettersson, L. J. (2003). A Brief History of Catalysis. *CATTech*, *7*(4), 130–138. doi:10.1023/A:1025001809516

Lotero, Liu, Lopez, Suwannakarn, Bruce, & Goodwin. (2005). *Synthesis of Biodiesel via Acid Catalysis*. doi:10.1021/IE049157G

Mcaloon, A., Taylor, F., Yee, W., Ibsen, K., & Wooley, R. (2000). *Determining the Cost of Producing Ethanol from Corn Starch and Lignocellulosic Feedstocks*. Retrieved from https://www.nrel.gov/docs/fy01osti/28893.pdf

Moniruzzaman, M., Dale, B. E., Hespell, R. B., & Bothast, R. J. (1997). Enzymatic hydrolysis of high-moisture corn fiber pretreated by afex and recovery and recycling of the enzyme complex. *Applied Biochemistry and Biotechnology*, *67*(1–2), 113–126. doi:10.1007/BF02787846

Narasimharao, K., Lee, A., & Wilson, K. (2007). Catalysts in Production of Biodiesel: A Review. *Journal of Biobased Materials and Bioenergy*, *1*(3), 301–314. doi:10.1166/jbmb.2007.002

Overend, R. P. (2004). *Thermochemical Biomass Gasification Technologies and Products. Stanford University: Global Climate and Energy Project (GCEP).* Battelle. Retrieved from https://web.stanford.edu/group/gcep/pdfs/energy_workshops_04_04/biomass_overend.pdf

Phillips, W. (1991). Critical Technologies: Panel selects 22 national priorities. *Chemical and Engineering News*, *69*(17), 4. doi:10.1021/cen-v069n017.p004

Pont, J., & Unnasch, S. (2014). *GranBio-Modified GREET Pathway for the Production of Ethanol from Sugarcane Straw*. Retrieved from https://www.arb.ca.gov/fuels/lcfs/2a2b/apps/gb-rpt-082514.pdf

Sagiroglu, A., Isbilir, Ş. S., Ozcan, H. M., Paluzar, H., & Toprakkiran, N. M. (2011). Comparison Of Biodiesel Productivities Of Different Vegetable Oils By Acidic Catalysis. *Chemical Industry & Chemical Engineering Quarterly*, *17*(1), 53–58. doi:10.2298/CICEQ100114054S

Schmidt, A. S., & Thomsen, A. B. (1998). Optimization of wet oxidation pretreatment of wheat straw. *Bioresource Technology*, *64*(2), 139–151. doi:10.1016/S0960-8524(97)00164-8

Sharma, Y. C., & Singh, B. (2008). Development of biodiesel from karanja, a tree found in rural India. *Fuel*, *87*(8–9), 1740–1742. doi:10.1016/j.fuel.2007.08.001

Streets, D. G., Yarber, K. F., Woo, J.-H., & Carmichael, G. R. (2003). Biomass burning in Asia: Annual and seasonal estimates and atmospheric emissions. *Global Biogeochemical Cycles, 17*(4). doi:10.1029/2003GB002040

Sun, Y., & Cheng, J. (2002). Hydrolysis of lignocellulosic materials for ethanol production: A review. *Bioresource Technology*, *83*(1), 1–11. doi:10.1016/S0960-8524(01)00212-7 PMID:12058826

Technology Vision 2020: The U.S. Chemical Industry. (1996). Retrieved from http://www.acs.org

Watling, L., & Thiel, M. (2013). *The natural history of the Crustacea*. Oxford University Press. Retrieved from https://global.oup.com/academic/product/lifestyles-and-feeding-biology-9780199797028#.WkoN_7fQqTQ.mendeley

Weil, J., Brewer, M., Hendrickson, R., Sarikaya, A., & Ladisch, M. R. (1998). Continuous pH monitoring during pretreatment of yellow poplar wood sawdust by pressure cooking in water. *Applied Biochemistry and Biotechnology*, *70–72*(1), 99–111. doi:10.1007/BF02920127

Werpy, T. G., & Petersen, G. (2004). *Top Value Added Chemicals from Biomass Volume I—Results of Screening for Potential Candidates from Sugars and Synthesis Gas Energy Efficiency and Renewable Energy*. Retrieved from https://www.nrel.gov/docs/fy04osti/35523.pdf

Wyman, C. E., Dale, B. E., Elander, R. T., Holtzapple, M., Ladisch, M. R., & Lee, Y. Y. (2005). Coordinated development of leading biomass pretreatment technologies. *Bioresource Technology*, *96*(18), 1959–1966. doi:10.1016/j.biortech.2005.01.010 PMID:16112483

Yunis & Oliveira. (2015). *Brazilian Modified GREET Pathway for the Production of Ethanol from Sugarcane -Method 2B: Pathway for Brazilian Sugarcane Ethanol from molasses, considering specifics mechanization level inputs and co-generation export credit from ABENGOA*. Retrieved from https://www.arb.ca.gov/fuels/lcfs/2a2b/apps/abg-brz-rpt-072915.pdf

Zerbe, J. I., & Baker, A. J. (1988). *Investigation of fundamentals of two-stage, dilute sulfuric acid hydrolysis of wood*. Forests Products Laboratory. Retrieved from https://catalogue.nla.gov.au/Record/4049848

Chapter 6
New Catalytic Approaches for Producing Alternative to MTBE Additives for Reformulation of Gasoline

Saud Aldrees
University of Oxford, UK

ABSTRACT

Due to the lead phase out that began in 1973, refiners had to replace the octane loss in gasoline. For this purpose, oxygenates and highly branched alkylates will play a major role as gasoline additives because they have relatively high octane ratings. Methyl tertiary-butyl ether (MTBE) is the oxygenate as a gasoline additive the most widely used nowadays in most of the countries in the world. It is used to raise octane levels, enhance engine performance, improve combustion efficiency and to reduce emissions of air pollutants such as carbon monoxide and hydrocarbons.

INTRODUCTION

Alkylates containing C_8 fractions or higher hydrocarbons are usually produced from alkylation of isobutane with olefin process in the presence of concentrated sulphuric acid or hydrogen fluoride that is used as liquid catalyst. The demand for branched C_8 fractions (isomers) increases sharply due to their uses as prime solvents and additives to gasoline. The present alkylation process suffers from inherent drawbacks such as corrosion, toxicity, and environmental problems. Therefore, oligomerization of light

DOI: 10.4018/978-1-5225-8033-1.ch006

olefins might be an attractive alternative to produce liquid hydrocarbon alkylates as components of gasoline and high value petrochemical products.

In addition, in view of the growing concern on environmental pollution, fuel reformulation is now carried out on a worldwide basis due to US and European legislations, which are focused on reducing evaporative emissions, lower sulphur content, lower aromatics and on complete fuel combustion. In this context, an interesting route for the production of environmentally friendly fuel is the dimerization, trimerization, tetramerization or oligomerization of light olefins

Oxygenates are hydrocarbons that contain one or more oxygen atoms. The primary oxygenates are alcohols and ethers, including: fuel ethanol, MTBE, ethyl tertiary butyl ether (ETBE), and tertiary amyl methyl ether (TAME).

The blending of (MTBE) into motor gasoline increased dramatically since it was first produced 37 years ago. MTBE has a blending octane in the range of 106-110 and make up approximately 5-15% of the gasoline pool. MTBE usage grew in the early 1980s in response to octane demand resulting initially from the phase out of lead from gasoline and later from rising demand for premium gasoline.

Methyl tertiary butyl ether use was mandated into united state automobile gasoline in the early 1990s. Due to leaking storage and distribution systems, MTBE contamination of water supplies has been discovered, causing state of California to completely banned MTBE in 2005. With other states proposing similar legislation. Refiners will replace MTBE in gasoline with ethanol. Some MTBE plant shut down and some will be converted to isooctane or alkylation in order to consume the excess isobutylene no longer consumed for MTBE.

Upgrading Gasoline in Oil Refining Processes

In view of MTBE phase out schedules adopted, various options for octane enhancement have been explored. The following main oil refining processes play key role in gasoline production:

- Crude oil distillation.
- Catalytic cracking conversion processes.
- Upgrading processes, such as catalytic reforming, isomerization, alkylation, polymerization of light olefins to form polygasoline, etherification processes, and isobutene dimerization.

In a breakthrough development, a research team at the National Center for Petrochemical Technology has developed a technology that has the potential to substantially expand the prospects for clean fuel. The technology involves a new nanocatalyst, clean fuel, additive and a new process to produce fuel distillates and additives. It also enables the use of different feedstock than commonly used at this time. And, if the current bench-scale results hold during production scale, a conversion rate of 90% or above may be achieved from feedstock into additive and distillates in the fuel range including gasoline, jet fuel and diesel. The end product will be nanocatalyst, additive and distillates free of sulfur, aromatics and nitrogen compounds. Besides being environment-friendly, the distillates and non-oxygenated alternative to MTBE as an anti-knock additive to gasoline, will generate significant market opportunities as for refineries in Saudi Arabia (SAUDI-ARAMCO) and around the world that choose to use this technology.

Composion of Gasoline as a Fuel

Gasoline is a mixture of relatively volatile hydrocarbons, including alkanes, cycloalkanes, alkenes, and aromatics. It is considered as one of the most important products of the oil refining industry; this is due to its extensive usage as a fuel. Because of this gasoline gains importance and is a subject of many of investigations that are aimed towards:

- Developing and improving its characteristics.
- Increasing the effectiveness of production and usage.
- Reduction of toxicity resulting from its use.
- Improving processes and the economics of production.

Gary et al. (1994) mentioned that gasolines are complex volatile mixtures of liquid hydrocarbons having a boiling point range from 38-205 °C at atmospheric pressure. Hamilton (1996) also mentioned that gasoline contains over 500 hydrocarbons that may have 5 to 12 carbon atoms. It is suitable as a fuel in spark-ignition internal combustion engines.

As Kettering (1944) discussed, hydrocarbons are any molecules that just contain hydrogen and carbon, both of which are fuel molecules that can be completely burnt to form water (H_2O), carbon dioxide (CO_2) and evolved large quantity of heat. But, If the combustion is not complete, carbon monoxide (CO) maybe formed.

The physical and chemical properties of gasoline differ by the difference of the percentage of its constituents, which typically present in traditional gasoline are shown in Table 1.

Table 1. Typical composition of gasoline

Fraction	Percentage
- Aliphatic – straight chain (n-paraffins) - Aliphatic branched (iso-paraffins)	30-50
- Aliphatic cyclic (cycloparaffins)	20-30
- Aromatic	20-35
- olefins	0-8

Gasoline Production

Gasoline is most often produced by the fractional distillation of crude oil. The crude oil is separated into fractions according to different boiling points of hydrocarbons of varying chain lengths. This fractional distillation process yields approximately 25% of straight-run gasoline from each barrel of crude oil.

Gasoline components can also be produced from several processing units in the crude oil refineries such as catalytic cracking unit, hydrocracking unit and reforming unit. In addition, the components can also be produced from isomerization of naphtha and alkylation of isooctane with 1-butene, processes besides naphtha that is produced from atmospheric distillation unit. These components form what is commonly called the "gasoline pool", and are mixed according to scientific methods and procedures to give the required type of gasoline meeting product specifications, conforming to scientific characteristics and legal standard and regulations. and in the required amounts with good economic benefits and return.

Thus, gasoline is constituted in oil refineries by blending the hydrocarbons which fall within its boiling range that are produced from different processing units of the refinery.

Upgrading Gasoline in Oil Refining Processes

The authors of Hobson (1992) and Kutler (1969) discuss in their research the modern refinery processes for gasoline components production include the following conversion processes.

1. Conversion Processes

a. Catalytic cracking: It converts heavy feeds into lighter materials for further processing or direct blending. Fluid catalytic cracking (FCC) is the most important process which breaks larger, higher-boiling hydrocarbons into more valuable gasoline range including:

- A moderate-quality, high-octane gasoline blend stock (91-93RON) called FCC gasoline, and
- About 30% aromatics, 20-30% olefins. and refinery gases, which may be sold for use as feed to alkylation and oxygenate production.

b. Hydrocracking Cracks and adds hydrogen to molecules, producing a more saturated, stable, gasoline fraction.

2. Upgrading Processes

Straight-run gasoline (directly from refinery distillation column) has an octane number of about 70. In other words, straight-run gasoline has some knocking properties as a mixture of 70% isooctane and 30% heptane. Many of these compounds are straight chain alkanes. Cracking, isomerization, and other refinery processes can be used to increase the rating of gasoline to about 90. Anti-knocking agents may be added to further increase the octane rating.The most important upgrading processes are:

3. Isomerisation

Isomerization raises gasoline fraction octane by converting straight chain hydrocarbons into branched isomers. The process is very important to upgrade light naphtha (70-78 RON) to isomerate blend stock (85-90 RON).

4. Reforming

Reforming is a refinery process to upgrade medium, heavy naphtha and saturated low octane hydrocarbons into higher octane product containing about 60% aromatics for gasoline blending. Reforming upgrades heavy naphtha (35-55 RON) (low octane components namely paraffins and naphthenes) to a prime gasoline blend stock (higher octane components such as isoparaffins and aromatics), called reformat. The refiner can vary the octane level of reformat over a wide range (90-120 RON) by controlling the "severity" of the former. There are four major types of reactions which occur during reforming processes:

a. Dehydrogenation of naphthenes to aromatics;
b. Dehydrocyclization of paraffins to aromatics;
c. Isomerization;
d. Hydrocracking.

5. Alkylation

Alkylation is the reaction between gaseous olefin (propene, n-butenes, and isobutene which are produced mainly by the FCC unit), and isobutane (which is produced by the FCC, hydrocracking, reforming and natural gas processing) to produce liquid high octane isoalkanes blends stock (92-97 RON). Alkylate is one of the best gasoline blending component because it is a clean sulphur component with no olefinic or aromatic compounds and low Reid Vapour Pressure (RVP).

6. Polymerization

Polymerization is formation of long, repeating organic polymer chains.This process converts light olefins typically the C_3 olefin to form polygasoline, an olefinic, high-octane blend stock (97 RON and 83 MON). Polymerization is a less-favoured process than alkylation because the products are also olefins, which may have to be converted to paraffins before they are blended to gasoline.

7. Dimerization

The authors of Al-Kinany et al. (2011), Al-Kinany et al. (2012), and Marchionna et al.(2001), mentioned in their research that the dimerization involves catalytic dimerization of butenes to isooctene (and other heavier iso-olefins) and then hydrogenation of the isooctene to isooctane. Very recently, environmental issues with MTBE have made it more desireable to demerize isobutene from the FCC unit, rather than etherify it. Fortunately, isobutene dimerization may be done with minimal modifications to existing MTBE plants and process conditions, using the same acidic catalysts, where olefin levels are not restricted; the extra blending octane boast of the diisobutylene can be retained. Where olefin levels are restricted, the diisobutylene can be hydrotreated to produce a relatively pure isooctane stream that can supplement alkylate for reducing olefins and aromatics in reformulated gasolines.

8. Etherification Processes

High oxygenates called ethers may be produced at the refinery by reacting suitable alcohols such as methanol and ethanol with branched olefins from the FCC, such as isobutene and isopentene, under the influence of acid catalysts.

These processes produce oxygenate blend stocks such as ethyltert-butyl ether (ETBE), tert-amyl methyl ether (TAME), and diisopropyl ether (DIPE). Of these blend stocks; MTBE is the most widely used in most of the countries in the world. It has exceptionally high octane (115 RON).

Octane Number

Many properties are considered essential for defining the quality of motor gasoline, most important of these is the octane number. It determines whether a fuel will knock in a given engine under given operating conditions, the higher the octane number, the higher the resistance to knock.

The octane rating of gasoline essentially tells us how much the air-fuel mixture can be compressed before it will spontaneously ignite. Gasoline with an optimal octane rating performs best in an engine designed to run on that octane level. Refiners aim to produce gasoline that has this optimal octane rating so it will meet the specifications for the most kinds of engines.

As we mentioned above, the Octane Number is considered an important measure in assessing the quality of the gasoline. There are two types of octane numbers,

1. **Research Octane Number (RON):** Is an index for anti-knock performance at low speeds, or the number simulates fuel performance under low severity engine operation. This number can be determine by "ASTM (the American Society for Testing and Materials) D-2699".
2. **Motor Octane Number (MON):** Is an index for anti-knock performance of high speeds, or the number simulates fuel performance under more severe operation.This can be determined according to a standard test method known as motor method (MON) "ASTM (the American Society for Testing and Materials) D-2700." Gibbs (1993) and Campbell et al. (1992) showed in their previous work that both methods use the same basic test engine but working under different operating conditions (higher engine speed, temperature and defined spark advance profiles. Simanaitis et al. (1989) mentioned that the numerical difference between the two octane numbers (RON - MON) is called octane sensitive. For any gasoline, RON is higher than MON, usually by 8 to 12 numbers. Most oil refineries produce gasoline in two different grades Regular and Premium the octane number (RON) for regular gasoline is about 94 but it maybe as high as 99 for the premium gasoline.
3. **Control Octane Number (CON):** In practice the octane of a gasoline is reported as the (CON). It is the arithmetic average of RON and MON or 1/2(RON+MON). If for example, a gasoline has an RON of 98 and a MON of 90, then the octane number would be the average of the two values or 94.

High octane gasoline does not outperform regular octane gasoline in preventing engine deposits from forming, in removing them, or in cleaning the engine. Consumers should select the lowest octane grade at which the car's engine runs without knocking.

Occasional light knocking or pinging won't harm the engine, and doesn't indicate a need for higher octane.

The Chevron Company review (2001) shows that gasoline blending is a mechanical mixing of the blending components to produce finished motor gasoline. The later maybe further mixed with other motor gasoline blending components or additives such as oxygenates, such as, teriarybutylmethyl ether or ethanol or other oxygenated compounds, up to about 8-12% to produce "oxygenated gasoline" and changing the components and compositions to produce reformulated gasoline ("RFG") with high octane number.

The octane number of a gasoline is a measure of its resistance to detonation "Knocking" which occurs when the octane number is too low for the engine and its operating conditions. When the spark plug fires, the flame moves through the air/fuel mixture, burning it very rapidly and if a portion of the unburned air/fuel mixture gets raised to a temperature and pressure it cannot tolerate and ignites before the flame front gets to it. Therefore, "Knocking" is a very good description of the sound heard from an engine using fuel of too low octane.

Full octane requirements for gasoline engines vary with the compression ratio of the engine, which is the relative volume of a cylinder from the bottom most position of the pistons stroke to the top most position of the piston's stroke. The higher an engine's compression ratio, the greater the amount of heat generated in the cylinder during the compression stroke.

Stickers (2002) mentioned that the efficiency of a spark-ignited gasoline engine can be related to the compression ratio, which is constrained by the unwanted "knock" that could rapidly destroy engines . For example, if fuel octane is too low for a given compression ratio, the fuel prematurely and spontaneously ignites too early and the fuel charge explodes rather than burns resulting in incomplete combustion. The net effect is loss of power, possible engine damage, and an audible "knock" or "ping" referred to as detonation. The higher a gasoline's octane the better is its antiknock performance.

Hobson (1984) and Han et al. (1984) discuss in their work the relationship between thermal efficiency, octane requirements and compression ratio is descibed in the following references. There are other factors affecting the antiknock ability, such as, Combustion chamber shape, Chemical structure of the fuel, Presence of antiknock additives, Stoichiometry, Number and position of spark plugs and Turbulence. In addition, there are some more factors which bsed on the motor manufacturer.

OXYGENATES ADDITIVES TO GASOLINE

Gasoline refiners began using oxygenates in the U.S. principally as octane boosters. Due to the lead phase out which began in 1973, refiners had to replace the octane loss in gasoline. Oxygenates were a natural choice to replace the octane loss because they have relatively high octane ratings.

Oxygenated gasoline, introduced in November 1992, contains additives that supplement the oxygen content of gasoline. They were first required to reduce exhaust emissions of carbon monoxide (CO) and, to a lesser extent, hydrocarbon emissions from motor vehicles. However, adding oxygenates generally increases emissions of oxides of nitrogen (NOx) from motor vehicles. Producers have flexibility in the choice of additives used to meet the requirements, including methyl tertiary butyl ether (MTBE) and ethanol.

Oxygenates are hydrocarbons that contain one or more oxygen atoms. The primary oxygenates are alcohols such as ethanol, and ethers such as, methyl tertiary butyl ether (MTBE), ethyl tertiary butyl ether (ETBE), tertiary amyl ethyl ether (TAEE) and tertiary amyl methyl ether (TAME) to produce reformulated gasoline ("RFG"). MTBE, the most widely used oxygenate in the market today, Table 2 shows typical properties of oxygenates.

While there are several oxygenates that can be used to meet oxygen requirements in gasoline, methyl tertiary butyl ether (MTBE) and ethanol are used most frequently. Other oxygenates such as tertiary amyl methyl ether (TAME) and ethyl tertiary butyl ether (ETBE) are used in much smaller quantities.

Methyltertiary Butyl Ether (MTBE)

MTBE (Methyl Tertiary Butyl Ether) is widely used as a gasoline additive. It is used to raise octane levels, enhance engine performance, improve combustion

Table 2. Typical properties of some oxygenates

Oxygenates Properties	Ethanol	MTBE	ETBE	TAME
Chemical formula	CH_3CH_2OH	$CH_4OC(CH_3)_3$	$CH_4CH_2OC(CH_3)_3$	$(CH_3)_3CCH_2OCH_3$
Oxygen content, wt%	34.73	18.15	15.66	15.66
Octane $(R+M)/_2$	115	110	111	105
Blending vapor pressure (RVP)	18	8	4	1.5

Source: National Petroleum Council, U.S. Petroleum Refining: Meeting Requirements for Cleaner Fuels and Refineries (Washington, DC, August 1993) Appendix L.

efficiency, and to reduce emissions of air pollutants such as carbon monoxide and hydrocarbons. Motor vehicles require a minimum octane level to operate properly. MTBE has been used as a replacement for tetraethyl lead when that substance was banned as a fuel additive to increase octane levels.

Nadim et al. (2001) shows the first of the three nationwide content regulations is the prohibition on lead-based additives effective in 1986. Prior to 1986, producers used lead-based additives to increase octane. In 1996, leaded gasoline was formally banned in the US. Oxygenated gasoline, introduced in 1992, contains additives that supplement the oxygen content of gasoline. In 1996, over 95 percent of the gasoline used in California was blended with MTBE. The remainder was blended with ethanol, ETBE and TAME.

The blending of MTBE into motor gasoline increased dramatically since it was fist produced 37 years ago. MTBE has a blending octane in the range of 106-110 and make up approximately 10-11% of the gasoline pool. MTBE usage grew in the early 1980s in response to octane demand resulting initially from the phase out of lead from gasoline and later from rising demand for premium gasoline.

Due to leaking storage and distribution systems, MTBE contamination of water supplies has been discovered, causing state of California to completely baned MTBE by the end of 2004. Refiners replaced MTBE in gasoline with ethanol, Some MTBE plants shutdown and some will be converted to isoocatane or alkylation in order to consume the excess isobutylenes no longer consumed for MTBE. Producers have flexibility in the choice of additives used to meet the requirements, including ethanol.

Ethanol

The key chemical property that differentiates ethanol from gasolines is the presence of oxygen. Ethanol can be used as an oxygenate to help control both carbon monoxide and ozone pollution. Whether ethanol is used in oxygenated fuels, reformulated gasoline, or conventional gasoline, there are certain technical parameters and issues that must be considered.

Ethanol has been added to gasoline since the late 1970s. Until the late 1980s ethanol's primary role in the fuels market was that of an octane enhancer and it was viewed as an environmentally sound alternative to the use of lead in gasoline. In the same year, some states in US began to use ethanol and other oxygenates in mandatory oxygenated fuel programs to reduce automobile exhaust emissions of carbon monoxide (CO). and, to a lesser extent, hydrocarbon emissions from motor vehicles. However, adding oxygenates generally increases emissions of oxides of nitrogen (NOx) from motor vehicles

Ethanol, a known product, has been accepted as a replacement for MTBE as a 10% oxygenating blend in gasoline. Ethanol is offered in some areas of the US as an 85% ethanol blend, and in some countries, such as Brazil, 100% ethanol has been promoted as a complete motor fuel in an effort to reduce pollution.

Ethanol is a high quality, high octane gasoline blend component that is derived from corn and other renewable resources. It is an oxygen-enhancing additive that helps gasoline burn cleaner, thus, reducing air pollution under current U.S. standards. It has been used for many years as a blending component for gasoline at a maximum of 10 volume percent. In 2009 allowable Ethanol Content of Gasoline to 15 percent with U.S. EPA.

Ethanol is blended with gasoline in order to produce an ethanol-gasoline blend that meets strict industry requirements for fuel quality and performance. For several years the EPA has mandated the use of ethanol-gasoline blends in certain cities having air pollution problems.The use of ethanol in the US is expanding with currently, as blended fuel represent more than 18% of US motor gasoline sales.

Adding ethanol to gasoline raises the vapour pressure (RVP) by approximately 1.0 psi, even at very low ethanol concentrations. Therefore, refiners have to make base gasoline at lower RVP to meet regulatory specifications that prevent gasoline from evaporating into atmosphere to form smoke and increase toxics. Lower RVP gasoline specifications reduce the quantity of gasoline that can be produced from a barrel of crude without additional processing steps.

David et al. (1999) showed in their previous work that Ethanol-gasoline blends are limited to a maximum of 10 percent ethanol by volume. This corresponds to an oxygen concentration of about 3.5 weight prevent-gasoline containing ethers, such as MTBE or ethyl tertiary butyl ether (ETBE), are limited to a 2.7 weight-percent oxygen concentration. This corresponds to 15% MTBE or 17.2% ETBE (7.7% ethanol), by volume.

At present, ethanol is the oxygenate most widely used in reformulated gasoline. This is in part due to the fact that use of the other oxygenate, MTBE as an gasoline additive., has been banned in 25 states in US in 2006 the petroleum industry switched to ethanol in their RFG due to concerns about ground water contamination from MTBE. More recently, motivated by concerns of MTBE related water pollution, California and twenty other states chose to ban MTBE as a gasoline additive.

The Driving Performance of Gasoline

Because gasoline almost always performs well, drivers forget what a sophisticated product it is. More thought would reveal a demanding set of performance expectations: gasoline plays an important role in meeting the following expectations (Chevron Company, 2001).

- An engine which starts easily when cold, warms up rapidly and runs smoothly under all conditions.
- An engine which delivers adequate power without knocking.
- A vehicle which provides good fuel economy and generates low emissions.
- Gasoline which does not add to engine deposits or contaminate or corrode the fuel system.

In the following gasoline characteristics, shows how they affect driving performance.

Volatility

Gasoline is composed of a variety of ingredients that evaporate at different temperatures. The faster vaporizing components evaporate at lower temperatures then the slower vaporizing components.

Gasoline with lower evaporation temperatures (more highly volatile) starts easier, warms up better and contribute less to deposits but would have more fuel losses and would be more likely to vapor lock. Gasoline with higher evaporation temperatures (decreased volatility) wouldn't have the losses or vapor lock problems but would start harder and not warm up as easily with increased deposits and dilution of engine oil. Exhaust emissions could also increase.

Driveability is the term used to describe how an engine starts, warms up and runs. It is the assessment of a vehicle's response to the accelerator, relative to what the driver expects. Driveability problems include: hard starting, backfiring, rough idle, poor throttle response and stalling at idle, under load or when decelerating.

The key gasoline characteristic for good driveability is volatility. Volatility is important because in the combustion chamber of an engine gasoline must be vaporized before it can burn. In cold weather, gasoline is blended to vaporize easily. This allows an engine to start quickly and run smoothly until it is warm. In warm weather, gasoline is blended to vaporize less easily to prevent vapor lock and minimize evaporation, which contributes to air pollution. It is important to note that there is no single best volatility for gasoline.

Volatility must be adjusted for altitude and seasonal temperature variations. Three properties are used to measure gasoline volatility: vapor pressure, distillation profile and vapor-liquid ratio. A fourth property, driveability index, is calculated from the distillation profile.

Vapor Pressure

The vapor pressure test is another measure of the fuel's volatility. More volatile fuels vaporize more readily and create more pressure when measured. Less volatile fuels exert less pressure and evaporate more slowly.

Vapor pressure is the most important property for cold-start and warmup driveability. Cold-start means that the engine is at ambient temperature, not that the ambient temperature is cold. When the gasolines vapor pressure is low, the engine may have to be cranked a long time before it starts. When it is extremely low, the engine may not start at all. Engines with port fuel injection appear to start more readily with low vapor pressure fuel than carbureted engines.

Distillation Profile

Since gasoline is a mixture of hundreds of hydrocarbons, many of which have different boiling points. It boils or distills over a range of temperatures, unlike a pure compound, that boils at a single temperature. To ensure the proper gasoline is being sold for the season, it must check the vapor pressure and the distillation temperatures. A gasolines distillation profile is the set of increasing temperatures at which it evaporates for a fixed series of increasing volume percentages (5%, 10%, 20%, 30%, 50%, ...90% and 100% (referred to as end point) under specific conditions. Alternatively, it maybe the set of increasing evaporation volume percents for a fixed series of increasing temperatures.

Vapor-Liquid Ratio

The vapor locking tendency of a gasoline is influenced both by the temperatures at the front end of its distillation profile and by its vapor pressure. But the property that correlates best with vapor lock is the temperature at which the gasoline forms a vapor-liquid ratio of twenty, the temperature at which it exists as twenty volumes of vapor in equilibrium with one volume of liquid at atmospheric pressure. This correlation was developed for vehicles with suction-type fuel pumps and carburetors. How well it applies to later-model fuel-injected cars with pressurized fuel systems needs investigations.

Vapor-Liquid Ratio - Driveability Index (DI)

While each range of the distillation profile is important, the gasoline represented by the entire profile is what the engine must distribute, vaporize and burn. To predict cold start and warmup driveability, a driveability index (DI) has been developed using the temperatures for the evaporated percentages of 10% (T10), 50% (T50) and 90% (T90):

The DI varies with gasoline grade and season; the normal range is 850 to 1300. Lower values of DI generally result in better cold-start and warmup performance, but once good driveability is achieved, there is no benefit to further lowering the DI. The equation was developed using data for conventional gasolines and has not been proven to be applicable to oxygenated gasolines. Work continues to improve the DI equation and to make it universally applicable.

Heating Values

The average fuel economics of vehicles is affected by many factors such as a vehicle's size, weight, aerodynamics, fuel delivery system, heating value of the gasoline,engine type, transmission type, weather conditions, air conditioner use, road conditions, the route driven, driving speed and driving style, the mechanical condition of the car, engine tune, wheel alignment and tire pressure.

As shown in Chevron Company report (2001), the most important factor is heating value of the gasoline. Chevron showed that the average fuel economics of vehicles are proportional to the heating values of the gasolines two fleets. Conventional fuels, such as gasoline jet fuel and diesel, have always varied in heating value. One cause is the formulation differences among batches and among refiners. The heating value also varies by grade and by season.

On average, the heating value of premium-grade gasoline is about 0.7% higher than regular-grade because premium-grade, in general, contains more aromatic hydrocarbons. The heating value of winter gasoline is about 1.5% lower than summer gasoline because winter gasoline contains more volatile, less dense hydrocarbons (n-butane).

Oxygenated gasolines, have lower heating values because the heating values of the oxygenate components are lower than those of the hydrocarbons they displace. The percent decrease in heating value is close to the mass percent oxygen in the gasoline.

Antiknock Performance

As we have seen above, Knock-free engine performance is as important as good driveability. Octane number is a measure of a gasoline's antiknock. There are two laboratory test methods to measure the octane number of a gasoline. One yields the Research octane number (RON), the other, the Motor octane number (MON). RON correlates best with low speed, mild-knocking conditions; MON correlates best with highspeed and high-temperature knocking conditions and with part-throttle operation. For a given gasoline, RON is always greater than MON.

Trends in Automotive Fuels Quality

The challenge for the internal combustion engine is the reduction of pollutant exhaust coupled with a lower CO_2 emission. A higher quality of fuels can reduce the exhaust either directly through an improved combustion or indirectly by enabling the use improved exhaust abatement technologies.

Ideal Gasoline

Gasoline for fuel engines needs to be of high octane number. This number describes the knocking resistance of the gasoline and, by definition, isooctane (2,2,4-trimethyl pentane) has octane number 100 and n-heptane 0. Lead components were once added to gasoline to increase the octane number, but in the 1980s, when they became recognised as environmentally harmful, they were gradually replaced by various ethers.

Methyl tert-butylether (MTBE) was one of these ethers. MTBE has a high blending octane number (RON/MON=118/100) and as an oxygen-containing compound it promotes clean burning of fuel. Although MTBE is an excellent gasoline component, one property has caused problems: its water solubility is 4.3 wt % and it has occasionally found its way into ground water from leaking storage tanks.

REFERENCES

Al-Kinany, M., Al-Khowaiter, S., Al-Drees, S., Alshehry, F., & Al Rasheed, R. (2018). *Nanocatalyst for conversion of monoolefins, process for conversion of monoolefins and process for preparing catalyst.* European Patent Office, EP2332647 B1.

Al-Kinany, M., Al-Khowaiter, S., Al-Drees, S., Alshehry, F., & Al Rasheed, R. (2019). *Supported nanocatalyst for conversion of monoolefins, process for conversion of monoolefins and process for preparing the nanocatalyst.* European Patent Office, EP2522424 A1.

Altun, S., Oztop, H. F., Oner, C., & Varol, Y. (2013). Exhaust Emissions of Methanol and Ethanol Unleaded Gasoline Blends in a Spark Ignition Engine. *Thermal Science, 17*(1), 291–297. doi:10.2298/TSCI111207034A

Andress, D., & Hadder, G. R. (1999). Ethanol Demand in U.S. Gasoline Production. Office of Fuels Dept. U.S. Dept. of Energy.

Awad, O. I., Mamat, R., Ibrahim, T. K., Hagos, F. Y., Noor, M. M., Yusri, I. M., & Leman, A. M. (2017). Calorific value enhancement of fusel oil by moisture removal and its effect on the performance and combustion of a spark ignition engine. *Energy Conversion and Management, 137*, 86–96. doi:10.1016/j.enconman.2017.01.009

Campbell, J. M., & Boyd, T. A. (1992). Methods of Knock Rating. 15. Measurement of the Knocking Characteristics of Automotive Fuels. In The Science of Petroleum (vol. 4). Oxford University Press.

Delgado, R. C. O. B., Araujo, A. S., & Fernandes, V. J. Jr. (2007). Properties of Brazilian gasoline mixed with hydrated ethanol for flex-fuel technology. *Fuel Processing Technology, 88*(4), 365–368. doi:10.1016/j.fuproc.2006.10.010

Di Girolamo, M., &Patrini, R. (2001). Light olefins dimerization to high quality gasoline components. *Catalysis Today, 65*, 397-403.

Gary, J. H., & Handwerk, G. E. (1994). *Petroleum Refining Technology and Economics* (3rd ed.). New York: Marcel Dekker, Inc.

Gibbs, L., Anderson, B., Barnes, K., Engeler, G., Freel, J., Horn, J., ... Tom, C. (2001). *Motor Gasolines Technical Review. In Gasoline and Driving Performance* (pp. 1–12). Chevron Company.

Gibbs, L., Anderson, B., Barnes, K., Engeler, G., Freel, J., Horn, J., ... Tom, C. (2001). *Motor Gasolines Technical Review. In Gasoline Refining and Testing* (pp. 31–52). Chevron Company.

Gibbs, L. M. (1993). *How Gasoline Has Changed.* Fuels and Lubricants Meeting and Expostion, SAE 932828.

Hamilton, B. (1996). *Gasoline FAQ Parts 1,2,3,4.* Internet FAQ Archives.

Han, D., Ickes, A. M., Bohac, S. V., Huang, Z., & Assanis, D. N. (2011). Premixed low-temperature combustion of blends of diesel and gasoline in a high speed compression ignition engine. *Proceedings of the Combustion Institute, 33*(2), 3039–3046. doi:10.1016/j.proci.2010.07.045

Hobson, G. D. (1984). Modern Petroleum Technology. In Fuels for Spark-ignition Engines (5th ed.). John Wiley.

Hobson, G. D. (1992). Hydrocarbon Processing. In *Refining Handbook.* Gulf Publishing Co.

Kettering, C. F. (1944). The Effect of the Molecular Structure of Fuels on the Power and Efficiency of Internal Combustion Engines. *Industrial & Engineering Chemistry, 36*(12), 1079–1085. doi:10.1021/ie50420a002

Kutler, A. A. (1969). *Crude distillation.* New York: Petro/Chem Engineering, John G. Simmonds & Co., Inc.

Lindsey, B. D., Ayotte, J. D., Jurgens, B. C., & Desimone, L. A. (2017). Using groundwater age distributions to understand changes in methyl tert-butyl ether (MTBE) concentrations in ambient groundwater, Northeastern United States. *The Science of the Total Environment, 579*, 579–587. doi:10.1016/j.scitotenv.2016.11.058 PMID:27884531

Mancini, E. R., Steen, A., Rausina, G. A., Wong, G. A., Arnold, D. C. L., Gostkowski, W. R., ... Errico, P. (2002). MTBE ambient water quality criteria development: A public/private partnership. *Environmental Science & Technology, 36*(2), 125–129. doi:10.1021/es002059b PMID:11831214

Methyl tertiary-Butyl Ether (MTBE) in drinking water. (2015). World Health Organization.

Nadim, F., Zack, P., Hoag, G. E., & Liu, S. (2001). United States experience with gasoline additives. *Energy Policy, 29*(1), 1–5. doi:10.1016/S0301-4215(00)00099-9

Nelson, W. L. (1958). *Petroleum Refinery Engineering* (4th ed.). New York: McGraw-Hill Book Co.

Pandey, R. K., Rehman, A., & Sarviya, R. M. (2012). Impact of alternative fuel properties on fuel spray behavior and atomization. *Renewable & Sustainable Energy Reviews, 16*(3), 1762–1778. doi:10.1016/j.rser.2011.11.010

Rizwanul Fattah, I. M., Masjuki, H. H., Kalam, M. A., Mofijur, M., & Abedin, M. J. (2014). Effect of antioxidant on the performance and emission characteristics of a diesel engine fuelled with palm biodiesel blends. *Energy Conversion and Management, 79*, 265–272. doi:10.1016/j.enconman.2013.12.024

Semiannual report of UST performance measures. (2015). Environmental Protection Agency.

Simanaitis & Kott. (1989). Octane Enhancers. *Road & Track*, 82-88.

Stickers, D.E. (2002). Octane and the environment. *The Science of the Total Environment, 299*, 37-56.

Yang, H. H., Liu, T. C., Chang, C.-F., & Lee, E. (2012). Effects of ethanol-blended gasoline on emissions of regulated air pollutants and carbonyls from motorcycles. *Applied Energy, 89*(1), 281–286. doi:10.1016/j.apenergy.2011.07.035

Zaharin, M. S. M., Abdullah, N. R., Najafi, G., Sharudin, H., & Yusaf, T. (2017). Effects of physicochemical properties of biodiesel fuel blends with alcohol on diesel engine performance and exhaust emissions: A review. *Renewable & Sustainable Energy Reviews, 79*, 475–493. doi:10.1016/j.rser.2017.05.035

Compilation of References

Abiev, R. (2019). Analysis of Hydrodynamics and Mass Transfer of Gas-Liquid and Liquid-Liquid Taylor Flows in Microchannels: Theoretical Approach and Experimental Proofs. In H. Santana, J. Lameu da Silva Jr, & O. Taranto (Eds.), *Process Analysis, Design, and Intensification in Microfluidics and Chemical Engineering* (pp. 1–49). Hershey, PA: IGI Global. doi:10.4018/978-1-5225-7138-4.ch001

Aboelazm, E. A., Ali, G. A. M., & Chong, K. F. (2018). Cobalt oxide supercapacitor electrode recovered from spent lithium-ion battery. *Chemistry of Advanced Materials*, *3*(4), 67–74.

Agency for Toxic Substances & Disease Registry (ATSDR). (1999). Toxicological Profile for Ethylbenzene (CAS# 100-41-4). Atlanta, GA: U. S. Department of health and human services, public health service.

Aguilar, J., Pergher, S. B. C., Detoni, C., Corna, A., Melo, F. V., & Sastre, E. (2008). Alkylation of biphenyl with propene using MCM-22 and ITQ-2 zeolites. *Catalysis Today*, 33–135, 667–672.

Al-Ayed, O., & Kunzru, D. (1988). Cyclohexane dehydrogenation on a nickel catalyst: Kinetics and catalyst fouling. *Journal of Chemical Technology and Biotechnology (Oxford, Oxfordshire)*, *43*(1), 23–38. doi:10.1002/jctb.280430104

Alba, M., Choi, J.-C., & Sakakura, T. (2001). Halogen-free process for the conversion of carbon dioxide to urethanes by homogeneous catalysis. *Chemical Communications (Cambridge)*, 2238–2239. PMID:12240129

Ali, G. A. M., Fouad, O. A., & Makhlouf, S. A. (2013a). Structural, optical and electrical properties of sol–gel prepared mesoporous Co3O4/SiO2 nanocomposites. *Journal of Alloys and Compounds*, *579*, 606–611. doi:10.1016/j.jallcom.2013.07.095

Ali, N. P., Riyahi, F., Housaindokht, M. R., Irani, M., Shahri, S. M. K., & Hatami, B. (2013b). Hydrocarbon production rates in Fischer-Tropsch synthesis over a Fe/Cu/La/Si catalyst. *Journal of Energy Chemistry*, *22*(1), 119–129. doi:10.1016/S2095-4956(13)60016-9

Al-Khattaf, S., Ali, M., & Čejka, J. (2010). Recent Development in Transformations of Aromatic Hydrocarbons over Zeolites. Zeolites and Catalysis, Synthesis, Reactions and Applications, 2, 623-648..

Al-Khattaf, S. (2003). The Influence of Alumina on the Performance of FCC Catalysts during Hydrotreated VGO Catalytic Cracking. *Energy & Fuels*, *17*(1), 62–68. doi:10.1021/ef020066a

Al-Khattaf, S., Ali, S., Aitani, A., Zilkova, N., Kubicka, D., & Čejka, J. (2014). Recent advances in reactions of alkylbenzenes over novel zeolites: The effects of zeolite structure and morphology. *Catalysis Reviews. Science and Engineering*, *56*(4), 333–402. doi:10.1080/01614940.2014.946846

Al-Kinany, M. C., & Al-Khowaiter, S. H. (1997). Ethylation and Isopropylation of Benzene using superacid catalyst. *15th World Petroleum Congress*, 2, 830–834.

Al-Kinany, M. C., & Al-Khowaiter, S. H. (1998). *Isomerization and disproportionation of diethylbenzenesin the presence of trifluoromethane-sulphonic acid as catalyst.*P resented at the Middle East PETROTECH'98, Manama, Bahrain.

Al-Kinany, M. C., Al-Khowaiter, S. H., Al-Rasheed, R. H., Al-Drees, S. A., & Al-Kharji, A. A. (2002). *Low temperature synthesis of Ethylbenzene by transalkylation of o-Diethylbenzene with benzene using triflic acid as catalyst.* The 5th International Conference and Exhibition on Chemistry and Industry, Manama, Bahrain.

Al-Kinany, M. C., Al-Megren, H. A., Aldrees, S., Al-Ghilan, E., Al-Dress, S., Al-Ghamdi, A., … Xiao, T. (2017). *Method of preparing a modified zeolite catalyst and preparing ethylbenzene using one cycle process.* US 0233308 A.

Al-Kinany, M. C., Al-Megren, H. A., Edwards, P., Xiao, T., Al-Drees, S., Alghilan, E., … Al-Ghamdi, K. (2013). *Isopropylbenzene (Cumene) Production by Alkylation of Benzene with Propylene Using Super acid Catalysts.* 3rd KACST-Oxford Petrochemicals Forum "Disruptive Innovations for Modern Petroleum Refining and Petrochemical Industry", Oxford, UK.

Al-Kinany, M. C., Al-Megren, H. A., Edwards, P., Xiao, T., Al-Shammari, A. S., Al-Drees, S. A., & Alghilan, E. (2012). *Synthesis of Ethylbenzene by Alkylation of Benzene with Ethylene over Modified Zeolite Catalyst in a Gas Phase Process.* 2ndKACST-Oxford Petrochemical Forum "Sustainability Through Innovation in the Moderen Petrochemical Industries", Riyadh, Saudi Arabia.

Al-Kinany, M. C., Al-Megren, H.A., Alghilan, E., Edwards, P., Xiao, T., Al-Drees, S., … Al-Ghamdi, K. (2016). *Solid Acid Zeolite Catalysts for Benzene/Ethylene Alkylation Reactions.* IRPC 2016, Milan, Italy.

Al-Kinany, M., Al-Khowaiter, S., Al-Drees, S., Alshehry, F., & Al Rasheed, R. (2018). *Nanocatalyst for conversion of monoolefins, process for conversion of monoolefins and process for preparing catalyst.* European Patent Office, EP2332647 B1.

Al-Kinany, M., Al-Khowaiter, S., Al-Drees, S., Alshehry, F., & Al Rasheed, R. (2019). *Supported nanocatalyst for conversion of monoolefins, process for conversion of monoolefins and process for preparing the nanocatalyst.* European Patent Office, EP2522424 A1.

Al-Kinany, M.C., Al-Megren, H. A., Edwards, P. P., Xiao, T., Al-Drees S. A., Alghilan, E. A., … Liu, B. (2015). *Isopropylation of Benzene over Modified Zeolite with Metal Oxide to Produce Isopropylbenzene (Cumene).* Pacifichem, Honolulu, HI.

Al-Kinany, M.C., Al-Megren, H.A., Edwards, P.P., Xiao, T., Al-Shammari, A.S., Al-Drees, S.A., & Al-Ghilan, E. A. (2011). *Selective Catalysis of Zeolite in Alkylation of Benzene with Ethylene to Produce Ethylbenzene.* The 1st KACST-Oxford Petrochemical Forum "Petroleum and Petrochemical Industries: Building the Future in a Sustainable Way", Oxford, UK.

Al-Kinany, M. C., & Al-Khowaiter, S. H. (1999). Synthesis of Ethylbenzene from Diethylbenzenes in the presence of Benzene using Trifilic Acid as Catalyst. *Studies in Surface Science and Catalysis*, *122*, 375–384. doi:10.1016/S0167-2991(99)80169-4

Al-Kinany, M. C., Al-Khowaiter, S. H., & Al-Malki, F. H. (2001). Synthesis of cumene (isopropylbenzene) from Diisopropylbenzenes in the presence of benzene using triflic acid as catalyst at room temperature. *Studies in Surface Science and Catalysis*, *133*, 459–464. doi:10.1016/S0167-2991(01)81995-9

Al-Kinany, M. C., Al-Megren, H. A., Al-Ghilan, E. A., Edwards, P. P., Xiao, T., Al-Shammari, A. S., & Al-Drees, S. A. (2012). Selective Zeolite Catalyst for Alkylation of Benzene with Ethylene to Produce Ethylbenzene. *Applied Petrochemical Research*, *2*(3-4), 73–83. doi:10.100713203-012-0022-6

Al-Kinany, M. C., Jibril, B. Y., Al-Khowaiter, S. H., Al-Dosari, M. A., Al-Megren, H. A., Al-Zahrani, S. M., & Al-Humaizi, K. I. (2005). Low temperature transalkylation of o-diethylbenzene with benzene to ethylbenzene using triflic acid as a catalyst. *Chemical Engineering and Processing*, *44*(8), 841–846. doi:10.1016/j.cep.2004.08.011

Al-mulali, U., & Mohammed, A. H. (2015). The relationship between energy consumption and GDP in emerging countries. *International Journal of Energy Sector Management*, *9*(1), 77–93. doi:10.1108/IJESM-04-2013-0006

Altun, S., Oztop, H. F., Oner, C., & Varol, Y. (2013). Exhaust Emissions of Methanol and Ethanol Unleaded Gasoline Blends in a Spark Ignition Engine. *Thermal Science*, *17*(1), 291–297. doi:10.2298/TSCI111207034A

Al-Zahrani, S. M., Al-Kinany, M. C., Al-Humaizi, K. I., & Al-Khowaiter, S. H. (2002). Transalkylation and isomerization of ortho-diethylbenzene with benzene using trifluoromethanesulphonic acid catalyst: Kinetic analysis. *Chemical Engineering and Processing*, *41*(4), 321–327. doi:10.1016/S0255-2701(01)00148-9

Al-Zahrani, S. M., Al-Kinany, M. C., Al-Humaizi, K. I., & Al-Khowaiter, S. H. (2003). Kinetics and Mechanisms of Transalkylation and disproportionation of meta-Diethylbenzene by Triflic Acid Catalyst. *International Journal of Chemical Kinetics*, *35*(11), 555–563. doi:10.1002/kin.10152

Amadi, L. A., & Igwe, P. I. (2018). Globalization and Rethinking of Environmental Consumption From a Sustainability Perspective. In S. Idris (Ed.), *Promoting Global Environmental Sustainability and Cooperation* (pp. 1–30). Hershey, PA: IGI Global. doi:10.4018/978-1-5225-3990-2.ch001

Ancheyta, J., Morales, P., Betancourt, G., Centeno, G., Marroquin, G., & Munoz, J. (2004). Individual Hydrotreating of FCC Feed Components. *Energy and Fuels, 181*, 1-4.

Andersson, S., & Myrstad, T. (1997). Optimum performance of residue FCC catalysts. *Oil Gas European, 23*, 19–22.

Andress, D., & Hadder, G. R. (1999). Ethanol Demand in U.S. Gasoline Production. Office of Fuels Dept. U.S. Dept. of Energy.

Añez, R., & Sierraalta, A. (2018). Advances in Theoretical Studies on Solid Catalysts for Renewable Energy Production. In S. González-Cortés & F. Imbert (Eds.), *Advanced Solid Catalysts for Renewable Energy Production* (pp. 1–32). Hershey, PA: IGI Global. doi:10.4018/978-1-5225-3903-2.ch001

Anwar, Z., Gulfraz, M., & Irshad, M. (2014). Agro-industrial lignocellulosic biomass a key to unlock the future bio-energy: A brief review. *Journal of Radiation Research and Applied Sciences, 7*(2), 163–173. doi:10.1016/j.jrras.2014.02.003

Aoki, U., Takaya, J., & Iwasawa, N. (2006). Rhodium(I)-Catalyzed Carboxylation of Aryl- and Alkenylboronic Esters with CO_2. *Journal of American Chemical Society, 128*, 8706-8707.

Aresta, M., Dibenedetto, A., & He, L.N. (2013). *Analysis of Demand for Captured CO_2 and Products from CO_2 Conversion*. TCGR Report.

Aresta, M. (2010). *Carbon dioxide Recovery and Utilization*. Boston, MA: Kluwer Academic Publishers.

Arsalanfar, M., Mirzaei, A. A., Bozorgzadeh, H. R., Samimi, A., & Ghobadi, R. (2014). Effect of support and promoter on the catalytic performance and structural properties of the Fe–Co–Mn catalysts for Fischer–Tropsch synthesis. *Journal of Industrial and Engineering Chemistry, 20*(4), 1313–1323. doi:10.1016/j.jiec.2013.07.011

Arteconi, A., Mazzarini, A., & Nicola, G. D. (2011). Emissions from Ethers and organic carbonate Fuel Additives: A Review. *Water, Air, and Soil Pollution, 221*(1-4), 405–423. doi:10.100711270-011-0804-y

Artyukhov, A., Artyukhova, N., & Krmela, J. (2019). Computer Simulation of the Aerodisperse Systems Hydrodynamics in Granulation and Drying Apparatus. In H. Santana, J. Lameu da Silva Jr, & O. Taranto (Eds.), *Process Analysis, Design, and Intensification in Microfluidics and Chemical Engineering* (pp. 277–321). Hershey, PA: IGI Global. doi:10.4018/978-1-5225-7138-4.ch010

Arzamendi, G., Campo, I., Arguiñarena, E., Sánchez, M., Montes, M., & Gandía, L. M. (2007). Synthesis of biodiesel with heterogeneous NaOH/alumina catalysts: Comparison with homogeneous NaOH. *Chemical Engineering Journal, 134*(1–3), 123–130. doi:10.1016/j.cej.2007.03.049

Ashrafiyan, O., Saremi, M., Pakseresht, A., & Ghasali, E. (2018). Oxidation-Protective Coatings for Carbon-Carbon Composites. In A. Pakseresht (Ed.), *Production, Properties, and Applications of High Temperature Coatings* (pp. 429–446). Hershey, PA: IGI Global. doi:10.4018/978-1-5225-4194-3.ch016

Aslam, S., & Ahmad, M. (2019). Nanoparticles for Degradation of Organic Pollutants. In R. Nazir (Ed.), *Nanotechnology Applications in Environmental Engineering* (pp. 241–267). Hershey, PA: IGI Global. doi:10.4018/978-1-5225-5745-6.ch011

Atadashi, I. M., Aroua, M. K., Abdul Aziz, A. R., & Sulaiman, N. M. N. (2013). The effects of catalysts in biodiesel production: A review. *Journal of Industrial and Engineering Chemistry*, *19*(1), 14–26. doi:10.1016/j.jiec.2012.07.009

Awad, O. I., Mamat, R., Ibrahim, T. K., Hagos, F. Y., Noor, M. M., Yusri, I. M., & Leman, A. M. (2017). Calorific value enhancement of fusel oil by moisture removal and its effect on the performance and combustion of a spark ignition engine. *Energy Conversion and Management*, *137*, 86–96. doi:10.1016/j.enconman.2017.01.009

Axe, W. N. (1946a). Alkenylation Process. Patent 2,403,963.

Axe, W. N. (1946b). Process for Reacton of Aromatic Hydorocarbons with normally Gaseous Unsaturated Hydrocarbons. Patent 2,412,595.

Azadi, M. (2018). Deposition of TiN, TiC, and DLC Coatings by PACVD. In A. Pakseresht (Ed.), *Production, Properties, and Applications of High Temperature Coatings* (pp. 381–402). Hershey, PA: IGI Global. doi:10.4018/978-1-5225-4194-3.ch014

Bach, G., Zimmermann, G., Kopinke, F., Barendregt, D., Oosterkamp, P., & Woerde, H. (1995). Transfer-Line Heat Exchanger Fouling during Pyrolysis of Hydrocarbons. 1. *Deposits from Dry Cracked Gases Indstrial and Engineering Chemistry Research*, *34*, 1132–1139.

Baechler, S. M., Gonzalez-Cortes, S., Xiao, T., Al-Megren, H. A., & Edwards, P. P. (2018). Perspective on the Deep Hydrotreating of Renewable and Non-Renewable Oils. In S. González-Cortés & F. Imbert (Eds.), *Advanced Solid Catalysts for Renewable Energy Production* (pp. 61–94). Hershey, PA: IGI Global. doi:10.4018/978-1-5225-3903-2.ch003

Bae, J. W., Park, S.-J., Kang, S.-H., Lee, Y.-J., Jun, K.-W., & Rhee, Y.-W. (2009). Effect of Cu content on the bifunctional Fischer–Tropsch Fe–Cu–K/ZSM5 catalyst. *Journal of Industrial and Engineering Chemistry*, *15*(6), 798–802. doi:10.1016/j.jiec.2009.09.002

Bahome, M. C., Jewell, L. L., Hildebrandt, D., Glasser, D., & Coville, N. J. (2005). Fischer–Tropsch synthesis over iron catalysts supported on carbon nanotubes. *Applied Catalysis A, General*, *287*(1), 60–67. doi:10.1016/j.apcata.2005.03.029

Bajaj, A., Lohan, P., Jha, P. N., & Mehrotra, R. (2010). Biodiesel production through lipase catalyzed transesterification: An overview. *Journal of Molecular Catalysis. B, Enzymatic*, *62*(1), 9–14. doi:10.1016/j.molcatb.2009.09.018

Bankole, M. T., Mohammed, I. A., Abdulkareem, A. S., Tijani, J. O., Ochigbo, S. S., Abubakre, O. K., & Afolabi, A. S. (2018). Optimization of supported bimetallic (Fe-Co/CaCO3) catalyst synthesis parameters for carbon nanotubes growth using factorial experimental design. *Journal of Alloys and Compounds*, *749*, 85–102. doi:10.1016/j.jallcom.2018.03.150

Bao, B., El-Halwagi, M. M., & Elbashir, N. O. (2010). Simulation, integration, and economic analysis of gas-to-liquid processes. *Fuel Processing Technology*, *91*(7), 703–713. doi:10.1016/j.fuproc.2010.02.001

Bao, Y. H., Jiang, H., Xing, W. H., Chen, R. Z., & Fan, Y. Q. (2015). Liquid phase hydroxylation of benzene to phenol over vanadylacetylacetonate supported on amine functionalized SBA-15. *Reaction Kinetics, Mechanisms and Catalysis*, *116*(2), 535–547. doi:10.100711144-015-0898-0

Bartholomew, C. H. (2001). Mechanisms of catalyst deactivation. *Applied Catalysis A, General*, *212*(1), 17–60. doi:10.1016/S0926-860X(00)00843-7

Bartholomew, C. H., & Farrauto, R. J. (2011). *Fundamentals of Industrial Catalytic Processes*. New York: John Wiley & Sons.

Bazarov, A. I. (1870). Obtaining urea directly from Carbon dioxide and Ammonia. *Journal für Praktische Chemie*, *2*, 283–312.

Beerala, A. K., Gobinath, R., Shyamala, G., & Manvitha, S. (2018). Water Quality Prediction Using Statistical Tool and Machine Learning Algorithm. *International Journal of Chemoinformatics and Chemical Engineering*, *7*(2), 43–58. doi:10.4018/IJCCE.2018070104

Behr, A., & Brehme, V. A. (2002). Homogeneous and heterogeneous catalyzed three-step synthesis of 2-ethylheptanoic acid from carbon dioxide, butadiene and hydrogen. *Journal of Molecular Catalysis A Chemical*, *187*(1), 69–80. doi:10.1016/S1381-1169(02)00190-5

Behr, A., & Heite, M. (2000). Telomerisation von Kohlendioxid und 1,3-Butadien: Verfahrensentwicklung via Miniplant-Technik. *Chemieingenieurtechnik (Weinheim)*, *72*(1-2), 58–61. doi:10.1002/1522-2640(200001)72:1/2<58::AID-CITE58>3.0.CO;2-F

Behrens, M., Zander, S., Kurr, P., Jacobsen, N., Senker, J., Koch, G., ... Schlogl, R. (2013). Methanol Steam Reforming, in Catalysis for alternative energy generation. *Journal of the American Chemical Society*, *135*, 6061–6068. doi:10.1021/ja310456f PMID:23488720

Bellussi, G. (2004). Zeolite catalysts for the production of chemical commodities: BTX derivatives. *Studies in Surface Science and Catalysis, 154*(Part A), 53-65.

Bellussi, G., Clerici, M. G., Carati, A., & Cavani, F. (1991). *US Patent 5 049 536*. US Patent Office.

Bentham, M.F., Gajda, G.J., Jensen, R.H., & Zinnen, H.A. (1996). *Development and commercialization of solid acid catalysts*. DGMKConference, Catalysis on Solid Acids and Bases, Berlin, Germany.

Bentham, M. F., Gadja, G. J., Jensen, R. H., & Zinner, H. A. (1997). Development and commercialization of solid acid catalysts. *ErdoelErdgasKohle*, *113*(2), 84–88.

Bhattacharyya, A., Woltermann, G., Yoo, J., Karch, J., & Cormier, W. (1988). Catalytic SOx abatement: The role of magnesium aluminate spinel in the removal of SOx from fluid catalytic cracking (FCC) flue gas. *Industrial & Engineering Chemistry Research*, *27*(8), 1356–1360. doi:10.1021/ie00080a004

Bhatt, R. (2017). Zero Tillage for Mitigating Global Warming Consequences and Improving Livelihoods in South Asia. In W. Ganpat & W. Isaac (Eds.), *Environmental Sustainability and Climate Change Adaptation Strategies* (pp. 126–161). Hershey, PA: IGI Global. doi:10.4018/978-1-5225-1607-1.ch005

Biasca, F., Dickenson, R., Chang, E., Johnson, H., Bailey, R., & Simbeck, D. (2003). *Upgrading Heavy Crude Oils and Residue to Transportation Fuel: Technology Economics and Outlook, Phase 7*. SAF Pacific Inc. Engineering and Economic Consultant.

Biodiesel Definition And Benefits. (n.d.). Retrieved from http://www.tristatebiodiesel.com/download/biodieselbenefits.pdf

Bioenergy, A. (2015). *Lifecycle GHG Emissions from Abengoa Bioenergy Biomass of Kansas, LLC Hugoton Cellulosic Ethanol Plant*. Retrieved from https://www.arb.ca.gov/fuels/lcfs/2a2b/apps/abbk-rpt-012915.pdf

Blasco, T., Corma, A., & Matinez-Triguero, J. (2006). Hydrothermal stabilization of ZSM-5 catalytic-cracking additives by phosphorus addition. *Journal of Catalysis*, *237*(2), 267–277. doi:10.1016/j.jcat.2005.11.011

Bloch, S. H., & Pat, U. S. (1972). hydrocarbon conversion catalyst and process. *Chemical Abstracts*, *77*.

Boerefijn, R., Gudde, N., & Ghadiri, M. (2000). A review of attrition of fluid cracking catalyst particles. *Advanced Powder Technology*, *11*(2), 145–174. doi:10.1163/156855200750172286

Bokade, V. V., & Yadav, G. D. (2008). Heteropolyacid supported on acidic clay: A novel efficient catalyst for alkylation of ethylbenzene with dilute ethanol to diethylbenzene in presence of C8 aromatics. *Journal of Molecular Catalysis A Chemical*, *285*(1-2), 155–161. doi:10.1016/j.molcata.2007.11.018

Bolleddu, V., Racherla, V., & Bandyopadhyay, P. P. (2018). Microstructural and Tribological Characteristics of Air Plasma Sprayed Alumina-Titania Coatings. In A. Pakseresht (Ed.), *Production, Properties, and Applications of High Temperature Coatings* (pp. 268–298). Hershey, PA: IGI Global. doi:10.4018/978-1-5225-4194-3.ch011

Bonura, G., Cannilla, C., Frusteri, L., Mezzapica, A., & Frusteri, F. (2017). DME production by CO_2 hydrogenation: Key factors affecting the behaviour of CuZnZr/ferrierite catalysts. *Catalysis Today*, *81*, 337–344. doi:10.1016/j.cattod.2016.05.057

Booth, B. L., Al-Kinany, M.C., & Khosrow, L. (1987). Trifluormethanesulphonic Acid as a catalyst for Ethylation of Benzene. *Chemical Society, Perkin Transactions*, 2049-2055.

Boot-Handford, M. E., Abanades, J. C., Anthony, E. J., Blunt, M. J., Brandani, S., Mac Dowell, N., ... Fennell, P. S. (2014). Carbon capture and storage update. *Energy & Environmental Science*, *7*(1), 130–189. doi:10.1039/C3EE42350F

Bose, G. K., Ghosh, P., & Pal, D. (2019). Analytical and Numerical Modelling of Liquid Penetration in a Closed Capillary. In H. Santana, J. Lameu da Silva Jr, & O. Taranto (Eds.), *Process Analysis, Design, and Intensification in Microfluidics and Chemical Engineering* (pp. 114–135). Hershey, PA: IGI Global. doi:10.4018/978-1-5225-7138-4.ch004

Bostanci, S. H., & Albayrak, A. N. (2017). The Role of Eco-Municipalities in Climate Change for a Sustainable Future. In W. Ganpat & W. Isaac (Eds.), *Environmental Sustainability and Climate Change Adaptation Strategies* (pp. 213–231). Hershey, PA: IGI Global. doi:10.4018/978-1-5225-1607-1.ch008

Botes, F. G., & Govender, N. S. (2007). Secondary reactions of ethylene as studied in a laboratory-scale recycle slurry reactor. *Energy & Fuels*, *21*(6), 3095–3101. doi:10.1021/ef7003464

Boveri, M., Marquez-Alvarez, C., Laborde, M. A., & Sastre, E. (2006). Steam and acid dealumination of mordenite: Characterization and influence on the catalytic performance in linear alkylbenzene synthesis. *Catalysis Today*, *114*(2-3), 217–225. doi:10.1016/j.cattod.2006.01.012

BP. (2017). *BP Energy Outlook Country and regional insights – India*. Retrieved from https://www.bp.com/content/dam/bp/pdf/energy-economics/energy-outlook-2017/bp-energy-outlook-2017-country-insight-india.pdf

Bridge, A., Scott, J., & Reed, E. (1975). *Hydrocarbon Processing*. Paper presented at the API Midyear Meeting.

Bridjanian, H., & Samimi, K. (2011). Bottom of The Barrel, An Important Challenge of the Petroleum Refining industry A. *Petroleum and Coal*, *53*, 13.

Broder, J. D., & Barrier, J. W. (1988). Producing ethanol and coproducts from multiple feedstocks. *American Society of Agricultural Engineers (Microfiche Collection) (USA)*. Retrieved from http://agris.fao.org/agris-search/search.do?recordID=US9023079

Bungay, H. R. (1981). *Energy, the biomass options. Energy, the biomass options*. New York: John Wiley & Sons. Retrieved from https://www.cabdirect.org/cabdirect/abstract/19840693586

Busca, G. (2007). Acid catalysts in industrial hydrocarbon chemistry. *Chemical Reviews*, *107*(11), 5366–5410. doi:10.1021/cr068042e PMID:17973436

Campbell, J. M., & Boyd, T. A. (1992). Methods of Knock Rating. 15. Measurement of the Knocking Characteristics of Automotive Fuels. In The Science of Petroleum (vol. 4). Oxford University Press.

Camp, C., Chatelain, L., Kefalidis, C. E., Pecaut, J., Maron, L., & Mazzanti, M. (2015). CO_2 conversion to isocyanate via multiple N–Si bond cleavage at a bulky uranium(III) complex.(2015) ChemComm. *Chemical Communications (Cambridge)*, *51*(84), 15454–15457. doi:10.1039/C5CC06707C PMID:26346380

Canakci, M., & Van Gerpen, J. (1999). Biodiesel production viaacid catalysis. *Transactions of the ASAE. American Society of Agricultural Engineers*, *42*(5), 1203–1210. doi:10.13031/2013.13285

Can, F., Travert, A., Ruaux, V., Gilson, J., Mauge, F., Hu, R., & Wormsbecher, R. (2007). FCC gasoline sulfur reduction additives: Mechanism and active sites. *Journal of Catalysis*, *249*(1), 79–92. doi:10.1016/j.jcat.2007.04.001

Cardoso, D., Lorena da Silva, L., & Zapelini, I. W. (2018). Mesoporous Silicas as Basic Heterogeneous Catalysts for the Formation of Biodiesel. In S. González-Cortés & F. Imbert (Eds.), *Advanced Solid Catalysts for Renewable Energy Production* (pp. 119–155). Hershey, PA: IGI Global. doi:10.4018/978-1-5225-3903-2.ch005

Catizzone, E., Bonura, G., Migliori, M., Frusteri, F., & Giordano, G. (2018). CO_2 Recycling to Dimethyl Ether: State-of-the-Art and Perspectives. *Molecules (Basel, Switzerland)*, *23*, 1–28.

Čejka, J., Corma, A., & Zones, S. I. (2010). *Zeolites and Catalysis: Synthesis, Reactions and Applications*. Wiley-VCH. doi:10.1002/9783527630295

Čejka, J., & Wichterlova, B. (2002). Acid-catalyzed synthesis of mono- and dialkyl benzenes over zeolites: Active sites, zeolite topology, and reaction mechanisms. *Catalysis Reviews. Science and Engineering*, *44*(3), 375–421. doi:10.1081/CR-120005741

Cejka, J., Wichterlova, B., & Bednarova, S. (1991). Alkylation of toluene with ethene over H-ZSM-5 zeolites. *Applied Catalysis A, General*, *79*(2), 215–226. doi:10.1016/0926-860X(91)80007-L

Chandel, A. K., & Sukumaran, R. K. (n.d.). *Sustainable biofuels development in India*. Academic Press.

Chang, & Grover, J. J., (1977). *US4058576A*. US Patent Office.

Chang, V. S., & Holtzapple, M. T. (2000). Fundamental Factors Affecting Biomass Enzymatic Reactivity. *Applied Biochemistry and Biotechnology*, *84–86*(1–9), 5–38. doi:10.1385/ABAB:84-86:1-9:5

Chang, C. D., Lang, W. H., & Smith, R. L. (1979). The Conversion of Methanol and Other O-Compounds to Hydrocarbons over Zeolite Catalysts. *Journal of Catalysis*, *56*(2), 169–173. doi:10.1016/0021-9517(79)90103-9

Chang, J., Roh, H., Park, M., & Park, S. (2002). Propane Dehydrogenation over a Hydrogen Permselective Membrane Reactor. *Bulletin of the Korean Chemical Society*, *23*(5), 674–678. doi:10.5012/bkcs.2002.23.5.674

Chang, V. S., Nagwani, M., & Holtzapple, M. T. (1998). Lime pretreatment of crop residues bagasse and wheat straw. *Applied Biochemistry and Biotechnology*, *74*(3), 135–159. doi:10.1007/BF02825962 PMID:9429297

Cheng, J. C., Degnan, T. F., Beck, J. S., Huang, Y. Y., Kalyanaraman, M. J., Kowalski, A., ... Mazzone, D. N. (1998). A comparison of zeolites MCM-22, beta, and USY for liquid phase alkylation of benzene with ethylene. *Studies in Surface Science and Catalysis*, *121*, 53–60. doi:10.1016/S0167-2991(99)80044-5

Cheng, W., Kim, G., Peters, A., Zhao, X., Rajagopalan, K., Ziebarth, M. S., & Pereira, C. J. (1998). Environmental Fluid Catalytic Cracking Technology. *Catalysis Reviews. Science and Engineering*, *40*(1-2), 39–79. doi:10.1080/01614949808007105

Chen, N. Y., Garwood, W. A., & Dwyer, F. G. (1996). *Shape Selective Catalysis in Industrial Applications*. New York: Marcel Dekker.

Chen, N. Y., & Garwood, W. E. (1986). Industrial application of shape-selective catalysis. *Catalysis Reviews. Science and Engineering*, *28*(2-3), 185–264. doi:10.1080/01614948608082251

Chen, Y. (2006). Recent advances in FCC technology. *Powder Technology*, *163*(1-2), 2–8. doi:10.1016/j.powtec.2006.01.001

Chianelli, R., & Daage, M. (1989). Structure/Function Relations in Transition Metal Sulfide. *Studies in Surface Science and Catalysis*, *50*, 1–19. doi:10.1016/S0167-2991(08)60486-3

Chisanga, T., & Mbale, J. (2019). Integration of ICTs in Radio Programs (II-RP) for Environmental Awareness for Peasant Farmers of Rural Zambia. In S. Narula, S. Rai, & A. Sharma (Eds.), *Environmental Awareness and the Role of Social Media* (pp. 15–42). Hershey, PA: IGI Global. doi:10.4018/978-1-5225-5291-8.ch002

Chopade, S. G., Kulkarni, K. S., & Topare, N. S. (2012). Solid Heterogeneous Catalysts For Production Of Biodiesel From Trans-Esterification Of Triglycerides With Methanol: A Review [Abstract]. *Acta Chimica and Pharmaceutica Indica*, 8–14. Retrieved from http://www.tsijournals.com/abstract/solid-heterogeneous-catalysts-for-production-of-biodiesel-from-transesterification-of-triglycerides-with-methanol-a-revi-11071.html

Choudhary, V. R., Mondal, K. C., & Mulla, S. A. R. (2005). Simultaneous Conversion of Methane and Methanol into Gasoline over Bifunctional Ga-, Zn-, In-, and/or Mo-Modified ZSM-5 Zeolites. *Angewandte Chemie International Edition*, *44*(28), 4381–4385. doi:10.1002/anie.200500694 PMID:15952225

Christensen, C.H., Johannsen, K., &Schmidt, I. (2003). Catalytic benzene alkylation over mesoporous zeolite single crystals: improving activity and selectivity with a new family of porous materials. *American Chemical Society, 125*(44), 13370-13371.

Clary, J. J. (2013). *The Toxicology of Methanol* (1st ed.). John Wiley & Sons,Inc. doi:10.1002/9781118353110

Coates, G. W., & Moore, D. R. (2004). Discrete metal-based catalysts for the copolymerization of CO_2 and epoxides: Discovery, reactivity, optimization, and mechanism. *Angewandte Chemie International Edition*, *43*(48), 6618–6639. doi:10.1002/anie.200460442 PMID:15558659

Cooper, C. (2009). *A Technical Basis for Carbon Dioxide Storage, CO_2 Capture Project*. available at: http://unfccc.int/resource/docs/2011/smsn/ngo/276.pdf

Corma, A., Corresa, E., Mathieu, Y., Sauvanaud, L., Al-Bogami, S., Al-Ghrami, M. S., & Bourane, A. (2017). Crude oil to chemicals: Light olefins from crude oil. *Catalysis Science & Technology*, *7*(1), 12–46. doi:10.1039/C6CY01886F

Cormier, W., Woltermann, G., Magee, J., Baars, F., & Upson, L. (1994). *Increasing metal-tolerance of FCC catalyst by sulfur oxide removal.* US Patent: 5,371,055.

Craciun, I., Reyniers, M.-F., Guy, B., & Marin, G. B. (2007). Effects of acid properties of Y zeolites on the liquid-phase alkylation of benzene with 1-octene: A reaction path analysis. *Journal of Molecular Catalysis A Chemical*, *277*(1-2), 1–14. doi:10.1016/j.molcata.2007.06.027

Cui, X., Zeng, S., Li, Z., Zheng, Q., Yu, X., & Han, B. (2018). Advanced Composites for Civil Engineering Infrastructures. In K. Kumar & J. Davim (Eds.), *Composites and Advanced Materials for Industrial Applications* (pp. 212–248). Hershey, PA: IGI Global. doi:10.4018/978-1-5225-5216-1.ch010

Dahl, I. M., & Kolboe, S. (1994). On the Reaction Mechanism for Hydrocarbon Formation from Methanol over SAPO-34: I. Isotopic Labeling Studies of the Co-Reaction of Ethene and Methanol. *Journal of Catalysis*, *149*(2), 458–464. doi:10.1006/jcat.1994.1312

Davey, S. (1999). *Proceedings of the European Refining Technology Conference.* Paris, France: Academic Press.

Davis, B. H., & Occelli, M. L. (2006). *Fischer-Tropsch synthesis, catalysts and catalysis* (Vol. 163). Elsevier.

Da, Z., Han, Z., Magnoux, P., & Guisnet, M. (2001). Liquid-phase alkylation of toluene with long-chain alkenes over HFAU and HBEA zeolites. *Applied Catalysis A, General*, *219*(1-2), 45–52. doi:10.1016/S0926-860X(01)00662-7

De Almeida, L. G., Dufaux, M., Taarit, Y. B., & Naccache, C. (1994). Liner alkylbenzene. *J.American Oil Chemists'. Society*, *71*(7), 675–694.

De Carvalho, M., Morgado, E., Cerqueira, H., De Resende, N., & Schmal, M. (2004). Behavior of Fresh and Deactivated Combustion Promoter Additives. *Industrial & Engineering Chemistry Research*, *43*(12), 3133–3136. doi:10.1021/ie034292z

De Stefanis, A., & Tomlinson, A. A. G. (2006). Towards designing pillared clays for catalysis. *Catalysis Today*, *114*(2-3), 126–141. doi:10.1016/j.cattod.2006.01.019

Degnan, T. F. Jr, Smith, C. M., & Venkat, C. R. (2001). Alkylation of aromatics with ethylene and propylene: Recent developments in commercial processes. *Applied Catalysis A, General*, *221*(1-2), 283–294. doi:10.1016/S0926-860X(01)00807-9

Degnan, T., Chitnis, G., & Schipper, P. (2000). History of ZSM-5 fluid catalytic cracking additive development at Mobil. *Microporous and Mesoporous Materials*, *35-36*, 245–252. doi:10.1016/S1387-1811(99)00225-5

Delgado, R. C. O. B., Araujo, A. S., & Fernandes, V. J. Jr. (2007). Properties of Brazilian gasoline mixed with hydrated ethanol for flex-fuel technology. *Fuel Processing Technology*, *88*(4), 365–368. doi:10.1016/j.fuproc.2006.10.010

Dell'Amico, D. B., Calderazzo, F., Labella, L., Marchetti, F., & Pampaloni, G. (2003). Converting carbon dioxide into carbamato derivatives. *Chemical Reviews*, *103*(10), 3857–3898. doi:10.1021/cr940266m PMID:14531715

Demirbaş, A. (2001). Biomass resource facilities and biomass conversion processing for fuels and chemicals. *Energy Conversion and Management*, *42*(11), 1357–1378. doi:10.1016/S0196-8904(00)00137-0

Demirbas, A. (2007). Converting biomass derived synthetic gas to fuels via Fisher-Tropsch synthesis. *Energy Sources. Part A, Recovery, Utilization, and Environmental Effects*, *29*(16), 1507–1512. doi:10.1080/15567030601003676

Deshmukh, S. R., Tonkovich, A. L. Y., McDaniel, J. S., Schrader, L. D., Burton, C. D., Jarosch, K. T., ... LeViness, S. (2011). Enabling cellulosic diesel with microchannel technology. *Biofuels*, *2*(3), 315–324. doi:10.4155/bfs.11.17

Detailed California-Modified GREET Pathway for Cellulosic Ethanol from Forest Waste. (2009). Retrieved from https://www.arb.ca.gov/fuels/lcfs/022709lcfs_forestw.pdf

Devassy, B. M., Lefebvre, F., & Halligudi, S. B. (2005). Zirconia Supported 12-Tungstophosphoric acid as Solid Catalyst for the Synthesis of Linear Alkyl Benzenes. *Journal of Catalysis*, *231*(1), 1–10. doi:10.1016/j.jcat.2004.09.024

Di Girolamo, M., &Patrini, R. (2001). Light olefins dimerization to high quality gasoline components. *Catalysis Today*, *65*, 397-403.

Di Girolamo, G. (2018). Current Challenges and Future Perspectives in the Field of Thermal Barrier Coatings. In A. Pakseresht (Ed.), *Production, Properties, and Applications of High Temperature Coatings* (pp. 25–59). Hershey, PA: IGI Global. doi:10.4018/978-1-5225-4194-3.ch002

Dickenson, R., Biasca, F., Schulman, B., & Johnson, H. (1997). Refiner options for converting and utilizing heavy fuel oil. *Hydrocarbon Processing*, *76*, 57–62.

Ding, M., Yang, Y., Wu, B., Wang, T., Xiang, H., & Li, Y. (2011). Effect of reducing agents on microstructure and catalytic performance of precipitated iron-manganese catalyst for Fischer–Tropsch synthesis. *Fuel Processing Technology*, *92*(12), 2353–2359. doi:10.1016/j.fuproc.2011.08.011

Dinus, R. J. (2000). *Genetic Modification of Short Rotation Poplar Biomass Feedstock for Efficient Conversion to Ethanol*. doi:10.2172/769297

Dipardo, J. (n.d.). *Outlook for Biomass Ethanol Production and Demand*. Retrieved from https://www.agmrc.org/media/cms/biomass_E6EE9065FD69D.pdf

Dixit, A. (2018). Application of Silica-Gel-Reinforced Aluminium Composite on the Piston of Internal Combustion Engine: Comparative Study of Silica-Gel-Reinforced Aluminium Composite Piston With Aluminium Alloy Piston. In K. Kumar & J. Davim (Eds.), *Composites and Advanced Materials for Industrial Applications* (pp. 63–98). Hershey, PA: IGI Global. doi:10.4018/978-1-5225-5216-1.ch004

Dobele, G., Dizhbite, T., Rossinskaja, G., Telysheva, G., Meier, D., Radtke, S., & Faix, O. (2003). Pre-treatment of biomass with phosphoric acid prior to fast pyrolysis: A promising method for obtaining 1,6-anhydrosaccharides in high yields. *Journal of Analytical and Applied Pyrolysis*, *68–69*, 197–211. doi:10.1016/S0165-2370(03)00063-9

Dry, M. E. (1990). Fischer-Tropsch synthesis over iron catalysts. *Catalysis Letters*, *7*(1-4), 241–251. doi:10.1007/BF00764506

Dry, M. E. (2002). The Fischer–Tropsch process: 1950–2000. *Catalysis Today*, *71*(3-4), 227–241. doi:10.1016/S0920-5861(01)00453-9

Dry, M., & Hoogendoorn, J. (1981). Technology of the Fischer-Tropsch process. *Catalysis Reviews. Science and Engineering*, *23*(1-2), 265–278. doi:10.1080/03602458108068078

Duruji, M. M., Olanrewaju, F. O., & Duruji-Moses, F. U. (2018). From Kyoto to Paris: An Analysis of the Politics of Multilateralism on Climate Change. In S. Idris (Ed.), *Promoting Global Environmental Sustainability and Cooperation* (pp. 31–56). Hershey, PA: IGI Global. doi:10.4018/978-1-5225-3990-2.ch002

Dutta, K., Daverey, A., & Lin, J.-G. (2014). Evolution retrospective for alternative fuels: First to fourth generation. *Renewable Energy*, *69*, 114–122. doi:10.1016/j.renene.2014.02.044

Dutta, M. M. (2019). Characterization of Nanomaterials. In R. Nazir (Ed.), *Nanotechnology Applications in Environmental Engineering* (pp. 44–80). Hershey, PA: IGI Global. doi:10.4018/978-1-5225-5745-6.ch003

Du, Y., Wang, H., & Chen, S. (2002). Study on alkylation of benzene with ethylene over β-zeolite catalyst to ethylbenzene by in situ IR.*Molecular Catalysis A. Chemical*, *179*(1-2), 253–261.

Dwyer, F. G., & Walker, O. (1981). Bargaining in an asymmetrical power structure *Catalysis in Organic Reactions. Chemical Industries, Marcel Dekker, New York*, *37*, 39–47.

Ebert, G. W., Juda, W. L., Kosakowski, R. H., Ma, B., Dong, L., Cummings, K. E., ... Luo, J. (2005). Carboxylation and Esterification of Functionalized Arylcopper Reagents. *The Journal of Organic Chemistry*, *70*(11), 4314–4317. doi:10.1021/jo047731s PMID:15903305

Ebrahimi, A. N., & Ashkan, S. (2011). Modification and optimization of benzene alkylation process for production of ethylbenzene. *Chemical Engineering and Processing: Process Intensification*, *50*(1), 31–36. doi:10.1016/j.cep.2010.10.011

Elewuwa, F. A., & Makkawi, Y. T. (2015). Hydrogen production by steam reforming of DME in a large scale CFB reactor. Part I: Computational model and predictions. *International Journal of Hydrogen Energy*, *40*(46), 15865–15876. doi:10.1016/j.ijhydene.2015.10.050

Elewuwa, F. A., & Makkawi, Y. T. (2016). A computational model of hydrogen production by steam reforming of dimethyl ether in a large scale CFB reactor. Part II: Parametric analysis. *International Journal of Hydrogen Energy*, *41*(44), 19819–19828. doi:10.1016/j.ijhydene.2016.08.072

ElMekawy, A., Hegab, H. M., Mohanakrishna, G., Elbaz, A. F., Bulut, M., & Pant, D. (2016). Technological advances in CO_2. conversion electro-biorefinery: A step toward commercialization. *Bioresource Technology, 215*, 357–370. doi:10.1016/j.biortech.2016.03.023 PMID:27020396

Engels, H.-W., Pirkl, H.-G., Albers, R., Albach, R. W., Krause, J., Hoffmann, A., ... Dormish, J. (2013). Polyurethanes: Versatile materials and sustainable problem solvers for today's challenges. *Angewandte Chemie International Edition, 52*(36), 9422–9441. doi:10.1002/anie.201302766 PMID:23893938

Erena, J., Vicente, J., Aguayo, A. T., Gayubo, A. G., Olazar, M., & Bilbao, J. (2013). Effect of combining metallic and acid functions ib CZA/HZSM-5 desilicated zeolite catalysts on the DME steam reforming in a fluidized bed. *International Journal of Hydrogen Energy, 38*(24), 10019–10028. doi:10.1016/j.ijhydene.2013.05.134

Espinoza, R., Steynberg, A., Jager, B., & Vosloo, A. (1999). Low temperature Fischer–Tropsch synthesis from a Sasol perspective. *Applied Catalysis A, General, 186*(1-2), 13–26. doi:10.1016/S0926-860X(99)00161-1

Eudoxie, G., & Roopnarine, R. (2017). Climate Change Adaptation and Disaster Risk Management in the Caribbean. In W. Ganpat & W. Isaac (Eds.), *Environmental Sustainability and Climate Change Adaptation Strategies* (pp. 97–125). Hershey, PA: IGI Global. doi:10.4018/978-1-5225-1607-1.ch004

European Commission. (2003). European Commission Regulation (EC) No 1831/2003 of the European Parliament and the Council of 22nd September 2003 on additives for use in animal nutrition. Off J Eur Union 2003. L268 18.10.

Faghihian, H., & Mohammadi, M. H. (2014). Acid activation effect on the catalytic performance of Al-pillared bentonite in alkylation of benzene with olefins. *Applied Clay Science, 93-94*, 1–7. doi:10.1016/j.clay.2014.02.026

Falco, M., Morgado, E., Amadeo, N., & Sedran, U. (2006). Accessibility in alumina matrices of FCC catalysts. *Applied Catalysis A, General, 315*, 29–34. doi:10.1016/j.apcata.2006.08.028

Fernandez-Baujin, J., Sundaram, K., & Chien, J. (1993). *Inhibition of coke formation during vaporization of heavy hydrocarbons.* US Patent: 5,190,634.

Fetisova, V. A., Ivashkina, E. D., Ivanchina, E. N., & Kravtsov, A. V. (2010). Mathematical model for the process of benzene alkylation by higher olefines. *Catalysis in Industry, 2*(1), 55–61. doi:10.1134/S2070050410010095

Feyzi, M., & Jafari, F. (2012). Study on iron-manganese catalysts for Fischer-Tropsch synthesis. *Journal of Fuel Chemistry and Technology, 40*(5), 550–557. doi:10.1016/S1872-5813(12)60021-8

Field, C. B., Barros, V. R., Dokken, D. J., Mach, K. J., Mastrandrea, M. D., Bilir, T. E., ... White, L. L. (2014). *IPCC, 2014: Climate Change 2014 Impacts, Adaptation, and Vulnerability. Part A: Global and Sectoral Aspects. Contribution ofWorking Group II to the Fifth Assessment Report of the Intergovernmental Panel on Climate Change.* Cambridge University Press.

Figueira, R. B. (2018). Hybrid Sol-Gel Coatings: Erosion-Corrosion Protection. In A. Pakseresht (Ed.), *Production, Properties, and Applications of High Temperature Coatings* (pp. 334–380). Hershey, PA: IGI Global. doi:10.4018/978-1-5225-4194-3.ch013

Fischer, F., & Tropsch, H. (1923). The preparation of synthetic oil mixtures (synthol) from carbon monoxide and hydrogen. *Brennstoff-Chem*, *4*, 276–285.

Fischer, F., Tropsch, H., & Dilthey, P. (1925). Reduction of carbon monoxide to methane in the presence of various metals. *Brennst. Chem*, *6*, 265–271.

Fouad, O. A., Ali, G. A. M., El-Erian, M. A. I., & Makhlouf, S. A. (2012). Humidity sensing properties of cobalt oxide/silica nanocomposites prepared via sol–gel and related routes. *Nano*, *07*(05), 1250038. doi:10.1142/S1793292012500385

Fouad, O. A., Makhlouf, S. A., Ali, G. A. M., & El-Sayed, A. Y. (2011). Cobalt/silica nanocomposite via thermal calcination-reduction of gel precursors. *Materials Chemistry and Physics*, *128*(1), 70–76. doi:10.1016/j.matchemphys.2011.02.072

Franck, H. G., & Stadelhofer, J. W. (1988). *Industrial Aromatic Chemistry, Raw Materials-Processes- Products*. Heidelberg, Germany: Springer-Verlag. doi:10.1007/978-3-642-73432-8

Freeman, D., Wells, R. P. K., & Hutchings, G. J. (2002, January). Ga2O3/H-ZSM-5 and Ga2O3/WO3 Catalysts. *Journal of Catalysis*, *205*(2), 358–365. doi:10.1006/jcat.2001.3446

Friedel, C., & Crafts, J. M. (1877). A New General Synthetical Method of Production Hydrocarbons. *Journal of the Chemical Society*, *32*, 725–791.

Frogneux, X., von Wolff, N., Thuery, P., Lefevre, G., & Cantat, T. (2016). CO_2 Conversion into Esters by Fluoride-Mediated Carboxylation of Organosilanes and Halide Derivatives. *Chemistry (Weinheim an der Bergstrasse, Germany)*, *22*(9), 2930–2934. doi:10.1002/chem.201505092 PMID:26712565

Frusteri, F., Migliori, M., Cannilla, C., Frusteri, L., Catizzone, E., Aloise, A., Giordano, G., & Bonura, G. (2017), Direct CO_2-to-DME hydrogenation reaction: New evidences of a superior behaviour of FER-based hybrid systems to obtain high DME yield. *Journal of CO_2 Utilization, 18,* 353–361.

Furimsky, E. (2000). Characterization of cokes from fluid/flexi-coking of heavy feeds. *Fuel Processing Technology*, *67*(3), 205–230. doi:10.1016/S0378-3820(00)00103-X

Gadde, B., Bonnet, S., Menke, C., & Garivait, S. (2009). Air pollutant emissions from rice straw open field burning in India, Thailand and the Philippines. *Environmental Pollution*, *157*(5), 1554–1558. doi:10.1016/j.envpol.2009.01.004 PMID:19201513

Gajda, Patton, & Wilson. (1995). Discrete Molecular Sieve and Use in Aromatic- Olefin Alkylation. UOP Inc. (US 5434326)

Galadima, A., & Muraza, O. (2015). Role of zeolite catalysts for benzene removal from gasoline via alkylation: A review. *Microporous and Mesoporous Materials*, *213*, 169–180. doi:10.1016/j.micromeso.2015.02.029

Gao, J., Zhang, L., Hu, J., Li, W., & Wang, J. (2009). Effect of zinc salt on the synthesis of ZSM- 5 for alkylation of benzene with ethanol. *Catalysis Communications*, *10*(12), 1615–1619. doi:10.1016/j.catcom.2009.04.029

Gao, L. (2011). Well-to-Wheels Analysis of Energy Use and Greenhouse Gas Emissions for Alternative Fuels. *International Journal of Applied Science and Technology*, *1*(6). Retrieved from http://www.ijastnet.com/journals/Vol_1_No_6_November_2011/1.pdf

Gao, P., Zhong, L. S., Zhang, L. N., Wang, H., Zhao, N., Wei, W., & Sun, Y. H. (2015). Yttrium oxide modified Cu/ZnO/Al_2O_3 catalysts *via*hydrotalcite-like precursors for CO_2hydrogenation to methanol. *Catalysis Science & Technology*, *5*(9), 4365–4377. doi:10.1039/C5CY00372E

Garrick, T. A., & Liburd, O. E. (2017). Impact of Climate Change on a Key Agricultural Pest: Thrips. In W. Ganpat & W. Isaac (Eds.), *Environmental Sustainability and Climate Change Adaptation Strategies* (pp. 232–254). Hershey, PA: IGI Global. doi:10.4018/978-1-5225-1607-1.ch009

Gary, J. H., & Handwerk, G. E. (1994). *Petroleum Refining Technology and Economics* (3rd ed.). New York: Marcel Dekker, Inc.

Ge, H., Liu, X., Wang, S., Yang, T., & Wen, X. (2017). *Innovative Applications of Mo(W)-Based Catalysts in the Petroleum and Chemical Industry: Emerging Research and Opportunities* (pp. 1–162). Hershey, PA: IGI Global. doi:10.4018/978-1-5225-2274-4

Gembicki, S. (2000). New solid acid based breakthrough technologies. *Studies in Surface Science and Catalysis*, *130*, 147–156. doi:10.1016/S0167-2991(00)80951-9

Ghani, F., & Qureshi, K. A. (2018). "Airpocalypse" or Tsar Economic China: Analysis of Unsustainable Environment and Reason Behind Increased Global Warming. In S. Idris (Ed.), *Promoting Global Environmental Sustainability and Cooperation* (pp. 137–161). Hershey, PA: IGI Global. doi:10.4018/978-1-5225-3990-2.ch006

Ghosh, A. K., & Brindisi, M. (2015). Organic carbamates in drug design and medicinal chemistry. *Journal of Medicinal Chemistry*, *58*(7), 2895–2940. doi:10.1021/jm501371s PMID:25565044

Gibbs, L. M. (1993). *How Gasoline Has Changed.* Fuels and Lubricants Meeting and Expostion, SAE 932828.

Gibbs, L., Anderson, B., Barnes, K., Engeler, G., Freel, J., Horn, J., ... Tom, C. (2001). *Motor Gasolines Technical Review. In Gasoline and Driving Performance* (pp. 1–12). Chevron Company.

Gibbs, L., Anderson, B., Barnes, K., Engeler, G., Freel, J., Horn, J., ... Tom, C. (2001). *Motor Gasolines Technical Review. In Gasoline Refining and Testing* (pp. 31–52). Chevron Company.

Girotti, G., & Cappellazzo, O. (1997). *Catalyst for the alkylation or transalkylation of aromatic compounds*. Eur. Pat. Appl. 687,500.

Goelzer, A. R., Ram, S., Hernandez, A., Chin, A. A., Harandi, M. N., & Morris, C. M. (1993). *Mobil-Badger Technologies for Benzene Reduction in Gasoline*. Paper presented at the 1993 NPRA Annual Meeting, San Antonio, TX.

Goeppert, A., Czaun, M., Jones, J.-P., Prakash, G. K. S., & Olah, G. A. (2014). Recycling of carbon dioxide to methanol and derived products - closing the loop. *Chemical Society Reviews*, *43*(23), 7995–8048. doi:10.1039/C4CS00122B PMID:24935751

Golden, S., Pulley, R., & Dean, C. (2004). Catalyst Changes, Downstream Improvements Increase FCC Propylene yields. *Oil & Gas Journal*, *102*, 44–46.

Gonzalez-Gil, R., Herrera, C., Larrubioa, M. A., Kowalik, P., Pieta, I. S., & Alemany, L. J. (2016). Hydrogen production by steam reforming of DME over Ni-based catalysts modified with vanadium. *International Journal of Hydrogen Energy*, *41*(43), 19781–19788. doi:10.1016/j.ijhydene.2016.05.074

Gooßen, L. J., Rodríguez, N., Manjolinho, F., & Lange, P. P. (2010). Synthesis of propiolic acids via copper-catalyzed insertion of carbon dioxide into the C–H bond of terminal alkynes. *Advanced Synthesis & Catalysis*, *352*(17), 2913–2917. doi:10.1002/adsc.201000564

Goundar, S., & Appana, S. (2017). Mainstreaming Development Policies for Climate Change in Fiji: A Policy Gap Analysis and the Role of ICTs. In W. Ganpat & W. Isaac (Eds.), *Environmental Sustainability and Climate Change Adaptation Strategies* (pp. 1–31). Hershey, PA: IGI Global. doi:10.4018/978-1-5225-1607-1.ch001

Graf, P. O., & Lefferts, L. (2009). Reactive separation of ethylene from the effluent gas of methane oxidative coupling via alkylation of benzene to ethylbenzene on ZSM-5. *Chemical Engineering Science*, *64*(12), 2773–2780. doi:10.1016/j.ces.2009.02.031

Grego, S., Aricò, F., & Tundo, P. (2012). Phosgene-free carbamoylation of aniline via dimethyl carbonate. *Pure and Applied Chemistry*, *84*(3), 695–705. doi:10.1351/PAC-CON-11-07-17

Groen, J. C., Sano, T., Moulijn, J. A., & Pérez-Ramírez, J. (2007). Alkaline-mediated mesoporousmordenite zeolites for acid-catalyzed conversions. *Journal of Catalysis*, *251*(1), 21–27. doi:10.1016/j.jcat.2007.07.020

Groode, T. A. (2008). *Biomass to ethanol: potential production and environmental impacts*. Massachusetts Institute of Technology. Retrieved from http://dspace.mit.edu/handle/1721.1/43144

Grove, J., Greig, C. R., Smart, S., & Lant, P. A. (2017). Producing a CO_2-neutral clean cooking fuel in India- where and at what cost. *International Journal of Hydrogen Energy*, *42*(30), 19076–19078. doi:10.1016/j.ijhydene.2017.06.070

Haas, A., Nee, J., & Kanz, B. (1997). *Proceedings of the DGMK Conference*. Aachen, Germany: Academic Press.

Hall, K. R. (2005). A new gas to liquids (GTL) or gas to ethylene (GTE) technology. *Catalysis Today*, *106*(1-4), 243–246. doi:10.1016/j.cattod.2005.07.176

Hamelinck, C. N., van Hooijdonk, G., & Faaij, A. P. (2005). Ethanol from lignocellulosic biomass: Techno-economic performance in short-, middle- and long-term. *Biomass and Bioenergy*, *28*(4), 384–410. doi:10.1016/j.biombioe.2004.09.002

Hamid, G., Ahari, S., Farshi, J.A., & Majid, K. (2004). Modelling and Simulation Of Benzene Alkylation Process Reactors For Production Of Ethylbenzene. *Petroleum and Coal*, *46*(1), 55–63.

Hamilton, B. (1996). *Gasoline FAQ Parts 1,2,3,4*. Internet FAQ Archives.

Hancock, E. G. (Ed.). (1975). *Benzene and its industrial derivatives*. Ernest Benn Limited.

Han, D., Ickes, A. M., Bohac, S. V., Huang, Z., & Assanis, D. N. (2011). Premixed low-temperature combustion of blends of diesel and gasoline in a high speed compression ignition engine. *Proceedings of the Combustion Institute*, *33*(2), 3039–3046. doi:10.1016/j.proci.2010.07.045

Han, M. H., Lin, S. X., & Roduner, E. (2003). Study on the alkylation of benzene with propylene over Hβ zeolite. *Applied Catalysis A, General*, *243*(1), 175–184. doi:10.1016/S0926-860X(02)00565-3

Han, M. Z., Cui, Z., Xu, C., Chen, W., & Jin, Y. (2003). Synthesis of linear alkylbenzene catalyzed by Hβ-zeolite. *Applied Catalysis A, General*, *238*(1), 99–107. doi:10.1016/S0926-860X(02)00334-4

Hansen, N., Bruggemannt, T., Bell, A., & Keil, F. (2008). Theoretical investigation of benzene alkylation with ethene over H-ZSM5. *The Journal of Physical Chemistry C*, *112*(39), 15402–15411. doi:10.1021/jp8036022

Harding, R., Peters, A., & Nee, J. (2001). New developments in FCC catalyst technology. *Applied Catalysis A, General*, *221*(1-2), 389–396. doi:10.1016/S0926-860X(01)00814-6

Harmer, M. A., Farneth, W. E., & Sun, Q. (1996). High Surface Area Nafion Resin/Silica Nanocomposites: A New Class of Solid Acid Catalyst. *Journal of the American Chemical Society*, *118*(33), 7708–7715. doi:10.1021/ja9541950

Harper, E. F., Ko, D. Y., Lese, H. K., Sabourin, E. T., & Williamson, R. C. (1977). *Industrial and Laboratory Alkylations*. American Chemical Society.

Hasdemir, T. A., & Çetin, S. (2019). Ecological Issues and Public Awareness in Social Media: Greenpeace Mediterranean's Turkey Case. In S. Narula, S. Rai, & A. Sharma (Eds.), *Environmental Awareness and the Role of Social Media* (pp. 1–14). Hershey, PA: IGI Global. doi:10.4018/978-1-5225-5291-8.ch001

Hassanzadeh, M., Saremi, M., Valefi, Z., & Pakseresht, A. H. (2018). Investigation on Improved-Durability Thermal Barrier Coatings: An Overview of Nanostructured, Multilayered, and Self-Healing TBCs. In A. Pakseresht (Ed.), *Production, Properties, and Applications of High Temperature Coatings* (pp. 60–78). Hershey, PA: IGI Global. doi:10.4018/978-1-5225-4194-3.ch003

Hellring, S. D., Schmitt, K. D., & Chang, C. D. (1987). Synthesis and decomposition of trimethyloxonium ZSM-5, a purported intermediate in methanol conversion into gasoline. *Journal of the Chemical Society. Chemical Communications*, (17): 1320–1322. doi:10.1039/c39870001320

Helwani, Z., Othman, M. R., Aziz, N., Fernando, W. J. N., & Kim, J. (2009). Technologies for production of biodiesel focusing on green catalytic techniques: A review. *Fuel Processing Technology*, *90*(12), 1502–1514. doi:10.1016/j.fuproc.2009.07.016

Hemler, C., & Upson, L. (1999). Maximising FCC propylene production. *Petroleum Technology Quarterly*, *4*, 31–35.

Henao, J., & Sotelo, O. (2018). Surface Engineering at High Temperature: Thermal Cycling and Corrosion Resistance. In A. Pakseresht (Ed.), *Production, Properties, and Applications of High Temperature Coatings* (pp. 131–159). Hershey, PA: IGI Global. doi:10.4018/978-1-5225-4194-3.ch006

Hernández-Cortez, J. G., Martinez, L., Soto, L., López, A., Navarrete, J., Manríquez, M., ... López-Salinas, E. (2010). Liquid phase alkylation of benzene with dec-1-ene catalyzed on supported 1.2-tungstophosphoric acid. *Catalysis Today*, *150*(3-4), 346–352. doi:10.1016/j.cattod.2009.12.003

Hernández-Romero, R. M., & Martínez, Y. J. (2018). Electrochemical Techniques as a Tool for Catalysts' Characterization: VIMP and SECM – A Provocative Approach. In S. González-Cortés & F. Imbert (Eds.), *Advanced Solid Catalysts for Renewable Energy Production* (pp. 386–428). Hershey, PA: IGI Global. doi:10.4018/978-1-5225-3903-2.ch012

Hertwich, E. G., Aaberg, M., Singh, B., & Strømman, A. H. (2008). Life-cycle Assessment of Carbon Dioxide Capture for Enhanced Oil Recovery. *Chinese Journal of Chemical Engineering*, *16*(3), 343–353. doi:10.1016/S1004-9541(08)60085-3

Hervert, G. L., Grove, D., & Linn, C. B. (1960). Alkylation OF Aromatic Hydrocarrons. US Patent 2,939,89.

Heyn, R. H., Jacobs, I., & Carr, R. H. (2014). *CO_2 Chemistry.* In M. Aresta & R. van Eldik (Eds.), *Advances in Inorganic Chemistry* (Vol. 66, pp. 83–115). Boca Raton, FL: Academic Press.

Hoberg, H., Peres, Y., & Milchereit, A. (1986). C-C-verknijpfung von alkenen mit co, an nickel(o); n-pentensauren aus ethen. *Journal of Organometallic Chemistry*, *307*, C38–C40. doi:10.1016/0022-328X(86)80487-9

Hoberg, H., Schefer, D., Burkhart, G., Kruger, C. M., & Romao, J. (1984). Nickel(0)-induzierte C-C-verknüpfung zwischen kohlendioxid und alkinen sowie alkenen. *Journal of Organometallic Chemistry*, *266*(2), 203–224. doi:10.1016/0022-328X(84)80129-1

Hobson, G. D. (1984). Modern Petroleum Technology. In Fuels for Spark-ignition Engines (5th ed.). John Wiley.

Hobson, G. D. (1992). Hydrocarbon Processing. In *Refining Handbook*. Gulf Publishing Co.

Hoeldrich, W. F. (2000). Environmentally benign manufacturing of fine and intermediate chemicals. *Catalysis Today*, *62*(1), 115–130. doi:10.1016/S0920-5861(00)00413-2

Honeywell, U. O. P. Aromatics. (n.d.). Retrieved from www.uop.com/processing-solutions/petrochemicals/benzene-para-xylene-production

Honeywell, U. O. P. Gasoline Desulfurization Process. (n.d.). Retrieved from www.uop.com/fcc-gasoline-hds-selectfining

Horsley, J. A. (1997). Producing bulk and fine chemicals using solid acids. *Chemtech*, *27*(10), 45–49.

Houk, B. (1996). Model predictive control of a delayed coking unit with [DMC]: Development issues. *Fuel Energy, 37,* 176.

Huang, C. C., Peng, J. J., Wu, S. H., Hou, H. Y., You, M. L., & Shu, C. M. (2010). Effects of cumenehydroperoxideon phenol and acetone manufacturing by DSC and VSP2. *Journal of Thermal Analysis and Calorimetry*, *102*(2), 579–585. doi:10.100710973-010-0953-z

Hu, C., Zhang, Y., Xu, L., & Peng, G. (1999). Continuous syntheses of octyl phenol or nonane phenol on supported heteropoly acid catalysts. *Applied Catalysis A, General*, *177*(2), 237–244. doi:10.1016/S0926-860X(98)00269-5

Huh, B., & Yan, T. (1990). *Catalytic cracking process with partial CO combustion.* US Patent: 4,980,050.

ICIS Chemical Business. (2006). *Kemira to expand Finnish formic acid capacity*. Author.

Idris, S. (2018). Water Security in Pakistan. In S. Idris (Ed.), *Promoting Global Environmental Sustainability and Cooperation* (pp. 58–84). Hershey, PA: IGI Global. doi:10.4018/978-1-5225-3990-2.ch003

IEA. (2015). *India Energy Outlook - World Energy Outlook Special Report 2015*. Retrieved from https://www.iea.org/publications/freepublications/publication/IndiaEnergyOutlook_WEO2015.pdf

Iglesia, E. (1997). Design, synthesis, and use of cobalt-based Fischer-Tropsch synthesis catalysts. *Applied Catalysis A, General*, *161*(1-2), 59–78. doi:10.1016/S0926-860X(97)00186-5

Iglesia, E., Reyes, S. C., Madon, R. J., & Soled, S. L. (1993). Selectivity control and catalyst design in the Fischer-Tropsch synthesis: Sites, pellets, and reactors. *Advances in Catalysis*, *39*, 221–302. doi:10.1016/S0360-0564(08)60579-9

Ince, F. (2018). International Businesses and Environmental Issues. In S. Idris (Ed.), *Promoting Global Environmental Sustainability and Cooperation* (pp. 86–111). Hershey, PA: IGI Global. doi:10.4018/978-1-5225-3990-2.ch004

Inoue, S., Koinuma, H., & Tsuruta, T. (1969). Copolymerization of carbon dioxide and epoxide. *Journal of Polymer Science. Part B: Polymer Letters*, *7*(4), 287–292. doi:10.1002/pol.1969.110070408

Ion, A., Van Doorslaer, C., Parvulescu, V., Jacobs, P., & De Vos, D. (2008). Green synthesis of carbamates from CO_2, amines and alcohols. *Green Chemistry*, *10*(1), 111–116. doi:10.1039/B711197E

Ipatieff, V. N. (1937). Treatment of Hydrorocarbons. US Patent 2,067,764.

Ipatieff, V. N. (1945). *US Patent 2,382,318*. US Patent Office.

Ipatieff, V. N., & Linn, C. B. (1947). Alkylation of Aromatics. US Patent 2,428,279.

Ipatieff, V. N., Corson, B. B., & Pines, H. J. (1936). Influence of Sulfuric Acid Concentration upon Reaction between Olefins and Benzene. *Journal of the American Chemical Society*, *58*(6), 919–922. doi:10.1021/ja01297a021

Ipatieff, V. N., & Grosse, A. V. (1936). Alkylation of aromatics with olefins in the presence of boron fluoride. *Journal of the American Chemical Society*, *58*(11), 2339. doi:10.1021/ja01302a505

Ipatieff, V. N., Pines, H., & Komarewsky, V. I. (1936). Phosphoric acid as the catalyst for alkylation of aromatic hydrocarbons. *Industrial & Engineering Chemistry*, *28*(2), 222–223. doi:10.1021/ie50314a019

Ipatieff, V. N., & Schmerling, L. (1946). Ethylation of Benzene in Presence of Solid Phosphoric Acid. *Industrial & Engineering Chemistry*, *38*(4), 400–402. doi:10.1021/ie50436a017

Ishutenko, D., Minaev, P., Anashkin, Yu., Nikulshina, M., Mozhaev, A., Maslakov, K., & Nikulshin, P. (2017). Potassium effect in K-Ni(Co)PW/Al2O3 catalysts for selective hydrotreating of model FCC gasoline. *Applied Catalysis B: Environmental*, *203*, 237–246. doi:10.1016/j.apcatb.2016.10.043

Jackson, N.B. (1999). *Research Teams and Partnerships: Trends in the Chemical Sciences*. Washington, DC: National Academies Press. doi:10.17226/9759

Jacobs, G., Das, T. K., Zhang, Y., Li, J., Racoillet, G., & Davis, B. H. (2002). Fischer–Tropsch synthesis: Support, loading, and promoter effects on the reducibility of cobalt catalysts. *Applied Catalysis A, General*, *233*(1), 263–281. doi:10.1016/S0926-860X(02)00195-3

Jacquin, Y., Gimenez-Coronado, M., & Dai-Nghia, H. (1984). *Process for solvent deasphalting heavy hydrocarbon fractions*. US Patent: 4,440,633.

Jayapalan, S. (2018). A Review of Chemical Treatments on Natural Fibers-Based Hybrid Composites for Engineering Applications. In K. Kumar & J. Davim (Eds.), *Composites and Advanced Materials for Industrial Applications* (pp. 16–37). Hershey, PA: IGI Global. doi:10.4018/978-1-5225-5216-1.ch002

Jiang, H., She, F., Du, Y., Chen, R. Z., & Xing, W. H. (2014). One-step continuous phenol synthesis technology via selective hydroxylation of benzene over ultrafine TS-1 in a submerged ceramic membrane reactor. *Chinese Journal of Chemical Engineering*, *22*(11–12), 1199–1207. doi:10.1016/j.cjche.2014.09.016

Johnson, L., Grant, J., & Low, P. L. (2014). Two degrees of separation: ambition and reality: Low Carbon Economy Index. *The Journal of Organic Chemistry*, *67*, 9070–9072.

Jones, D., & Pujado, P. (2006). *Handbook of Petroleum Processing*. Berlin: Springer. doi:10.1007/1-4020-2820-2

Jucha, S. O., & Moskal, G. J. (2018). Thermal Barrier Coatings on the Base of Samarium Zirconates. In A. Pakseresht (Ed.), *Production, Properties, and Applications of High Temperature Coatings* (pp. 79–106). Hershey, PA: IGI Global. doi:10.4018/978-1-5225-4194-3.ch004

Júnior, J. L. (2019). An Introduction to Computational Fluid Dynamics and Its Application in Microfluidics. In H. Santana, J. Lameu da Silva Jr, & O. Taranto (Eds.), *Process Analysis, Design, and Intensification in Microfluidics and Chemical Engineering* (pp. 50–78). Hershey, PA: IGI Global. doi:10.4018/978-1-5225-7138-4.ch002

Júnior, J. L., & Santana, H. S. (2019). Experimental and Numerical Analyses of a Micro-Heat Exchanger for Ethanol Excess Recovery From Biodiesel. In H. Santana, J. Lameu da Silva Jr, & O. Taranto (Eds.), *Process Analysis, Design, and Intensification in Microfluidics and Chemical Engineering* (pp. 167–194). Hershey, PA: IGI Global. doi:10.4018/978-1-5225-7138-4.ch006

Kaeding, W. W., & Young, L. B. (1978). *U.S. Patent 4,086,287*. US Patent Office.

Kaeding, W. W., & Holland, R. E. (1988). Shape-selective reactions with zeolite catalysts: VI. Alkylation of benzene with propylene to produce cumene. *Journal of Catalysis*, *109*(1), 212–216. doi:10.1016/0021-9517(88)90200-X

Kaeding, W., Barile, G., & Wu, M. (1984). Mobil Zeolite Catalysts for Monomers. *Catalysis Reviews. Science and Engineering*, *26*(3-4), 597–612. doi:10.1080/01614948408064727

Kais, S. M. (2017). Climate Change: Vulnerability and Resilience in Commercial Shrimp Aquaculture in Bangladesh. In W. Ganpat & W. Isaac (Eds.), *Environmental Sustainability and Climate Change Adaptation Strategies* (pp. 162–187). Hershey, PA: IGI Global. doi:10.4018/978-1-5225-1607-1.ch006

Kang, S., Ju, M., & Kim, J. (2018). The Extraction of Metal Contaminants Using Supercritical Carbon Dioxide. *International Journal of Chemoinformatics and Chemical Engineering*, *7*(1), 52–60. doi:10.4018/IJCCE.2018010104

Kapilan, N. (2018). Catalyst for Biodiesel: An Overview of Catalysts Used in Biodiesel Production. In S. González-Cortés & F. Imbert (Eds.), *Advanced Solid Catalysts for Renewable Energy Production* (pp. 156–175). Hershey, PA: IGI Global. doi:10.4018/978-1-5225-3903-2.ch006

Kartal, O. E., & Onal, I. (2008). Synthesis of ZSM-5 from modified clinoptilolite and its catalytic activity in alkylation of benzene to ethylbenzene. *Chemical Engineering Communications*, *195*(8), 1043–1057. doi:10.1080/00986440801907078

Kathalewar, M. S., Joshi, P. B., Sabnis, A. S., & Malshe, V. C. (2013). Non-isocyanate polyurethanes: From chemistry to applications. *RSC Advances*, *3*(13), 4110–4129. doi:10.1039/c2ra21938g

Kavil, Y. N., Shaban, Y. A., Al Farawati, R. K., Orif, M. I., Zobidi, M., & Khan, S. U. M. (2017). Photocatalytic conversion of CO_2 into methanol over Cu-C/TiO_2 nanoparticles under UV light and natural sunlight. *Journal of Photochemistry and Photobiology A Chemistry*, *347*, 244–253. doi:10.1016/j.jphotochem.2017.07.046

KBR. (n.d.). *Petrochemical technologies, Olefins*. Retrieved from https://www.kbr.com/en/solutions/technologies/process-technologies/petrochemical-technologies

Keshavamurthy, R., Naveena, B. E., & Sekhar, N. (2018). Thermal Spray Coatings for Erosion-Corrosion Protection. In A. Pakseresht (Ed.), *Production, Properties, and Applications of High Temperature Coatings* (pp. 246–267). Hershey, PA: IGI Global. doi:10.4018/978-1-5225-4194-3.ch010

Keshav, T. R., & Basu, S. (2007). Gas-to-liquid technologies: India's perspective. *Fuel Processing Technology*, *88*(5), 493–500. doi:10.1016/j.fuproc.2006.12.006

Kettering, C. F. (1944). The Effect of the Molecular Structure of Fuels on the Power and Efficiency of Internal Combustion Engines. *Industrial & Engineering Chemistry*, *36*(12), 1079–1085. doi:10.1021/ie50420a002

Khanna, B. K. (2017). Indian National Strategy for Climate Change Adaptation and Mitigation. In W. Ganpat & W. Isaac (Eds.), *Environmental Sustainability and Climate Change Adaptation Strategies* (pp. 32–63). Hershey, PA: IGI Global. doi:10.4018/978-1-5225-1607-1.ch002

Khanna, B. K. (2017). Vulnerability of the Lakshadweep Coral Islands in India and Strategies for Mitigating Climate Change Impacts. In W. Ganpat & W. Isaac (Eds.), *Environmental Sustainability and Climate Change Adaptation Strategies* (pp. 64–96). Hershey, PA: IGI Global. doi:10.4018/978-1-5225-1607-1.ch003

Kharlampidi, K.E., Izmailov, I.R., & Batyrshin, N.N. (2000). Alkylation of benzene with 1-butene in the presence of $AlCl_3$. *Russian Journal of General Chemistry*, *70*(2), 278-280.

Khelfi, A. (2019). Toxicocinetic and Mechanisms of Action of Nanoparticles. In R. Nazir (Ed.), *Nanotechnology Applications in Environmental Engineering* (pp. 344–368). Hershey, PA: IGI Global. doi:10.4018/978-1-5225-5745-6.ch014

Kim, S. H., Kim, K. D., & Lee, Y. K. (2017). Effects of dispersed MoS2 catalysts and reaction conditions on slurry phase hydrocracking of vacuum residue. *Journal of Catalysis*, *347*, 127–137. doi:10.1016/j.jcat.2016.11.015

Kolbe, H. (1874). Ueber eine neue Darstellungsmethode und einige bemerkenswerte Eigenschaften der Salicylsäure. *Journal für Praktische Chemie*, *10*(1), 89–107. doi:10.1002/prac.18740100106

Kolboe, S., Kubelkova, L., Haag,W. O., Rooney, J. J., Tsyganenko, A. A., Hall, W.K., Bielanski A. & Kolesnikov, I.M. (n.d.). *Catalysis in the oil and gas industry*. Gubkin Russian State University of Oil and Gas.

Kolboe, S. (1993). Hydrocarbon Formation from Methanol/Dimethyl Ether Over Protonated Zeolites and Molecular Sieves. New Insights from Recent Experiments. *Studies in Surface Science and Catalysis*, *75*, 449–461. doi:10.1016/S0167-2991(08)64030-6

Komarewsky, V. I., & Ruther, W. E. (1950). Reactions of Paraffin Hydrocarbons in the Presence of Sulfuric Acid. *Journal of the American Chemical Society*, *72*(12), 5501–5503. doi:10.1021/ja01168a034

Kornas, A., Grabowski, R., Sliwa, M., Samson, K., Ruggiero-Mikołajczyk, M., & Zelazny, A. (2017). Dimethyl ether synthesis from CO_2 hydrogenation over hybrid catalysts: Effects of preparation methods. *Reaction Kinetics, Mechanisms and Catalysis*, *121*(1), 317–327. doi:10.100711144-017-1153-7

Kothandaraman, J., Goeppert, A., Czaun, M., Olah, G. A., & Surya Prakash, G. K. (2016). Conversion of CO_2 from Air into Methanol Using a Polyamine and a Homogeneous Ruthenium Catalyst. *Journal of the American Chemical Society*, *138*(3), 778–781. doi:10.1021/jacs.5b12354 PMID:26713663

Kothari, B. L., & Mishra, N. (2017). *CHAPTER VI | Department of Animal Husbandry, Dairying & Fisheries*. Retrieved January 10, 2018, from http://dahd.nic.in/related-links/chapter-vi

Kouzu, M., Kasuno, T., Tajika, M., Sugimoto, Y., Yamanaka, S., & Hidaka, J. (2008). Calcium oxide as a solid base catalyst for transesterification of soybean oil and its application to biodiesel production. *Fuel*, *87*(12), 2798–2806. doi:10.1016/j.fuel.2007.10.019

Krässig, H. A. (1993). *Cellulose: structure, accessibility, and reactivity*. Yverdon: Gordon and Breach Science. Retrieved from http://www.worldcat.org/title/cellulose-structure-accessibility-and-reactivity/oclc/876452167?referer=di&ht=edition

Kulkarni, M. G., & Dalai, A. K. (2006). Waste Cooking OilAn Economical Source for Biodiesel: A Review. *Industrial & Engineering Chemistry Research*, *45*(9), 2901–2913. doi:10.1021/ie0510526

Kumar, S., Yadav, R.K., Ram, K., Aguiar, A., Koh, J., & Sobral, A.J.F.N., (2018), Graphene oxide modified cobalt metallated porphyrin photocatalyst for conversion of formic acid from carbon dioxide. *Journal of CO_2 Utilization, 27*,107-114.

Kumari, N., & Kumar, K. (2018). Fabrication of Orthotic Calipers With Epoxy-Based Green Composite. In K. Kumar & J. Davim (Eds.), *Composites and Advanced Materials for Industrial Applications* (pp. 157–176). Hershey, PA: IGI Global. doi:10.4018/978-1-5225-5216-1.ch008

Kuppusamy, R. R. (2018). Development of Aerospace Composite Structures Through Vacuum-Enhanced Resin Transfer Moulding Technology (VERTMTy): Vacuum-Enhanced Resin Transfer Moulding. In K. Kumar & J. Davim (Eds.), *Composites and Advanced Materials for Industrial Applications* (pp. 99–111). Hershey, PA: IGI Global. doi:10.4018/978-1-5225-5216-1.ch005

Kutler, A. A. (1969). *Crude distillation*. New York: Petro/Chem Engineering, John G. Simmonds & Co., Inc.

Laccino, L. (1999). *Heteroatom removal through countercurrent sorption*. US Patent: 5,928,497.

Lagarde, C. (2013). *A New Global Economy for a New Generation*. International Monetary Fund.

Lalancette, J. M., Fournier-Breault, M. J., & Thiffault, R. (1974). Metals Intercalated in Graphite. II. The Friedel–Crafts Reactions with AlCl3–Graphite. *Canadian Journal of Chemistry*, *52*(4), 589–591. doi:10.1139/v74-091

Lalancette, J. M., & Lafontaine, J. (1973). Intercalation of antimony pentafluoride in the lattice of graphite. *Journal of the Chemical Society. Chemical Communications*, (21): 815. doi:10.1039/c39730000815

Lallo, C. H., Smalling, S., Facey, A., & Hughes, M. (2017). The Impact of Climate Change on Small Ruminant Performance in Caribbean Communities. In W. Ganpat & W. Isaac (Eds.), *Environmental Sustainability and Climate Change Adaptation Strategies* (pp. 296–321). Hershey, PA: IGI Global. doi:10.4018/978-1-5225-1607-1.ch011

Lamani, M. (2017). Rice Straw Extracted Cellulose Biocompatible Nanofiber. *International Journal of Chemoinformatics and Chemical Engineering*, *6*(2), 1–20. doi:10.4018/IJCCE.2017070101

Lam, M. K., Lee, K. T., & Mohamed, A. R. (2010). Homogeneous, heterogeneous and enzymatic catalysis for transesterification of high free fatty acid oil (waste cooking oil) to biodiesel: A review. *Biotechnology Advances*, *28*(4), 500–518. doi:10.1016/j.biotechadv.2010.03.002 PMID:20362044

Lapidus, A. L., Pirozhkov, S. D., & Zelinsky, N. D. (1988). *Proc. 9th Int. Congr. Catal.*, *3*, 1028.

Lateef, A., & Nazir, R. (2019). Application of Magnetic Nanomaterials for Water Treatment. In R. Nazir (Ed.), *Nanotechnology Applications in Environmental Engineering* (pp. 201–219). Hershey, PA: IGI Global. doi:10.4018/978-1-5225-5745-6.ch009

Lee, K., Lee, S., Cho, H., Jeong, S., Kim, W., Lee, S., & Lee, D. C. (2017). Cu^+ -incorporated TiO_2 overlayer on Cu_2O nanowire photocathodes for enhanced photoelectrochemical conversion of CO_2 to methanol. *Journal of Energy Chemistry*.

Lee, E.-H., Ahn, J.-Y., Dharman, M. M., Park, D.-W., Park, S.-W., & Kim, I. (2008). Synthesis of cyclic carbonate from vinyl cyclohexene oxide and CO_2 using ionic liquids as catalysts. *Catalysis Today*, *131*(1-4), 130–134. doi:10.1016/j.cattod.2007.10.012

Lei, Z. G., Dai, C. N., Wang, Y. L., & Chen, B. H. (2009). Process optimization on alkylation of benzene with propylene. *Energy & Fuels*, *23*(6), 3159–3166. doi:10.1021/ef900052j

Leung, D. Y. C., Wu, X., & Leung, M. K. H. (2010). A review on biodiesel production using catalyzed transesterification. *Applied Energy*, *87*(4), 1083–1095. doi:10.1016/j.apenergy.2009.10.006

Li, Z., Shi, W., Wang, X., & Jiang, F. (1994). Deep Catalytic Cracking III. In Process for Light-olefins Production (pp. 33-42). American Chemical Society.

Li, C., Yuan, X., & Fujimoto, K. (2014). Development of highly stable catalyst for. methanol synthesis from carbon dioxide. *Applied Catalysis A, General*, *469*, 306–311. doi:10.1016/j.apcata.2013.10.010

Lifecycle GHG Emissions from Poet-DSM Project Liberty Cellulosic Ethanol Plant. (2015). Retricved from https://www.arb.ca.gov/fuels/lcfs/2a2b/apps/poet-lib-rpt-121715.pdf

Lindsey, B. D., Ayotte, J. D., Jurgens, B. C., & Desimone, L. A. (2017). Using groundwater age distributions to understand changes in methyl tert-butyl ether (MTBE) concentrations in ambient groundwater, Northeastern United States. *The Science of the Total Environment*, *579*, 579–587. doi:10.1016/j.scitotenv.2016.11.058 PMID:27884531

Lindström, B., & Pettersson, L. J. (2003). A Brief History of Catalysis. *CATTech*, *7*(4), 130–138. doi:10.1023/A:1025001809516

Li, P., Zhang, W., Han, X., & Bao, X. (2010). Conversion of Methanol to Hydrocarbons over Phosphorus-modified ZSM-5/ZSM-11 Intergrowth Zeolites. *Catalysis Letters*, *134*(1-2), 124–130. doi:10.100710562-009-0214-6

Li, S., Krishnamoorthy, S., Li, A., Meitzner, G. D., & Iglesia, E. (2002). Promoted iron-based catalysts for the Fischer–Tropsch synthesis: Design, synthesis, site densities, and catalytic properties. *Journal of Catalysis*, *206*(2), 202–217. doi:10.1006/jcat.2001.3506

Liu, C., Gao, X., Zhang, Z., Zhang, H., Sun, S., & Deng, Y. (2004). Surface modification of zeolite Y and mechanism for reducing naphtha olefin formation in catalytic cracking reaction. *Applied Catalysis A, General*, *264*(2), 225–228. doi:10.1016/j.apcata.2003.12.048

Liu, R., Qin, Z., Ji, H., & Su, T. (2013). Synthesis of Dimethyl Ether from CO_2 and H_2 Using a Cu–Fe–Zr/HZSM-5 Catalyst System. *Industrial & Engineering Chemistry Research*, *52*(47), 16648–16655. doi:10.1021/ie401763g

Liu, X.-M., Lu, G. Q., Yan, Z.-F., & Beltramini, J. (2003). Recent Advances in Catalysts for Methanol Synthesis via Hydrogenation of CO and CO_2. *Industrial & Engineering Chemistry Research*, *4225*(25), 6518–6530. doi:10.1021/ie020979s

Liu, Y., Long, J., & Tian, H. (2010). *Innovations in DCC catalyst for propylene production from heavy oils.* Beijing: Technical Progress in Refining and Petrochemical Industry China Petrochemical Press.

Liu, Y., Long, J., Tian, H., Xu, Y., & Zhao, L. (2011). Advances in DCC Process and Catalyst for Propylene Production from Heavy Oils. *China Petroleum Processing and Petrochemical Technology*, *13*, 1–5.

Li, Y., Junge, K., & Beller, M. (2013). Improving the efficiency of the hydrogenation of carbonates and carbon dioxide to methanol. *ChemCatChem*, *5*(5), 1072–1074. doi:10.1002/cctc.201300013

Li, Y., Wang, J., Wang, S., Li, D., Song, S., Zhang, P., ... Yuan, H. (2019). Immiscible Two-Phase Parallel Microflow and Its Applications in Fabricating Micro- and Nanomaterials. In H. Santana, J. Lameu da Silva Jr, & O. Taranto (Eds.), *Process Analysis, Design, and Intensification in Microfluidics and Chemical Engineering* (pp. 136–166). Hershey, PA: IGI Global. doi:10.4018/978-1-5225-7138-4.ch005

Li, Z., Xie, C., Luo, Y., Zhao, L., & Shu, X. (2007). MMC-High Propylene Selectivity DCC Catalyst. *China Petroleum Processing and Petrochemical Technology*, *1*, 13–18.

Loi, N. K., Huyen, N. T., Tu, L. H., Tram, V. N., Liem, N. D., Dat, N. L., ... Minh, D. N. (2017). Sustainable Land Use and Watershed Management in Response to Climate Change Impacts: Case Study in Srepok Watershed, Central Highland of Vietnam. In W. Ganpat & W. Isaac (Eds.), *Environmental Sustainability and Climate Change Adaptation Strategies* (pp. 255–295). Hershey, PA: IGI Global. doi:10.4018/978-1-5225-1607-1.ch010

Long, J., Tian, H., Zhang, W., Zhu, Y., Liu, Y., Chen, Z., . . . He, M. (2006). *Cracking catalyst for hydrocarbon containing molecular sieve and preparation process thereof.* Chinese Patent Application: 03147987.1.

Long, J., Tian, H., Liu, Y., Xie, C., & Li, J. (2007). Study on the Performance and Commercial Application of New Generation DMMC-1 Type Catalyst for Deep Catalytic Cracking. *China Petroleum Processing and Petrochemical Technology*, *4*, 1–6.

Lotero, Liu, Lopez, Suwannakarn, Bruce, & Goodwin. (2005). *Synthesis of Biodiesel via Acid Catalysis.* doi:10.1021/IE049157G

Lu, B., Yakushi, Y., Oumi, Y., Itabashi, K., & Sano, T. (2006). Control of crystal size of high-silica mordenite by quenching in the course of crystallization process. *Microporous and Mesoporous Materials*, *95*(1-3), 141–145. doi:10.1016/j.micromeso.2006.05.017

Luo, D., Ma, W., Wu, L., & Nakata, K. (2018). CO2 Reduction by Photocatalysis on TiO2. In S. González-Cortés & F. Imbert (Eds.), *Advanced Solid Catalysts for Renewable Energy Production* (pp. 268–299). Hershey, PA: IGI Global. doi:10.4018/978-1-5225-3903-2.ch010

Lu, S.-M., Wang, Z., Li, J., Xiao, J., & Li, C. (2016). Base-free hydrogenation of CO_2 to formic acid in water with an iridium complex bearing a *N,N'*-diimine ligand. *Green Chemistry*, *18*(16), 4553–4558. doi:10.1039/C6GC00856A

Lu, X.-B., & Darensbourg, D. J. (2012). Cobalt catalysts for the coupling of CO_2 and epoxides to provide polycarbonates and cyclic carbonates. *Chemical Society Reviews*, *41*(4), 1462–1484. doi:10.1039/C1CS15142H PMID:21858339

Lyu, L., Jin, F., Zhong, H., Chen, H., & Yao, G. (2015). A novel approach to reduction of CO_2 into methanol by water splitting with aluminum over a copper catalyst. *RSC Advances*, *5*(40), 31450–31453. doi:10.1039/C5RA02872H

Madhavi, T. P. (2019). Importance of Social Media to Create Environmental Awareness in Sustainable Higher Education System. In S. Narula, S. Rai, & A. Sharma (Eds.), *Environmental Awareness and the Role of Social Media* (pp. 72–81). Hershey, PA: IGI Global. doi:10.4018/978-1-5225-5291-8.ch004

Maier, W. H., & Ward, J. D. (1969). *United States Patent No. US3478119A*. US Patent Office.

Maihom, T., Wannakao, S., Boekfa, B., & Limtrakul, J. (2013). Production of Formic Acid via Hydrogenation of CO_2 over a Copper- Alkoxide-Functionalized MOF: A Mechanistic Study. *The Journal of Physical Chemistry C*, *117*(34), 17650–17658. doi:10.1021/jp405178p

Ma, J., Sun, N. N., Zhang, X. L., Zhao, N., Mao, F. K., Wei, W., & Sun, Y. H. (2009). A short review of catalysis for CO2 conversion. *Catalysis Today*, *148*(3-4), 221–231. doi:10.1016/j.cattod.2009.08.015

Mallick, R., & Bajpai, S. P. (2019). Impact of Social Media on Environmental Awareness. In S. Narula, S. Rai, & A. Sharma (Eds.), *Environmental Awareness and the Role of Social Media* (pp. 140–149). Hershey, PA: IGI Global. doi:10.4018/978-1-5225-5291-8.ch007

Mancini, E. R., Steen, A., Rausina, G. A., Wong, G. A., Arnold, D. C. L., Gostkowski, W. R., ... Errico, P. (2002). MTBE ambient water quality criteria development: A public/private partnership. *Environmental Science & Technology*, *36*(2), 125–129. doi:10.1021/es002059b PMID:11831214

Manne, A., & Myneni, M. B. (2019). Social Media for Disaster Awareness and Management. In S. Narula, S. Rai, & A. Sharma (Eds.), *Environmental Awareness and the Role of Social Media* (pp. 82–116). Hershey, PA: IGI Global. doi:10.4018/978-1-5225-5291-8.ch005

Marafi, A., Al-Bazzaz, H., Al-Marri, M., Maruyama, F., Absi-Halabi, M., & Stanislaus, A. (2003). Residual-Oil Hydrotreating Kinetics for Graded Catalyst Systems: Effect of Original and Treated Feedstocks. *Energy & Fuels*, *17*(5), 1191–1197. doi:10.1021/ef020282j

Markewitz, P., Kuckshinrichs, W., Leitner, W., Linssen, J., Zapp, P., Bongartz, R., ... Mgller, T. E. (2012). Worldwide innovations in the development of carbon capture technologies and the utilization of CO_2. *Energy & Environmental Science*, *5*(6), 7281–7305. doi:10.1039/c2ee03403d

Marques, J., Guillaume, D., Merrignac, I., Espinat, D., & Brunet, S. (2011). Effect of catalysts acidity on residues hydrotreatment. *Applied Catalysis B: Environmental*, *101*(3-4), 727–737. doi:10.1016/j.apcatb.2010.11.015

Marquez, O., & Jairo Márquez, P. (2018). Synthesis of Electrocatalysts for Electrochemistry in Energy. In S. González-Cortés & F. Imbert (Eds.), *Advanced Solid Catalysts for Renewable Energy Production* (pp. 300–385). Hershey, PA: IGI Global. doi:10.4018/978-1-5225-3903-2.ch011

Marsh, G., Hill, N., & Sully, J. (2000). *Consultation on the Need to Reduce the Sulphur Content of Petrol and Diesel Fuels Below 50 PPM: - A Policy Makers Summary*. Restricted – Commercial AEAT/ENV/R/*0372* issue 5.

Martinez-Guerra, E., Muppaneni, T., Gude, V. G., & Deng, S. (2018). Non-Conventional Feedstock and Technologies for Biodiesel Production. In S. González-Cortés & F. Imbert (Eds.), *Advanced Solid Catalysts for Renewable Energy Production* (pp. 96–118). Hershey, PA: IGI Global. doi:10.4018/978-1-5225-3903-2.ch004

Mattox, W. J. (1945). Conversion of Hyorocarbons. US Patent 2,385,524.

Mcaloon, A., Taylor, F., Yee, W., Ibsen, K., & Wooley, R. (2000). *Determining the Cost of Producing Ethanol from Corn Starch and Lignocellulosic Feedstocks*. Retrieved from https://www.nrel.gov/docs/fy01osti/28893.pdf

McClure, J. D. (1977). Ethylbenzene Process using an Unsupported Perfluorinated Polymer Catalyst. US Patent 4,041,090.

McClure, J. D. (1977). Hydrocarbon Conversion Process Usingan Unsupported Perfluornated Polymer Catalyst. US Patent 4,022, 847.

McGhee, W. D., & Riley, D. P. (1992). Palladium-mediated synthesis of urethanes from amines, carbon dioxide, and cyclic diolefins. *Organometallics*, *11*(2), 900–907. doi:10.1021/om00038a060

McGhee, W., Riley, D., Christ, K., Pan, Y., & Parnas, B. (1995). Carbon Dioxide as a Phosgene Replacement: Synthesis and Mechanistic Studies of Urethanes from Amines, CO_2, and Alkyl Chlorides. *The Journal of Organic Chemistry*, *60*(9), 2820–2830. doi:10.1021/jo00114a035

McKetta. (1992). *Petroleum Processing Handbook*. New York: Marcel Dekkar.

Mehlberg, R., Rosser, F., Fei, Z., & Stevens, C. (2011). *Low NOx FCC catalyst regeneration process*. US Patent: 7,914,666.

Meima, G. R. (1998). Advances in cumene production. *CATTech*, *2*(1), 5–12.

Meima, G. R., Van der Aalst, M. J. M., Samson, M. S. U., Garces, J. M., & Lee, J. G. (1996). Recent developments in the production of cumene with zeolite catalysts. *EdrolErdgasKohle*, *112*(7), 3156–3318.

Methyl tertiary-Butyl Ether (MTBE) in drinking water. (2015). World Health Organization.

Metz, B., Davidson, O., de Coninck, H., Loos, M., & Meyer, L. (2005). *IPCC Special Report on Carbon Dioxide Capture and Storage*. Intergovernmental Panel on Climate Change. Available at: http://www.ipcc.ch/pdf/special-reports/srccs/ srccs_wholereport.pdf

Meyers, R. A. (2005). Handbook of Heterogeneous Catalysis. McGraw-Hill.

Michael Frenkel, G. J. K., Marsh, K. N., Roganov, G. N., & Wilhoit, R. C. (1994). *Thermodynamics of Organic Compounds in the Gas State* (Vol. 2). College Station, TX: Academic Press.

Ming-Yuan, H., Zhonghui, L., & Enze, M. (1988). Acidic and hydrocarbon catalytic properties of pillared clay. *Catalysis Today*, *2*(2-3), 321–338. doi:10.1016/0920-5861(88)85013-2

Mita, T., Higuchi, Y., & Sato, Y. (2014). Carboxylation with CO_2 via Brook Rearrangement: Preparation of α-Hydroxy Acid Derivatives. *Organic Letters*, *16*(1), 7–14. doi:10.1021/ol403099f PMID:24313846

Mitchell, S., Michels, N.-L., Kunze, K., & Ramirez-Perez, J. (2012). Visualization of hierarchically structured zeolite bodies from macro to nano length scales. *Nature Chemistry*, *4*(10), 825–831. doi:10.1038/nchem.1403 PMID:23000996

Mitsubishi Hitachi Power Systems. (n.d.). *Integrated Coal Gasification Combined Cycle (IGCC) Power Plants*. Retrieved from https://www.mhps.com/products/igcc/

Mitsutani, A. (2002). Future possibilities of recently commercialized acid/base-catalyzed chemical processes. *Catalysis Today*, *73*(1-2), 57–63. doi:10.1016/S0920-5861(01)00518-1

Miyano, M., Zhang, H., Yoshiba, M., & Izumi, Y. (2017). Selective Photoconversion of Carbon Dioxide into Methanol Using Layered Double Hydroxides at 0.40 MPa. *Energy Technology (Weinheim)*, *5*(6), 892–900. doi:10.1002/ente.201600578

Mkhize, N., & Ellis, D. (2018). Consumer Cooperation in Sustainability: The Green Gap in an Emerging Market. In S. Idris (Ed.), *Promoting Global Environmental Sustainability and Cooperation* (pp. 112–135). Hershey, PA: IGI Global. doi:10.4018/978-1-5225-3990-2.ch005

Mohajeri, A., Zamani, Y., & Shirazi, L. (2014). Investigation of products distribution in Fischer-Tropsch synthesis by nano-sized iron-based catalyst. *Journal of Nanoanalysis*, *1*(2), 58–64.

Moniruzzaman, M., Dale, B. E., Hespell, R. B., & Bothast, R. J. (1997). Enzymatic hydrolysis of high-moisture corn fiber pretreated by afex and recovery and recycling of the enzyme complex. *Applied Biochemistry and Biotechnology*, *67*(1–2), 113–126. doi:10.1007/BF02787846

Montazeri, A., & Karjibani, M. (2017). Preparation and Characterization of Chitosan - Double Walled Carbon Nanotubes Hydrogels. *International Journal of Chemoinformatics and Chemical Engineering*, *6*(2), 21–30. doi:10.4018/IJCCE.2017070102

Monteiro, W. F., Vieira, M. O., Aquino, A. S., de Souza, M. O., Lima, J. D., Einloft, S., & Ligabue, R. (2017). CO_2 conversion to propylene carbonate catalyzed by Ionic liquid containing organosilane groups supported on titanate nanotubes/nanowires. *Applied Catalysis A, General*, *544*, 46–54. doi:10.1016/j.apcata.2017.07.011

Morel, F., Kressmann, S., Harle, V., & Kasztelan, S. (1997). Processes and catalysts for hydrocracking of heavy oil and residues. *Studies in Surface Science and Catalysis*, *106*, 1–16. doi:10.1016/S0167-2991(97)80003-1

Moreno, A., Cohen, L., & Berna, J. L. (1988). Influence of unsulfonated material and its sulfone content on the physical properties of linear alkylbenzenesulfonates. *Surfactant Detergent*, *25*, 216.

Motokura, K., Kashiwame, D., Miyaji, A., & Baba, T. (2012). Copper-Catalyzed Formic Acid Synthesis from CO_2 with Hydrosilanes and H_2O. *Organic Letters*, *14*(10), 2642–2645. doi:10.1021/ol301034j PMID:22540994

Mubarak, S. M. (2019). Nanotechnology for Air Remediation. In R. Nazir (Ed.), *Nanotechnology Applications in Environmental Engineering* (pp. 121–142). Hershey, PA: IGI Global. doi:10.4018/978-1-5225-5745-6.ch006

Mujere, N., & Moyce, W. (2017). Climate Change Impacts on Surface Water Quality. In W. Ganpat & W. Isaac (Eds.), *Environmental Sustainability and Climate Change Adaptation Strategies* (pp. 322–340). Hershey, PA: IGI Global. doi:10.4018/978-1-5225-1607-1.ch012

Mukhopadhyay, A., Barman, T. K., & Sahoo, P. (2018). Electroless Nickel Coatings for High Temperature Applications. In K. Kumar & J. Davim (Eds.), *Composites and Advanced Materials for Industrial Applications* (pp. 297–331). Hershey, PA: IGI Global. doi:10.4018/978-1-5225-5216-1.ch013

Murzin, D. Y. (2009). Thermodynamic analysis of nanoparticle size effect on catalytic kinetics. *Chemical Engineering Science*, *64*(5), 1046–1052. doi:10.1016/j.ces.2008.10.066

Nadim, F., Zack, P., Hoag, G. E., & Liu, S. (2001). United States experience with gasoline additives. *Energy Policy*, *29*(1), 1–5. doi:10.1016/S0301-4215(00)00099-9

Nagashima, S., Miyagawa, T., Matsumoto, M., Suzuki, S., Komaki, H., Takagi, M., & Murai, S. (2011). *Life Cycle Assessment Performed on a CCS Model Case in Japan and Evaluation of Improvement Facilitated by Heat Integration.* Academic Press.

Naghiyev, F. B. (2018). Amplitude-Frequency Characteristics of the Oscillations of Methane Gas Bubbles in Oil. *International Journal of Chemoinformatics and Chemical Engineering*, *7*(2), 16–28. doi:10.4018/IJCCE.2018070102

Naik, S. P., Vy Bui, T. R., Miller, J. D., Drinnan, N. B., & Zmierczak, W. (2011). Synthesis of DME from CO2/H2 gas mixture. *Chemical Engineering Journal*, *167*, 362–368.

Najim, M., Feddaoui, M., Alla, A. N., & Charef, A. (2019). Heat and Mass Transfer Characteristics of Evaporating Falling Films: Application to Thermal Protection and Desalination. In H. Santana, J. Lameu da Silva Jr, & O. Taranto (Eds.), *Process Analysis, Design, and Intensification in Microfluidics and Chemical Engineering* (pp. 251–276). Hershey, PA: IGI Global. doi:10.4018/978-1-5225-7138-4.ch009

Namuangruk, S., Pantu, P., & Limtrakul, J. (2004). Alkylation of benzene with ethylene over faujasite zeolite investigated by the ONIOM method. *Journal of Catalysis*, *225*(2), 523–530. doi:10.1016/j.jcat.2004.04.016

Narasimharao, K., Lee, A., & Wilson, K. (2007). Catalysts in Production of Biodiesel: A Review. *Journal of Biobased Materials and Bioenergy*, *1*(3), 301–314. doi:10.1166/jbmb.2007.002

Nasreen, S., & Rafique, U. (2019). Nanotechnology for Water Environmental Application: Functionalized Silica Hybrids as Nano-Sorbents. In R. Nazir (Ed.), *Nanotechnology Applications in Environmental Engineering* (pp. 171–200). Hershey, PA: IGI Global. doi:10.4018/978-1-5225-5745-6.ch008

Nawaz, Z., Zhu, J., & Wei, F. (2009). Drastic enhancement of propene yield from 1-hexene catalytic cracking using a shape intensified MESO-SAPO-34 catalyst. *Journal of Engineering Science and Technology*, *4*, 409–418.

Nazir, R. (2019). Introduction to Environmental Nanotechnology: E-Nano. In R. Nazir (Ed.), *Nanotechnology Applications in Environmental Engineering* (pp. 1–27). Hershey, PA: IGI Global. doi:10.4018/978-1-5225-5745-6.ch001

Nee, J., Harding, R., Yaluris, G., Cheng, W., Zhao, X., Dougan, T., & Riley, J. (2004). *Kirk-Othmer Encyclopedia of Chemical Technology. Fluid Catalytic Cracking (FCC)* Catalysts and Additives.

Nelson, W. L. (1958). *Petroleum Refinery Engineering* (4th ed.). New York: McGraw-Hill Book Co.

Ngubevana, L. (2018). Process Synthesis and the G-H Space for Analysing and Optimising Biomass Gasification. *International Journal of Chemoinformatics and Chemical Engineering*, *7*(1), 13–41. doi:10.4018/IJCCE.2018010102

Nichols Roth, G. (2015). *G7 Climate Change The New Economy*. World News -Climate Change The New Economy Ltd.

Noman, D., & Owens, G. (2019). Wastewater Treatment and Role of Green Synthesized Metal Oxide Nanocomposites. In R. Nazir (Ed.), *Nanotechnology Applications in Environmental Engineering* (pp. 268–307). Hershey, PA: IGI Global. doi:10.4018/978-1-5225-5745-6.ch012

Nomura, R., Hasegawa, Y., Ishimoto, M., Toyosaki, T., & Matsuda, H. (1992). Carbonylation of Amines by Carbon Dioxide in the Presence of an Organoantimony Catalyst'. *The Journal of Organic Chemistry*, *57*(26), 7339–7342. doi:10.1021/jo00052a060

Nosheen, S. (2019). Nanomembrane Applications in Environmental Engineering. In R. Nazir (Ed.), *Nanotechnology Applications in Environmental Engineering* (pp. 103–120). Hershey, PA: IGI Global. doi:10.4018/978-1-5225-5745-6.ch005

O'Kelly, A. A., Kellet, J., & Plucker, J. (1947). Monoalkyibenzenes by Vapor-Phase Alkylation with Silica-alumina Catalyst. *Journal of Industrial and Engineering Chemistry*, *39*(2), 154–158. doi:10.1021/ie50446a016

Oar-Arteta, L., Aguayo, A. T., Remiro, M., Arandia, A., Bilbao, J., & Gayubo, A. G. (2016). Kinetics of the steam reforming of dimethyl ether over $CuFe_2O_4$/g-Al_2O_3. *Chemical Engineering Journal*, *306*, 401–412. doi:10.1016/j.cej.2016.07.075

Occelli, M., & Anthony, R. (1989). *Advances in Hydrotreating Catalysts*. Amsterdam: Elsevier.

Occelli, M., & O'Connor, P. (1994). *III, Materials and Processes*. American Chemical Society.

Occelli, M., & O'Connor, P. (1998). *Fluid Cracking Catalysts*. New York: Marcel Dekker.

Odedairo, T., & Al-Khattaf, S. (2011). Alkylation and transalkylation of alkylbenzenes in cymene production over zeolite catalysts. *Chemical Engineering Journal*, *167*(1), 240–254. doi:10.1016/j.cej.2010.12.043

Odetola, P. I., Popoola, P. A., & Oladijo, P. (2018). Thin Coating Deposition by Magnetron Sputtering. In A. Pakseresht (Ed.), *Production, Properties, and Applications of High Temperature Coatings* (pp. 403–428). Hershey, PA: IGI Global. doi:10.4018/978-1-5225-4194-3.ch015

Ohmiya, H., Tanabe, M., & Sawamura, M. (2011). Copper-Catalyzed Carboxylation of Alkylboranes with Carbon Dioxide: Formal Reductive Carboxylation of Terminal Alkenes. *Organic Letters*, *13*(5), 1086–1088. doi:10.1021/ol103128x PMID:21275416

Okuhara, T., Mizuno, N., & Misono, M. (1996). Catalytic Chemistry of Heteropoly Compounds. *Advances in Catalysis*, *41*, 113–252. doi:10.1016/S0360-0564(08)60041-3

Olah, G. A. (1973). *Friedel-Crafts Chemistry*. Wiley-Interscience.

Olah, G. A., Kaspi, J., & Bukala, J. (1977). Heterogeneous catalysis by solid superacids. 3. Alkylation of benzene and transalkylation of alkylbenzenes over graphite-intercalated Lewis acid halide and perfluorinated resin sulfonic acid (Nafion-H) catalysts. *The Journal of Organic Chemistry*, *42*(26), 4187–4191. doi:10.1021/jo00862a003

Olah, G. A., Prakash, G. K. S., & Sommer, J. (1979). Superacids. *Science*, *206*(4414), 13–20. doi:10.1126cience.206.4414.13 PMID:17812427

Olah, G. A., & Ruff, J. K. (1964). Friedel-Crafts and related reactions. Volume I: General Aspects. *Inorganic Chemistry*, *3*(8), 1205–1206. doi:10.1021/ic50018a043

Otto, A., Grube, T., Schiebahn, S., & Stolten, D. (2015). Closing the loop: Captured CO_2 as a feedstock in the chemical industry. *Energy & Environmental Science*, *8*(11), 3283–3297. doi:10.1039/C5EE02591E

Overend, R. P. (2004). *Thermochemical Biomass Gasification Technologies and Products. Stanford University: Global Climate and Energy Project (GCEP)*. Battelle. Retrieved from https://web.stanford.edu/group/gcep/pdfs/energy_workshops_04_04/biomass_overend.pdf

Pandey, J. (2019). Recent Progresses in Membranes for Proton Exchange Membrane Fuel Cell (PEMFC) for Clean and Environmentally Friendly Applications. In R. Nazir (Ed.), *Nanotechnology Applications in Environmental Engineering* (pp. 308–343). Hershey, PA: IGI Global. doi:10.4018/978-1-5225-5745-6.ch013

Pandey, R. K., Rehman, A., & Sarviya, R. M. (2012). Impact of alternative fuel properties on fuel spray behavior and atomization. *Renewable & Sustainable Energy Reviews*, *16*(3), 1762–1778. doi:10.1016/j.rser.2011.11.010

Peng, W., Zhao, N., Xiao, F., Wei, W., & Sun, Y. (2012). Recent progress in phosgene-free methods for synthesis of dimethyl carbonate. *Pure and Applied Chemistry*, *84*(3), 603–620. doi:10.1351/PAC-CON-11-06-02

Pennwell. (n.d.). *DCC reactor and generator*. Retrieved from http://images.pennwellnet.com/ogj/images/ogj2/9602jfufu02.gif

Perego, C., Amarilli, S., Carati, A., Flego, C., Pazzuconi, G., Rizzo, C., & Bellusi, G. (1999). Mesoporous silica-aluminas as catalysts for the alkylation of aromatic hydrocarbons with olefins. *Microporous and Mesoporous Materials*, *27*(2-3), 345–354. doi:10.1016/S1387-1811(98)00267-4

Perego, C., & Ingallina, P. (2002). Recent advances in the industrial alkylation of aromatics: New catalysts and new processes. *Catalysis Today*, *73*(1-2), 3–22. doi:10.1016/S0920-5861(01)00511-9

Perez-Uriarte, P., Ateka, A., Gamero, M., Aguayo, A. T., & Bilbao, J. (2016). Effect of the operating conditions in the transformation of DME to olefins over a HZSM-5 zeolite catalyst. *Industrial & Engineering Chemistry Research*, *55*(23), 6569–6578. doi:10.1021/acs.iecr.6b00627

Peters, A., Rakiewicz, E., Weatherbee, G., & Zhao, X. (2002). *Reduced NOx combustion promoter for use in FCC processes.* US Patent: 6,358,881 B.

Peters, M., Kchler, B., Kuckshinrichs, W., Leitner, W., Markewitz, P., & Muller, T. E. (2011). Chemical technologies for exploiting and recycling carbon dioxide into the value chain. *ChemSusChem*, *4*(9), 1216–1240. doi:10.1002/cssc.201000447 PMID:21866580

Phillips, W. (1991). Critical Technologies: Panel selects 22 national priorities. *Chemical and Engineering News*, *69*(17), 4. doi:10.1021/cen-v069n017.p004

Phosgene was used as a chemical weapon during World War I. (n.d.). Retrieved from https://www.opcw.org/chemical-weapons-convention/annexes/ annex-on-chemicals/schedule-3/

Poels, E., van Beek, W., den Hoed, W., & Visser, C. (1995). Deactivation of fixed-bed nickel hydrogenation catalysts by sulfur. *Fuel*, *74*(12), 1800–1805. doi:10.1016/0016-2361(95)80011-6

Pont, J., & Unnasch, S. (2014). *GranBio-Modified GREET Pathway for the Production of Ethanol from Sugarcane Straw*. Retrieved from https://www.arb.ca.gov/fuels/lcfs/2a2b/apps/gb-rpt-082514.pdf

Pour, A. N., Housaindokht, M. R., Irani, M., & Shahri, S. M. K. (2014). Size-dependent studies of Fischer–Tropsch synthesis on iron based catalyst: New kinetic model. *Fuel*, *116*, 787–793. doi:10.1016/j.fuel.2013.08.080

Pour, A. N., Housaindokht, M. R., Tayyari, S. F., Zarkesh, J., & Alaei, M. R. (2010). Deactivation studies of Fischer–Tropsch synthesis on nano-structured iron catalyst. *Journal of Molecular Catalysis A Chemical*, *330*(1-2), 112–120. doi:10.1016/j.molcata.2010.07.010

Pour, A. N., Zamani, Y., Tavasoli, A., Shahri, S. M. K., & Taheri, S. A. (2008). Study on products distribution of iron and iron–zeolite catalysts in Fischer–Tropsch synthesis. *Fuel*, *87*(10-11), 2004–2012. doi:10.1016/j.fuel.2007.10.014

Prez-Uriarte, P., Ateka, A., Aguayo, A. T., Gayubo, A. G., & Bilbao, J. (2016). Kinetic model for the reaction of DME to olefins over a HZSM-5 zeolite catalyst. *Chemical Engineering Journal*, *302*, 801–810. doi:10.1016/j.cej.2016.05.096

Progue, R. F. (1996). *World Patent 96/34843*. Dow Chemical Company.

Pujado, P. R., Salazar, J. R., & Berger, C. V. (1976). Cheapest Route to Phenol. *Hydrocarbon Processing*, *55*(3), 91–96.

Purohit, P., & Vagge, S. T. (2018). Oxidation Behavior of LTA/YSZ Intermixed Layer Coating. In A. Pakseresht (Ed.), *Production, Properties, and Applications of High Temperature Coatings* (pp. 107–130). Hershey, PA: IGI Global. doi:10.4018/978-1-5225-4194-3.ch005

Qn, L., & Hill, C. (1992). *US Patent 5 118 894*. US Patent Office.

Quaranta, E., & Aresta, M. (2010). The Chemistry of N−CO_2 Bonds: Synthesis of Carbamic Acids and Their Derivatives, Isocyanates and Ureas. In M. Aresta (Ed.), *Carbon Dioxide as Chemical Feedstock* (pp. 121–167). Weinheim: Wiley-VCH. doi:10.1002/9783527629916.ch6

Ramdani, N., & Azibi, M. (2018). Polymer Composite Materials for Microelectronics Packaging Applications: Composites for Microelectronics Packaging. In K. Kumar & J. Davim (Eds.), *Composites and Advanced Materials for Industrial Applications* (pp. 177–211). Hershey, PA: IGI Global. doi:10.4018/978-1-5225-5216-1.ch009

Ramesh, D., Muniraj, I. K., Thangavelu, K., & Karthikeyan, S. (2018). Pretreatment of Lignocellulosic Biomass Feedstocks for Biofuel Production. In S. González-Cortés & F. Imbert (Eds.), *Advanced Solid Catalysts for Renewable Energy Production* (pp. 33–60). Hershey, PA: IGI Global. doi:10.4018/978-1-5225-3903-2.ch002

Ramona, B. (2018). Global Implications of Sustainability and E-Society Infrastructure in Developing Economies. In S. Idris (Ed.), *Promoting Global Environmental Sustainability and Cooperation* (pp. 162–183). Hershey, PA: IGI Global. doi:10.4018/978-1-5225-3990-2.ch007

Rana, M., Samano, V., Ancheyta, J., & Diaz, J. (2007). A review of recent advances on process technologies for upgrading of heavy oils and residua. *Fuel*, *86*(9), 1216–1231. doi:10.1016/j.fuel.2006.08.004

Rangel, M. D., Pereira, A. L., Marchetti, G., Querino, P. S., & Albornoz, A. (2018). Case Study: Effect of Zirconium on the Textural and Catalytic Properties of Magnetite. In S. González-Cortés & F. Imbert (Eds.), *Advanced Solid Catalysts for Renewable Energy Production* (pp. 216–241). Hershey, PA: IGI Global. doi:10.4018/978-1-5225-3903-2.ch008

Reinoso, F., Santana, P., Palazon, E., Diez, M., & Marsh, H. (1998). Delayed coking: Industrial and laboratory aspects. *Carbon*, *36*(1-2), 105–116. doi:10.1016/S0008-6223(97)00154-1

Ren, T., Patel, M., & Blok, K. (2006). Olefins from conventional and heavy feedstocks: Energy use in steam cracking and alternative processes. *Energy*, *31*(4), 425–451. doi:10.1016/j.energy.2005.04.001

Ren, Y., & Sophie, A. L. (2018). Metal-Free Synthesis of Unsymmetrical Ureas and Carbamates from CO_2 and Amines via Isocyanate Intermediates. *The Journal of Organic Chemistry*, *83*(2), 913–920. doi:10.1021/acs.joc.7b02905 PMID:29226676

Reutemann, W., & Kieczka, H. (2012). Formic acid. In *Ullmann's encyclopedia of industrial chemistry* (7th ed.). Wiley-VCH. Retrieved from https://en.wikipedia.org/wiki/Formic_acid

Ribeiro, A. P. C., Martins, L. M. D. R. S., & Pombeiro, A. J. L. (2013). Carbon dioxide-to-methanol single-pot conversion using a C-scorpionate iron(II) catalyst. *Green Chemistry, 00*, 1–3.

Rizwanul Fattah, I. M., Masjuki, H. H., Kalam, M. A., Mofijur, M., & Abedin, M. J. (2014). Effect of antioxidant on the performance and emission characteristics of a diesel engine fuelled with palm biodiesel blends. *Energy Conversion and Management, 79*, 265–272. doi:10.1016/j.enconman.2013.12.024

Rodewald, P. G. (1975). Hydrocarbon Conversion in the Presence of a Catalyst of a Lewis acid Intercalated in Graphite. US Patent 3,925,495.

Rodríguez-Sulbarán, P. J., Lugo, C. A., Perez, M. A., Gonzalez-Cortes, S. L., D'Angelo, R., Rondon, J., ... Del Castillo, H. L. (2018). Dry Reforming of Methane on LaSrNiAl Perovskite-Type Structures Synthesized by Solution Combustion. In S. González-Cortés & F. Imbert (Eds.), *Advanced Solid Catalysts for Renewable Energy Production* (pp. 242–266). Hershey, PA: IGI Global. doi:10.4018/978-1-5225-3903-2.ch009

Rofer-DePoorter, C. K. (1981). A comprehensive mechanism for the Fischer-Tropsch synthesis. *Chemical Reviews, 81*(5), 447–474. doi:10.1021/cr00045a002

Rosa, R. L., & Seabra, A. C. (2019). Design Techniques for Microfluidic Devices Implementation Applicable to Chemical Analysis Systems: Theoretical and Experimental Analysis of Microcomponents. In H. Santana, J. Lameu da Silva Jr, & O. Taranto (Eds.), *Process Analysis, Design, and Intensification in Microfluidics and Chemical Engineering* (pp. 195–222). Hershey, PA: IGI Global. doi:10.4018/978-1-5225-7138-4.ch007

Rudesill, J., & Peters, A. (1996). The Effect of Oxidation State of Vanadium on the Selectivity of Fluid Catalytic Cracking Catalysts. *ACS Division of Petroleum Chemistry, Inc. Preprints, 41*, 407–409.

Sabharwal, D. (2019). Study of Consumers' Attitudes Towards Marketing of Environmentally Friendly Products. In S. Narula, S. Rai, & A. Sharma (Eds.), *Environmental Awareness and the Role of Social Media* (pp. 150–168). Hershey, PA: IGI Global. doi:10.4018/978-1-5225-5291-8.ch008

Sadegh, H., Helmi, H., Hamdy, A. S., Masjedi, A., Dastjerdi, M. J. H.& Shahryari-ghoshekandi, R. (2016). A developed simple, facile, economic and fast technique for preparation of high quality Fe_3O_4 magnetic nanoparticles. *Chemistry of Advanced Materials, 1*(1), 27-32.

Sadegh, H., Ali, G. A. M., Makhlouf, A. S. H., Chong, K. F., Alharbi, N. S., Agarwal, S., & Gupta, V. K. (2018). MWCNTs-Fe_3O_4 nanocomposite for Hg (II) high adsorption efficiency. *Journal of Molecular Liquids, 258*, 345–353. doi:10.1016/j.molliq.2018.03.012

Sadegh, H., Ali, G. A., Nia, H. J., & Mahmoodi, Z. (2019). Nanomaterial Surface Modifications for Enhancement of the Pollutant Adsorption From Wastewater: Adsorption of Nanomaterials. In R. Nazir (Ed.), *Nanotechnology Applications in Environmental Engineering* (pp. 143–170). Hershey, PA: IGI Global. doi:10.4018/978-1-5225-5745-6.ch007

Sadegh, H., Shahryari-ghoshekandi, R., & Kazemi, M. (2014). Study in synthesis and characterization of carbon nanotubes decorated by magnetic iron oxide nanoparticles. *International Nano Letters*, *4*(4), 129–135. doi:10.100740089-014-0128-1

Sadri, E., & Ashrafizadeh, F. (2018). High Temperature Nanocomposite Coatings by Plasma Spraying for Friction and Wear Applications. In A. Pakseresht (Ed.), *Production, Properties, and Applications of High Temperature Coatings* (pp. 216–245). Hershey, PA: IGI Global. doi:10.4018/978-1-5225-4194-3.ch009

Sagiroglu, A., Isbilir, Ş. S., Ozcan, H. M., Paluzar, H., & Toprakkiran, N. M. (2011). Comparison Of Biodiesel Productivities Of Different Vegetable Oils By Acidic Catalysis. *Chemical Industry & Chemical Engineering Quarterly*, *17*(1), 53–58. doi:10.2298/CICEQ100114054S

Sahoo, S. (2018). Laminated Composite Hypar Shells as Roofing Units: Static and Dynamic Behavior. In K. Kumar & J. Davim (Eds.), *Composites and Advanced Materials for Industrial Applications* (pp. 249–269). Hershey, PA: IGI Global. doi:10.4018/978-1-5225-5216-1.ch011

Sahu, R., Song, B. J., Im, J. S., Jeon, Y.-P., & Lee, C. W. (2015). A review of recent advances in catalytic hydrocracking of heavy residues. *Journal of Industrial and Engineering Chemistry*, *27*, 12–24. doi:10.1016/j.jiec.2015.01.011

Sahu, R., Song, B.-J., Jeon, Y. P., & Lee, C. W. (2016). Upgrading of vacuum residue in batch type reactor using Ni-Mo supported on goethite catalyst. *Journal of Industrial and Engineering Chemistry*, *35*, 115–122. doi:10.1016/j.jiec.2015.12.017

Sakakura, T., Choi, J.-C., & Yasuda, H. (2007). Transformation of Carbon Dioxide. *Chemical Reviews*, *107*(6), 2365–2387. doi:10.1021/cr068357u PMID:17564481

Salar, H. (2006). Analysis of metal ions in crude oil by reversed-phase high performance liquid chromatography using short column. *Journal of Chromatography. A*, *1118*(1), 82–84. doi:10.1016/j.chroma.2006.04.068 PMID:16723133

Salvatore, R. N., Shin, S. I., Nagle, A. S., & Jung, K. W. (2001). Efficient carbamate synthesis via a three-component coupling of an amine, CO2, and alkyl halides in the presence of Cs2CO3 and tetrabutylammonium iodide. *The Journal of Organic Chemistry*, *66*(3), 1035–1037. doi:10.1021/jo001140u PMID:11430069

Sambi, I., Khulbe, K., & Mann, R. (1982). Catalytic hydrotreatment of heavy gas oil. *Industrial & Engineering Chemistry Product Research and Development*, *21*(4), 575–580. doi:10.1021/i300008a012

Sanchez-Claros, M., Ortega-Casanova, J., & Galindo-Rosales, F. J. (2019). 2D Numerical Study of a Micromixer Based on Blowing and Vortex Shedding Mechanisms. In H. Santana, J. Lameu da Silva Jr, & O. Taranto (Eds.), *Process Analysis, Design, and Intensification in Microfluidics and Chemical Engineering* (pp. 79–113). Hershey, PA: IGI Global. doi:10.4018/978-1-5225-7138-4.ch003

Sanchez, J. M., & Toral, M. R. (2011). Model for Gasification of Residual Fuels from Petroleum Refineries Using the Equation Oriented (EO) Approach. *Industrial & Engineering Chemistry Research, 50*(5), 2628–2640. doi:10.1021/ie100663w

Santhoshkumar, R., Hong, Y.-C., Luo, C.-Z., Wu, Y.-C., Hung, C.-H., Hwang, K.-Y., ... Cheng, C.-H. (2016). Synthesis of Vinyl Carboxylic Acids using Carbon Dioxide as a Carbon Source by Iron-Catalyzed Hydromagnesiation. *ChemCatChem, 8*(13), 1–5. doi:10.1002/cctc.201600279

Sattarin, M., Modarresi, H., Talachi, H., & Teymori, M. (2006). Solvent deasphalting of vacuum residue in a bench-scale unit. *Petroleum and Coal, 48*, 14–19.

Satterfield, C. N. (1991). *Heterogeneous Catalysis in Industrial Practice* (2nd ed.). New York: McGraw-Hill.

Saylik, D., Horvath, M. J., Elmes, P. S., Jackson, W. R., Lovel, C. G., & Moody, K. (1999). Preparation of Isocyanates from Primary Amines and Carbon Dioxide Using Mitsunobu Chemistry. *The Journal of Organic Chemistry, 64*(11), 3940–3946. doi:10.1021/jo982362j

Schmerling, L. (1945). Alkylation of Aromatic Compounds. US Patent 2,385,303.

Schmidt, A. S., & Thomsen, A. B. (1998). Optimization of wet oxidation pretreatment of wheat straw. *Bioresource Technology, 64*(2), 139–151. doi:10.1016/S0960-8524(97)00164-8

Schulz, H., vein Steen, E., & Claeys, M. (1994). Selectivity and mechanism of Fischer-Tropsch synthesis with iron and cobalt catalysts. In Studies in Surface Science and Catalysis (Vol. 81, pp. 455-460). Elsevier.

Schulz, H. (1999). Short history and present trends of Fischer–Tropsch synthesis. *Applied Catalysis A, General, 186*(1-2), 3–12. doi:10.1016/S0926-860X(99)00160-X

Sebulsky, R. T., & Henke, A. M. (1971). Alkylation of Benzene with Dodecene-1 Catalyzed by Supported Silicotungstic Acid. *Industrial & Engineering Chemistry Process Design and Development, 10*(2), 272–279. doi:10.1021/i260038a022

Semelsberger, T. A., Borup, R. L., & Greene, H. L. (2006). Dimethyl ether (DME) as an alternative fuel. *Journal of Power Sources, 156*(2), 497–511. doi:10.1016/j.jpowsour.2005.05.082

Semiannual report of UST performance measures. (2015). Environmental Protection Agency.

Seyedin, S., Maghsoodloo, S., & Mottaghitalab, V. (2017). A Quantitative Study on Simultaneous Effects of Governing Parameters in Electrospinning of Nanofibers using Modified Neural Network using Genetic Algorithm. *International Journal of Chemoinformatics and Chemical Engineering, 6*(1), 20–38. doi:10.4018/IJCCE.2017010102

Sezgin, H., & Berkalp, O. B. (2018). Textile-Reinforced Composites for the Automotive Industry. In K. Kumar & J. Davim (Eds.), *Composites and Advanced Materials for Industrial Applications* (pp. 129–156). Hershey, PA: IGI Global. doi:10.4018/978-1-5225-5216-1.ch007

Shahien, M. (2018). Reactive Plasma Spray: A Method for Nitride Coatings Deposition in Thermal Spray. In A. Pakseresht (Ed.), *Production, Properties, and Applications of High Temperature Coatings* (pp. 299–332). Hershey, PA: IGI Global. doi:10.4018/978-1-5225-4194-3.ch012

Shahzadi, P. (2019). Properties of Nanomaterials and Environment. In R. Nazir (Ed.), *Nanotechnology Applications in Environmental Engineering* (pp. 28–43). Hershey, PA: IGI Global. doi:10.4018/978-1-5225-5745-6.ch002

Shaikh, A. A. G., & Sivaram, S. (1996). Organic carbonates. *Chemical Reviews*, *96*(3), 951–976. doi:10.1021/cr950067i PMID:11848777

Sharma, N., & Kumar, K. (2018). Fabrication of Porous NiTi Alloy Using Organic Binders. In K. Kumar & J. Davim (Eds.), *Composites and Advanced Materials for Industrial Applications* (pp. 38–62). Hershey, PA: IGI Global. doi:10.4018/978-1-5225-5216-1.ch003

Sharma, Y. C., & Singh, B. (2008). Development of biodiesel from karanja, a tree found in rural India. *Fuel*, *87*(8–9), 1740–1742. doi:10.1016/j.fuel.2007.08.001

Sheikholeslami, M. (2019). *Applications of Nanofluid Transportation and Heat Transfer Simulation* (pp. 1–692). Hershey, PA: IGI Global. doi:10.4018/978-1-5225-7595-5

Shi, F., Deng, Y. Q., SiMa, T. L., Peng, J. J., Gu, Y. L., & Qiao, B. T. (2003). Alternatives to Phosgene and Carbon Monoxide: Synthesis of Symmetric Urea Derivatives with Carbon Dioxide in Ionic Liquids. Angewandte Chemie International Edition, 42, 3257−3260.

Shi, Y., Xing, E., Xie, W., Zhang, F., Mu, X., & Shu, X. (2015). Size-controlled synthesis of MCM-49 zeolites and their application in liquid-phase alkylation of benzene with ethylene. *The Royal Society of Chemistry's*, *5*, 13420–13429.

Simanaitis & Kott. (1989). Octane Enhancers. *Road & Track*, 82-88.

Singh-Ackbarali, D., & Maharaj, R. (2017). Mini Livestock Ranching: Solution to Reducing the Carbon Footprint and Negative Environmental Impacts of Agriculture. In W. Ganpat & W. Isaac (Eds.), *Environmental Sustainability and Climate Change Adaptation Strategies* (pp. 188–212). Hershey, PA: IGI Global. doi:10.4018/978-1-5225-1607-1.ch007

Singh, R., & Dutta, S. (2018). Visible Light Active Nanocomposites for Photocatalytic Applications. In K. Kumar & J. Davim (Eds.), *Composites and Advanced Materials for Industrial Applications* (pp. 270–296). Hershey, PA: IGI Global. doi:10.4018/978-1-5225-5216-1.ch012

Sinopec Corp. (n.d.). *Oil Refining*. Retrieved from http://www.sinopec.com/listco/en/products_service/License/Oil_Refining/

Six, C., & Richter, F. (2003). Isocyanates, Organic. In *Ullmann's Encyclopedia of Industrial Chemistry*. Weinheim: Wiley-VCH. doi:10.1002/14356007.a14_611

Slaughter, J. I., & McCaulay, D. A. (1959). *United States Patent No. US2908728A*. US Patent Office.

Smith, J., Sy, A., Chen, J., & Dautzenberg, F. (1993, January). Catalytic Distillation Route for Cumene. DeWitt Petrochemical Review.

Sneeden, R. P. A. (1982). Article. In G. Wilkinson, F. A. G. Stone, & A. W. Abel (Eds.), Comprehensive Organometallic Chemistry (Vol. 8, p. 225). New York: Pergamon Press. doi:10.1016/B978-008046518-0.00109-4

Socha, R. F., Chang, C. D., Gould, R. M., Kane, S. E., & Avidan, A. A. (1987). Fluid-Bed Studies of Olefin Production from Methanol. Industrial Chemicals via C Processes: American Chemical Society, 328, 34-41. doi:10.1021/bk-1987-0328.ch003

Songa, H., Zhanga, N., Zhonga, C., Liua, Z., Xiaoa, M., & Gai, H. (2017). Hydrogenation of CO_2 into Formic Acid over Palladium Catalyst on Chitin. *New Journal of Chemistry*, *41*(17), 9170–9177. doi:10.1039/C7NJ00460E

Song, J., Feng, X., Yamamoto, Y., Almansour, A. I., Arumugam, N., Suresh Kumar, R., & Bao, M. (2016). Selective synthesis of δ-lactone via palladium nanoparticles-catalyzed telomerization of CO_2 with 1,3-butadiene. *Tetrahedron Letters*, *57*(29), 3163–3166. doi:10.1016/j.tetlet.2016.06.024

Song, Q.-W., Zhao, Y.-N., He, L.-N., Gao, J., & Yang, Z.-Z. (2012). Synthesis of Oxazolidinones/Polyurethanes from Aziridines and CO_2. *Current Catalysis*, *1*(2), 107–124. doi:10.2174/2211544711201020107

Sopena, S., Fiorani, G., Martin, C., & Kleij, A. W. (2015). Highly Efficient Organocatalyzed Conversion of Oxiranes and CO_2 into Organic Carbonates. *ChemSusChem*, *8*(19), 3248–3254. doi:10.1002/cssc.201500710 PMID:26331287

Srirattnai, K., Damronglerd, S., Omi, S., Roengsumran, S., Petsom, A., & Ma, G. H. (2002). Encapsulated $AlCl_3$: A convenient catalyst for the alkylation of benzene with dodecene. *Tetrahedron Letters*, *43*(25), 4555–4557. doi:10.1016/S0040-4039(02)00702-5

Srivastava, R., Srinivas, D., & Ratnasamy, P. (2005). CO_2 activation and synthesis of cyclic carbonates and alkyl/aryl carbamates over adenine-modified Ti-SBA-15 solid catalysts. *Journal of Catalysis*, *233*(1), 1–15. doi:10.1016/j.jcat.2005.03.023

Srivastava, R., Srinivas, D., & Ratnasamy, P. (2006). Syntheses of polycarbonate and polyurethane precursors utilizing CO_2 over highly efficient, solid as-synthesized MCM-41 catalyst. *Tetrahedron Letters*, *47*(25), 4213–4217. doi:10.1016/j.tetlet.2006.04.057

Steinhauser, G. (2008). Cleaner production in the Solvay process: General strategies and recent developments. *Journal of Cleaner Production*, *16*(7), 833–841. doi:10.1016/j.jclepro.2007.04.005

Steynberg, A. P. (2004). Introduction to Fischer-Tropsch Technology. In A. Steynberg & M. Dry (Eds.), *Studies in Surface Science and Catalysis* (Vol. 152, pp. 1–63). Elsevier.

Steynberg, A., Dry, M., Davis, B., & Breman, B. (2004). Fischer-tropsch reactors. *Studies in Surface Science and Catalysis*, *152*, 64–195. doi:10.1016/S0167-2991(04)80459-2

Steynberg, A., Espinoza, R., Jager, B., & Vosloo, A. (1999). High temperature Fischer–Tropsch synthesis in commercial practice. *Applied Catalysis A, General, 186*(1-2), 41–54. doi:10.1016/S0926-860X(99)00163-5

Stickers, D.E. (2002). Octane and the environment. *The Science of the Total Environment, 299*, 37-56.

Stocker, M. (1999). Methanol-to-hydrocarbons: Catalytic materials and their behavior1. *Microporous and Mesoporous Materials, 29*(1-2), 3–48. doi:10.1016/S1387-1811(98)00319-9

Stone, R. J. (2017). Modelling the Frequency of Tropical Cyclones in the Lower Caribbean Region. In W. Ganpat & W. Isaac (Eds.), *Environmental Sustainability and Climate Change Adaptation Strategies* (pp. 341–349). Hershey, PA: IGI Global. doi:10.4018/978-1-5225-1607-1.ch013

Storace, L., Anzalone, L., Confalone, P. N., Davis, W. P., Fortunak, J. M., Giangiordano, M., ... Wood, C. C. (2002). For the use of phosgene in the multikilogram-scale preparation of a medicinally important compound, see. *Organic Process Research & Development, 6*, 54–63. doi:10.1021/op015507o

Storch, H. H. (1951). *The Fischer-Tropsch and related syntheses: including a summary of theoretical and applied contact catalysis*. Wiley.

Streets, D. G., Yarber, K. F., Woo, J.-H., & Carmichael, G. R. (2003). Biomass burning in Asia: Annual and seasonal estimates and atmospheric emissions. *Global Biogeochemical Cycles, 17*(4). doi:10.1029/2003GB002040

Styring, P., Jansen, D., de Coninck, H., Reith, H., & Armstrong, K. (2011). *Carbon Capture and Utilisation in the Green Economy*. Centre for Low Carbon Futures. Available at: http://co2chem.co.uk/wpcontent/uploads/2012/06/CCU%20in%20the%20green%20economy%20report.pdf

Sun, Y., & Cheng, J. (2002). Hydrolysis of lignocellulosic materials for ethanol production: A review. *Bioresource Technology, 83*(1), 1–11. doi:10.1016/S0960-8524(01)00212-7 PMID:12058826

Svelle, S., Joensen, F., Nerlov, J., Olsbye, U., Lillerud, K. P., Kolboe, S., & Bjorgen, M. (2006). Conversion of methanol into hydrocarbons over zeolite H-ZSM-5: Ethene formation is mechanistically separated from the formation of higher alkenes. *Journal of the American Chemical Society, 128*(46), 14770–14771. doi:10.1021/ja065810a PMID:17105263

Tahir, L. (2019). Green Approaches to Environmental Sustainability. In R. Nazir (Ed.), *Nanotechnology Applications in Environmental Engineering* (pp. 81–101). Hershey, PA: IGI Global. doi:10.4018/978-1-5225-5745-6.ch004

Tai, C. C., Huck, M. J., McKoon, E. P., Woo, T., & Jessop, P. G. (2002). *Low-temperature synthesis of tetraalkylureas from secondary amines and carbon dioxide*. Academic Press.

Tao, J.-L., Jun, K.-W., & Lee, K.-W. (2001). Co-production of dimethyl ether and methanol from CO_2 hydrogenation: Development of a stable hybrid catalyst. *Applied Organometallic Chemistry, 15*(2), 105–108. doi:10.1002/1099-0739(200102)15:2<105::AID-AOC100>3.0.CO;2-B

Tavakoli, A., Sohrabi, M., & Kargari, A. (2008). Application of Anderson–Schulz–Flory (ASF) equation in the product distribution of slurry phase FT synthesis with nanosized iron catalysts. *Chemical Engineering Journal, 136*(2-3), 358–363. doi:10.1016/j.cej.2007.04.017

Technology Vision 2020: The U.S. Chemical Industry. (1996). Retrieved from http://www.acs.org

Tecza, W., & Brzozowski, R. (2003). Modern industrial processes for making ethylbenzene and cumene. *Przemysl Chemiczny, 82*(11), 1428–1434.

Teketel, S. Skistad, W., Benard, S., Olsbye, U., Lillerud, K. P., Beato, P., & Svelle, S. (2012). Shape selectivity in the conversion of methanol to hydrocarbons: the catalytic performance of one-dimensional 10-ring zeolite: ZSM-22,MZSM-23,ZSM-48 and EU-1. *American Chemical Society, 2*(1), 26-37.

Teketel, S., Svelle, S., Lillerud, K. P., & Olsbye, U. (2009). Shape-Selective Conversion of Methanol to Hydrocarbons Over 10-Ring Unidirectional-Channel Acidic H-ZSM-22. *ChemCatChem, 1*(1), 78–81. doi:10.1002/cctc.200900057

Tian, J.-S., Cai, F., Wang, J.-Q., Du, Y., & He, L.-N. (2008). Environmentally Benign Chemical Conversion of CO_2 into Organic Carbonates Catalyzed by Phosphonium Salts. *Phosphorus, Sulfur, and Silicon and the Related Elements, 183*(2-3), 494–498. doi:10.1080/10426500701761540

Tilwankar, V., Rai, S., & Bajpai, S. P. (2019). Role of Social Media in Environment Awareness: Social Media and Environment. In S. Narula, S. Rai, & A. Sharma (Eds.), *Environmental Awareness and the Role of Social Media* (pp. 117–139). Hershey, PA: IGI Global. doi:10.4018/978-1-5225-5291-8.ch006

Tsakoumis, N. E., Rønning, M., Borg, Ø., Rytter, E., & Holmen, A. (2010). Deactivation of cobalt based Fischer–Tropsch catalysts: A review. *Catalysis Today, 154*(3-4), 162–182. doi:10.1016/j.cattod.2010.02.077

Tsuda, T., Chujo, Y., & Saegusa, T. (1976). Copper(I) cyanoacetate as a carrier of activated carbon dioxide. *Journal of the Chemical Society. Chemical Communications*, (11): 415–416. doi:10.1039/c39760000415

Tsuda, T., Yamamoto, T., & Saegusa, T. (1992). Palladium-catalyzed cycloaddition of carbon dioxide with methoxyallene. *Journal of Organometallic Chemistry, 429*(3), C46–C48. doi:10.1016/0022-328X(92)83194-M

Tsuda, T., Yasukawa, H., & Komori, K. (1995). Nickel(0)-Catalyzed Cycloaddition Copolymerization of Ether Diynes with Carbon Dioxide to Poly(2-pyrone)s. *Macromolecules, 28*(5), 1356–1359. doi:10.1021/ma00109a004

Tsujino, T. (1996). Recent development of Texaco gasification technology and its applications. *Fuel and Energy Abstracts, 37*(3), 183. doi:10.1016/0140-6701(96)88538-9

Tuan Le, Q. A., Gon Kim, H., & Kim, Y. H. (2018). Electrochemical synthesis of formic acid from CO_2 catalyzed by Shewanella oneidensis MR-1 whole-cell biocatalyst. *Enzyme and Microbial Technology, 116*, 1–5. doi:10.1016/j.enzmictec.2018.05.005 PMID:29887011

Twigg, M. V. (1996). *The Catalyst Handbook* (2nd ed.). London: Manson Publishing.

Uitti, K. D. (1977). Process for Production and Dehydrogenation of Ethylbenzene. US Patent 4,009,217.

Universal Oil Products Co. (1970). *Preparation of monoalkylated aromatic compounds*. Patent 2,032,048.

Use of Process Analytics for Cracking Processes in Refineries. (2007). *Case Study*. Siemens.

Valange, S., Chatel, G., Amaniampong, P. N., Behling, R., & Jérôme, F. (2018). Ultrasound-Assisted Synthesis of Nanostructured Oxide Materials: Basic Concepts and Applications to Energy. In S. González-Cortés & F. Imbert (Eds.), *Advanced Solid Catalysts for Renewable Energy Production* (pp. 177–215). Hershey, PA: IGI Global. doi:10.4018/978-1-5225-3903-2.ch007

Valli, V. L. K., & Alper, H. (1995). A Simple, Convenient, and Efficient Method for the Synthesis of Isocyanates from Urethanes. *The Journal of Organic Chemistry*, *60*(1), 257–258. doi:10.1021/jo00106a044

van Santen, R. A., Ghouri, M. M., Shetty, S., & Hensen, E. M. (2011). Structure sensitivity of the Fischer–Tropsch reaction; molecular kinetics simulations. *Catalysis Science & Technology*, *1*(6), 891–911. doi:10.1039/c1cy00118c

Ventuto, P. B., & Hamilton, L. A. (1966). Organic Reaction Catalized by crystalline aluminosilicates:II. Alkylation Reactions: Mechanistic and Aging Consideration. *Journal of Catalysis*, *5*(3), 484–493. doi:10.1016/S0021-9517(66)80067-2

Verma, R., Suri, N. M., & Kant, S. (2018). Slurry Sprayed Mullite Coatings and Their Corrosion Performances. In A. Pakseresht (Ed.), *Production, Properties, and Applications of High Temperature Coatings* (pp. 187–214). Hershey, PA: IGI Global. doi:10.4018/978-1-5225-4194-3.ch008

Vezirov, R. (2011). Visbreaking – technologies tested by practice and time. *Chemistry and Technology of Fuels and Oils*, *46*(6), 367–374. doi:10.100710553-011-0236-9

Vincent, A. W., Kevin, J. F., & Gelbke, H-P. (2005). *Ethylbenzene*. Wiley-VCH.

Vincent, W. A., Kevin, F. J., & Gelbke, H-P. (2005). Ethylbenzene. In *Ullman's Encyclopedia of Industrial Chemistry*. Weinheim: Wiley.

Vincenzo, C., Nacci, A., Monopoli, A., & Fanizzi, A. (2002). Cyclic Carbonate Formation from Carbon Dioxide and Oxiranes in Tetrabutylammonium Halides as Solvents and Catalysts. *Organic Letters*, *4*(15), 2561–2563. doi:10.1021/ol026189w PMID:12123376

Vogt, E. T., Whiting, C., Dutta, G. T., Chowdhury, A., & Weckhuysen, B. M. (2015). Zeolites and zeotypes for oil and gas conversion. *Advances in Catalysis*, *58*, 143–314. doi:10.1016/bs.acat.2015.10.001

Vora, B. V., Marker, T. L., Barger, P. T., Nilsen, H. R., Kvisle, S., & Fuglerud, T. (1997). Economic route for natural gas conversion to ethylene and propylene. *Studies in Surface Science and Catalysis*, *107*, 87–98. doi:10.1016/S0167-2991(97)80321-7

Waldman, T. E., & McGhee, W. D. (1994). Isocyanates from primary amines and carbon dioxide: 'dehydration' of carbamate anions. *Journal of the Chemical Society. Chemical Communications*, *0*(8), 957–958. doi:10.1039/C39940000957

Wang, J., & Zhang, Y. (2016). Boronic Acids as Hydrogen Bond Donor Catalysts for Efficient Conversion of CO_2 into Organic Carbonate in Water. *ACS Catalysis*, *6*(8), 4871–4876. doi:10.1021/acscatal.6b01422

Wang, S. H. (1993). *Styrene, PEB Report 33C, Supplement C, Process Economic Program. Papers presented at SRI International Conference.*

Wang, T., Tang, X. P., Huang, X. F., Qian, W., Cui, Y., Hui, X., ... Wei, F. (2014). Conversion of methanol to aromatics in fluidized bed reactor. *Catalysis Today*, *233*, 8–13. doi:10.1016/j.cattod.2014.02.007

Ward, J.D. (1971). *Alkylation-transalkylation process*. Patent 1,246,728/19.

Watling, L., & Thiel, M. (2013). *The natural history of the Crustacea*. Oxford University Press. Retrieved from https://global.oup.com/academic/product/lifestyles-and-feeding-biology-9780199797028#.WkoN_7fQqTQ.mendeley

Weber, M., & Klenine-Boymann, M. (2004) Phenol. In Ullmann's Encylopedia of Industrial Chemistry. Wiley-VCH Verlag GmbH & Co.

Wei Chen, X., Zhao, C., Li, S., Song, Y., Ge, R., Wei, W., & Sun, Y. (2017). Exclusive formation of formic acid from CO_2 electro reduction by tunable Pd-Sn alloy. *Angewandte Chemie*, *129*(40), 12387–12391. doi:10.1002/ange.201707098

Weilhard, A., Qadir, M. I., Sans, V., & Dupont, J. (2018). Selective CO_2 Hydrogenation to Formic Acid with Multifunctional Ionic Liquids. *ACS Catalysis*, *8*(3), 1628–1634. doi:10.1021/acscatal.7b03931

Weil, J., Brewer, M., Hendrickson, R., Sarikaya, A., & Ladisch, M. R. (1998). Continuous pH monitoring during pretreatment of yellow poplar wood sawdust by pressure cooking in water. *Applied Biochemistry and Biotechnology*, *70–72*(1), 99–111. doi:10.1007/BF02920127

Weisser, D. (2007). A guide to life-cycle greenhouse gas (GHG) emissions from electric supply technologies. *Energy*, *32*(9), 1543–1559. doi:10.1016/j.energy.2007.01.008

Welch, V. A., Fallon, K. J., & Gelbke, H-P. (2005). Ethylbenzene. In Ullman's Encyclopedia of Industrial Chemistry. Wiley-VCH.

Werpy, T. G., & Petersen, G. (2004). *Top Value Added Chemicals from Biomass Volume I—Results of Screening for Potential Candidates from Sugars and Synthesis Gas Energy Efficiency and Renewable Energy*. Retrieved from https://www.nrel.gov/docs/fy04osti/35523.pdf

Whiteoak, C. J., Kielland, N., Laserna, V., Escudero-Adan, E. C., Martin, E., & Kleij, A. W. (2013). A Powerful Aluminum Catalyst for the Synthesis of Highly Functional Organic Carbonates. *Journal of the American Chemical Society*, *135*(4), 1228–1231. doi:10.1021/ja311053h PMID:23302007

Wight, C. G. (1979). *US Patent No. 4,169,111*. US Patent Office.

Williams, C.M., Johnson, J.B., & Rovis, T. (2008), Nickel-catalyzed reductive carboxylation of styrenes using CO_2. *Journal of American Chemical Society, 130*, 14936 – 14937.

Wu, C., Cheng, H., Liu, R., Wang, Q., Hao, Y., Yu, Y., & Zhao, F. (2010). Synthesis of urea derivatives from amines and CO_2 in the absence of catalyst and solvent. *Green Chemistry*, *12*(10), 1811–1816. doi:10.1039/c0gc00059k

Wu, L., Liu, Q., Fleischer, I., Jackstell, R., & Beller, M. (2014). Ruthenium-catalysed alkoxycarbonylation of alkenes with carbon dioxide. *Nature Communications*, *5*, 1–6. PMID:24518431

Wunderly, H. L., Sowa, J. F., & Nieuwland, J. A. (1936). The Olefin—Benzene Condensation with the Sulfuric Acid Catalyst. *Journal of the American Chemical Society*, *58*, 1007. doi:10.1021/ja01297a050

Wyman, C. E., Dale, B. E., Elander, R. T., Holtzapple, M., Ladisch, M. R., & Lee, Y. Y. (2005). Coordinated development of leading biomass pretreatment technologies. *Bioresource Technology*, *96*(18), 1959–1966. doi:10.1016/j.biortech.2005.01.010 PMID:16112483

Xiang, X., Guo, L., Wu, X., Ma, X., & Xia, Y. (2012). Urea formation from carbon dioxide and ammonia at atmospheric pressure. *Environmental Chemistry Letters*, *10*(3), 295–300. doi:10.100710311-012-0366-2

Xiaohe, J., Ningning, Z., Jian-Gong, M., & Peng, C. (2017). Conversion of CO_2 into cyclic carbonates by a Co (II) metal-organic framework and the improvement of catalytic activity via nanocrystallization. *Dalton Transactions (Cambridge, England)*, *00*, 1–3.

Xie, W. (2018). Thermal Spray Methods and Splat Formation. In A. Pakseresht (Ed.), *Production, Properties, and Applications of High Temperature Coatings* (pp. 1–24). Hershey, PA: IGI Global. doi:10.4018/978-1-5225-4194-3.ch001

Xing, E., Shi, Y., Xie, W., Zhang, F., Mu, X., & Shu, X. (2016). Size-controlled synthesis of MCM-49 zeolite from NaX for liquid-phase alkylation of benzene with ethylene. *Microporous and Mesoporous Materials*, *236*, 54–62. doi:10.1016/j.micromeso.2016.08.026

Yadav, G. D., & Asthana, N. S. (2003). Selective decomposition of cumene hydroperoxide into phenol and acetone by a novel cesium substituted heteropolyacid on clay. *Applied Catalysis A, General*, *244*(2), 341–357. doi:10.1016/S0926-860X(02)00605-1

Yang, H. H., Liu, T. C., Chang, C.-F., & Lee, E. (2012). Effects of ethanol-blended gasoline on emissions of regulated air pollutants and carbonyls from motorcycles. *Applied Energy*, *89*(1), 281–286. doi:10.1016/j.apenergy.2011.07.035

Yang, H., Xu, Z., Fan, M., Gupta, R., Slimane, R. B., Bland, A. E., & Wright, I. (2008). Progress in carbon dioxide separation and capture: A review. *Journal of Environmental Sciences (China)*, *20*(1), 14–27. doi:10.1016/S1001-0742(08)60002-9 PMID:18572517

Yang, K. C., Kang, Y., & Wang, R. (2019). Integrating Social and Mobile Media in Environmental Marketing Communications in China: Opportunities and Challenges. In S. Narula, S. Rai, & A. Sharma (Eds.), *Environmental Awareness and the Role of Social Media* (pp. 43–71). Hershey, PA: IGI Global. doi:10.4018/978-1-5225-5291-8.ch003

Yang, W., Wang, Z., Sun, H., & Zhang, B. (2016). Advances in development and industrial applications of ethylbenzene processes. *Chinese Journal of Catalysis*, *37*(1), 16–26. doi:10.1016/S1872-2067(15)60965-2

Yaseen, T., & Yaseen, A. (2019). Wastewater Treatment in Removal of Heavy Metals: Nanotechnology Applications in Environmental Engineering. In R. Nazir (Ed.), *Nanotechnology Applications in Environmental Engineering* (pp. 220–240). Hershey, PA: IGI Global. doi:10.4018/978-1-5225-5745-6.ch010

Yeung, C. S., & Dong, V. M. (2008). Beyond Aresta's Complex: Ni- and Pd-Catalyzed Organozinc Coupling with CO_2. *Journal of the American Chemical Society*, *130*(25), 7826–7827. doi:10.1021/ja803435w PMID:18510323

Yokogawa. (n.d.). *Top Gas Density Measurement in a Fluid Catalytic Cracker (FCC)*. Retrieved from https://www.yokogawa.com/library/resources/application-notes/top-gas-density-measurement-in-a-fluid-catalytic-cracker-fcc/

Yonemoto-Kobayashi, M., Inamoto, K., Tanaka, Y., & Kondo, Y. (2013). Carboxylation of alkynylsilanes with carbon dioxide mediated by cesium fluoride in DMSO. *Organic & Biomolecular Chemistry*, *11*(23), 3773–3775. doi:10.1039/c3ob40760h PMID:23674078

Yoo, W.-J., Capdevila, M. G., Du, X., & Kobayashi, S. (2012). Base-Mediated Carboxylation of Unprotected Indole Derivatives with Carbon Dioxide. *Organic Letters*, *14*(20), 5326–5329. doi:10.1021/ol3025082 PMID:23035967

Yoshida, M., Hara, N., & Okuyama, S. (2000). Catalytic production of urethanes from amines and alkyl halides in supercritical carbon dioxide. *Chemical Communications (Cambridge)*, *0*(2), 151–152. doi:10.1039/a908819i

Young, L. B., & Skillman, N. J. (1981). *US Patent 4 301 317*. US Patent Office.

Young, L. B., & Skillman, N.J. (1979). *US Patent 4 301 316*. US Patent Office.

Yu, B., Diao, Z.-F., Guo C.-X., & He, L.-N. (2013). Carboxylation of olefins/alkynes with CO_2 to industrially relevant acrylic acid derivatives. *Journal of CO_2 Utilization, 1*, 60–68.

Yuan, G., Qi, C., Wu, W., & Jiang, H. (2017). Curr. Opin. *Green and Sustainable Chemistry*, *3*, 22–27. doi:10.1016/j.cogsc.2016.11.006

Yuan, Y., Chen, C., Zeng, C., Mousavi, B., Chaemchuen, S., & Verpoort, F. (2017). Carboxylation of Terminal Alkynes with Carbon Dioxide Catalyzed by an *in-situ* Ag_2O/ N-Heterocyclic Carbene Precursor System. *ChemCatChem*, *9*(5), 882–887. doi:10.1002/cctc.201601379

Yu, D.-Y., & Zhang, Y.-G. (2011). The direct carboxylation of terminal alkynes with carbon dioxide. *Green Chemistry*, *13*(5), 1275–1279. doi:10.1039/c0gc00819b

Yunis & Oliveira. (2015). *Brazilian Modified GREET Pathway for the Production of Ethanol from Sugarcane -Method 2B: Pathway for Brazilian Sugarcane Ethanol from molasses, considering specifics mechanization level inputs and co-generation export credit from ABENGOA*. Retrieved from https://www.arb.ca.gov/fuels/lcfs/2a2b/apps/abg-brz-rpt-072915.pdf

Yurchak, S., Voltz, S. E., & Warner, J. P. (1979). Process Aging Studies in the Conversion of Methanol to Gasoline in a Fixed Bed Reactor. *Industrial & Engineering Chemistry Process Design and Development*, *18*(3), 527–534. doi:10.1021/i260071a030

Zaharin, M. S. M., Abdullah, N. R., Najafi, G., Sharudin, H., & Yusaf, T. (2017). Effects of physicochemical properties of biodiesel fuel blends with alcohol on diesel engine performance and exhaust emissions: A review. *Renewable & Sustainable Energy Reviews*, *79*, 475–493. doi:10.1016/j.rser.2017.05.035

Zaher, A. A. (2019). Smart Temperature Sensors: History, Current Practices, and Future Trends. In H. Santana, J. Lameu da Silva Jr, & O. Taranto (Eds.), *Process Analysis, Design, and Intensification in Microfluidics and Chemical Engineering* (pp. 223–250). Hershey, PA: IGI Global. doi:10.4018/978-1-5225-7138-4.ch008

Zaidi, H. A., & Pant, K. K. (2008). Activity of Oxalic Acid Treated ZnO/CuO/HZSM-5 Catalyst for the Transformation of Methanol to Gasoline Range Hydrocarbons. *Industrial & Engineering Chemistry Research*, *47*(9), 2970–2975. doi:10.1021/ie071339y

Zamani, Y., Rahimizadeh, M., & Seyedi, S. M. (2016). The Effect of Temperature on Product Distribution over Fe-Cu-K Catalyst in Fischer-Tropsch Synthesis. *Journal of Petroleum Science and Technology*, *6*(1), 46–52.

Zapp, P., Schreiber, J., Marx, A., Haines, M., Hake, J. F., & Gale, J. (2012). Overall environmental impacts of CCS technologies—A life cycle approach. *International Journal of Greenhouse Gas Control*, *8*, 12–21. doi:10.1016/j.ijggc.2012.01.014

Zare, K., Sadegh, H., Shahryari-ghoshekandi, R., Asif, M., Tyagi, I., Agarwal, S., & Gupta, V. K. (2016). Equilibrium and kinetic study of ammonium ion adsorption by Fe_3O_4 nanoparticles from aqueous solutions. *Journal of Molecular Liquids*, *213*, 345–350. doi:10.1016/j.molliq.2015.08.045

Zerbe, J. I., & Baker, A. J. (1988). *Investigation of fundamentals of two-stage, dilute sulfuric acid hydrolysis of wood*. Forests Products Laboratory. Retrieved from https://catalogue.nla.gov.au/Record/4049848

Zhang, W., Wang, S., Zhao, Y., & Ma, X. (2018). Hydrogenation of CO_2 to formic acid catalyzed by heterogeneous Ru-PPh_3/Al_2O_3 catalysts. *Fuel Processing Technology, 178*, 98–103. doi:10.1016/j.fuproc.2018.05.024

Zhang, Y. (2018). Covalent Radii and New Applications. *International Journal of Chemoinformatics and Chemical Engineering, 7*(1), 42–51. doi:10.4018/IJCCE.2018010103

Zhang, Z., Fan, F., Xing, H., Yang, Q., Bao, Z., & Ren, Q. (2017). Efficient Synthesis of Cyclic Carbonates from Atmospheric CO_2 Using a Positive Charge Delocalized Ionic Liquid Catalyst. *ACS Sustainable Chemistry & Engineering, 5*(4), 2841–2846. doi:10.1021/acssuschemeng.7b00513

Zhao, L., Pang, H., You, B., Fang, X., Wang, J., Duan, W., & Liu, J. (2007). *Gasoline hydrofinishing startup method and gasoline hydrofinishing operation method.* Chinese Patent: 101376835.

Zhao, X., Roberie, T., & Rajagopalan, K. (1996). p-Xylene yield in fluid catalytic cracking products. *Applied Catalysis A, General, 145*(1-2), 407–418. doi:10.1016/0926-860X(96)00180-9

Zhao, Y., Luo, Q., Gao, W., Gao, S., & Song, Y. (2003). Industrial application test of a FCC gasoline olefin reducing catalyst RFG-FS3. *Petroleum and Petrochemicals, 1*, 33–37.

Zhizhin, G. V. (2017). Systematics of Anomalies in the Filling of Electron Orbitals of Chemical Elements. *International Journal of Chemoinformatics and Chemical Engineering, 6*(1), 48–62. doi:10.4018/IJCCE.2017010104

Zhizhin, G. V. (2018). *Chemical Compound Structures and the Higher Dimension of Molecules: Emerging Research and Opportunities*. Hershey, PA: IGI Global. doi:10.4018/978-1-5225-4108-0

Zhizhin, G. V. (2018). The Metric Structure of Nucleic Acids and the Higher Dimension of Their Constituents. *International Journal of Chemoinformatics and Chemical Engineering, 7*(2), 1–15. doi:10.4018/IJCCE.2018070101

Zhou, X., Di, X., Yu, G., Lu, R., & Li, C. (2010). Simulation of Delayed Coking Reaction in Coke Drum. *Petroleum Science and Technology, 28*(3), 277–285. doi:10.1080/10916460802706505

Zhu, S., & Wang, F. (2018). Nanocrystalline, Enamel and Composite Coatings for Superalloys. In A. Pakseresht (Ed.), *Production, Properties, and Applications of High Temperature Coatings* (pp. 160–186). Hershey, PA: IGI Global. doi:10.4018/978-1-5225-4194-3.ch007

Zhu, X. X., & Chen, C. F., An, J., Zeng, P, & Xu, Y. (2011). Development and Industrialization of the Ethylbenzene Production Technologies from Dilute Ethylene in FCC Dry Gas. *Scientific Net Publisher in Materials Science & Engineering, 233*(235), 1708–1713.

Zindani, D., & Kumar, K. (2018). Industrial Applications of Polymer Composite Materials. In K. Kumar & J. Davim (Eds.), *Composites and Advanced Materials for Industrial Applications* (pp. 1–15). Hershey, PA: IGI Global. doi:10.4018/978-1-5225-5216-1.ch001

Zindani, D., Maity, S. R., & Bhowmik, S. (2018). A Decision-Making Approach for Material Selection of Polymeric Composite Bumper Beam. In K. Kumar & J. Davim (Eds.), *Composites and Advanced Materials for Industrial Applications* (pp. 112–128). Hershey, PA: IGI Global. doi:10.4018/978-1-5225-5216-1.ch006

Related Readings

To continue IGI Global's long-standing tradition of advancing innovation through emerging research, please find below a compiled list of recommended IGI Global book chapters and journal articles in the areas of chemical engineering, environmental sustainability, and oxidation. These related readings will provide additional information and guidance to further enrich your knowledge and assist you with your own research.

Abiev, R. (2019). Analysis of Hydrodynamics and Mass Transfer of Gas-Liquid and Liquid-Liquid Taylor Flows in Microchannels: Theoretical Approach and Experimental Proofs. In H. Santana, J. Lameu da Silva Jr, & O. Taranto (Eds.), *Process Analysis, Design, and Intensification in Microfluidics and Chemical Engineering* (pp. 1–49). Hershey, PA: IGI Global. doi:10.4018/978-1-5225-7138-4.ch001

Amadi, L. A., & Igwe, P. I. (2018). Globalization and Rethinking of Environmental Consumption From a Sustainability Perspective. In S. Idris (Ed.), *Promoting Global Environmental Sustainability and Cooperation* (pp. 1–30). Hershey, PA: IGI Global. doi:10.4018/978-1-5225-3990-2.ch001

Añez, R., & Sierraalta, A. (2018). Advances in Theoretical Studies on Solid Catalysts for Renewable Energy Production. In S. González-Cortés & F. Imbert (Eds.), *Advanced Solid Catalysts for Renewable Energy Production* (pp. 1–32). Hershey, PA: IGI Global. doi:10.4018/978-1-5225-3903-2.ch001

Artyukhov, A., Artyukhova, N., & Krmela, J. (2019). Computer Simulation of the Aerodisperse Systems Hydrodynamics in Granulation and Drying Apparatus. In H. Santana, J. Lameu da Silva Jr, & O. Taranto (Eds.), *Process Analysis, Design, and Intensification in Microfluidics and Chemical Engineering* (pp. 277–321). Hershey, PA: IGI Global. doi:10.4018/978-1-5225-7138-4.ch010

Ashrafiyan, O., Saremi, M., Pakseresht, A., & Ghasali, E. (2018). Oxidation-Protective Coatings for Carbon-Carbon Composites. In A. Pakseresht (Ed.), *Production, Properties, and Applications of High Temperature Coatings* (pp. 429–446). Hershey, PA: IGI Global. doi:10.4018/978-1-5225-4194-3.ch016

Aslam, S., & Ahmad, M. (2019). Nanoparticles for Degradation of Organic Pollutants. In R. Nazir (Ed.), *Nanotechnology Applications in Environmental Engineering* (pp. 241–267). Hershey, PA: IGI Global. doi:10.4018/978-1-5225-5745-6.ch011

Azadi, M. (2018). Deposition of TiN, TiC, and DLC Coatings by PACVD. In A. Pakseresht (Ed.), *Production, Properties, and Applications of High Temperature Coatings* (pp. 381–402). Hershey, PA: IGI Global. doi:10.4018/978-1-5225-4194-3.ch014

Baechler, S. M., Gonzalez-Cortes, S., Xiao, T., Al-Megren, H. A., & Edwards, P. P. (2018). Perspective on the Deep Hydrotreating of Renewable and Non-Renewable Oils. In S. González-Cortés & F. Imbert (Eds.), *Advanced Solid Catalysts for Renewable Energy Production* (pp. 61–94). Hershey, PA: IGI Global. doi:10.4018/978-1-5225-3903-2.ch003

Beerala, A. K., Gobinath, R., Shyamala, G., & Manvitha, S. (2018). Water Quality Prediction Using Statistical Tool and Machine Learning Algorithm. *International Journal of Chemoinformatics and Chemical Engineering*, *7*(2), 43–58. doi:10.4018/IJCCE.2018070104

Bhatt, R. (2017). Zero Tillage for Mitigating Global Warming Consequences and Improving Livelihoods in South Asia. In W. Ganpat & W. Isaac (Eds.), *Environmental Sustainability and Climate Change Adaptation Strategies* (pp. 126–161). Hershey, PA: IGI Global. doi:10.4018/978-1-5225-1607-1.ch005

Bolleddu, V., Racherla, V., & Bandyopadhyay, P. P. (2018). Microstructural and Tribological Characteristics of Air Plasma Sprayed Alumina-Titania Coatings. In A. Pakseresht (Ed.), *Production, Properties, and Applications of High Temperature Coatings* (pp. 268–298). Hershey, PA: IGI Global. doi:10.4018/978-1-5225-4194-3.ch011

Bose, G. K., Ghosh, P., & Pal, D. (2019). Analytical and Numerical Modelling of Liquid Penetration in a Closed Capillary. In H. Santana, J. Lameu da Silva Jr, & O. Taranto (Eds.), *Process Analysis, Design, and Intensification in Microfluidics and Chemical Engineering* (pp. 114–135). Hershey, PA: IGI Global. doi:10.4018/978-1-5225-7138-4.ch004

Bostanci, S. H., & Albayrak, A. N. (2017). The Role of Eco-Municipalities in Climate Change for a Sustainable Future. In W. Ganpat & W. Isaac (Eds.), *Environmental Sustainability and Climate Change Adaptation Strategies* (pp. 213–231). Hershey, PA: IGI Global. doi:10.4018/978-1-5225-1607-1.ch008

Cardoso, D., Lorena da Silva, L., & Zapelini, I. W. (2018). Mesoporous Silicas as Basic Heterogeneous Catalysts for the Formation of Biodiesel. In S. González-Cortés & F. Imbert (Eds.), *Advanced Solid Catalysts for Renewable Energy Production* (pp. 119–155). Hershey, PA: IGI Global. doi:10.4018/978-1-5225-3903-2.ch005

Chisanga, T., & Mbale, J. (2019). Integration of ICTs in Radio Programs (II-RP) for Environmental Awareness for Peasant Farmers of Rural Zambia. In S. Narula, S. Rai, & A. Sharma (Eds.), *Environmental Awareness and the Role of Social Media* (pp. 15–42). Hershey, PA: IGI Global. doi:10.4018/978-1-5225-5291-8.ch002

Cui, X., Zeng, S., Li, Z., Zheng, Q., Yu, X., & Han, B. (2018). Advanced Composites for Civil Engineering Infrastructures. In K. Kumar & J. Davim (Eds.), *Composites and Advanced Materials for Industrial Applications* (pp. 212–248). Hershey, PA: IGI Global. doi:10.4018/978-1-5225-5216-1.ch010

Di Girolamo, G. (2018). Current Challenges and Future Perspectives in the Field of Thermal Barrier Coatings. In A. Pakseresht (Ed.), *Production, Properties, and Applications of High Temperature Coatings* (pp. 25–59). Hershey, PA: IGI Global. doi:10.4018/978-1-5225-4194-3.ch002

Dixit, A. (2018). Application of Silica-Gel-Reinforced Aluminium Composite on the Piston of Internal Combustion Engine: Comparative Study of Silica-Gel-Reinforced Aluminium Composite Piston With Aluminium Alloy Piston. In K. Kumar & J. Davim (Eds.), *Composites and Advanced Materials for Industrial Applications* (pp. 63–98). Hershey, PA: IGI Global. doi:10.4018/978-1-5225-5216-1.ch004

Duruji, M. M., Olanrewaju, F. O., & Duruji-Moses, F. U. (2018). From Kyoto to Paris: An Analysis of the Politics of Multilateralism on Climate Change. In S. Idris (Ed.), *Promoting Global Environmental Sustainability and Cooperation* (pp. 31–56). Hershey, PA: IGI Global. doi:10.4018/978-1-5225-3990-2.ch002

Dutta, M. M. (2019). Characterization of Nanomaterials. In R. Nazir (Ed.), *Nanotechnology Applications in Environmental Engineering* (pp. 44–80). Hershey, PA: IGI Global. doi:10.4018/978-1-5225-5745-6.ch003

Eudoxie, G., & Roopnarine, R. (2017). Climate Change Adaptation and Disaster Risk Management in the Caribbean. In W. Ganpat & W. Isaac (Eds.), *Environmental Sustainability and Climate Change Adaptation Strategies* (pp. 97–125). Hershey, PA: IGI Global. doi:10.4018/978-1-5225-1607-1.ch004

Figueira, R. B. (2018). Hybrid Sol-Gel Coatings: Erosion-Corrosion Protection. In A. Pakseresht (Ed.), *Production, Properties, and Applications of High Temperature Coatings* (pp. 334–380). Hershey, PA: IGI Global. doi:10.4018/978-1-5225-4194-3.ch013

Garrick, T. A., & Liburd, O. E. (2017). Impact of Climate Change on a Key Agricultural Pest: Thrips. In W. Ganpat & W. Isaac (Eds.), *Environmental Sustainability and Climate Change Adaptation Strategies* (pp. 232–254). Hershey, PA: IGI Global. doi:10.4018/978-1-5225-1607-1.ch009

Ge, H., Liu, X., Wang, S., Yang, T., & Wen, X. (2017). *Innovative Applications of Mo(W)-Based Catalysts in the Petroleum and Chemical Industry: Emerging Research and Opportunities* (pp. 1–162). Hershey, PA: IGI Global. doi:10.4018/978-1-5225-2274-4

Ghani, F., & Qureshi, K. A. (2018). "Airpocalypse" or Tsar Economic China: Analysis of Unsustainable Environment and Reason Behind Increased Global Warming. In S. Idris (Ed.), *Promoting Global Environmental Sustainability and Cooperation* (pp. 137–161). Hershey, PA: IGI Global. doi:10.4018/978-1-5225-3990-2.ch006

Goundar, S., & Appana, S. (2017). Mainstreaming Development Policies for Climate Change in Fiji: A Policy Gap Analysis and the Role of ICTs. In W. Ganpat & W. Isaac (Eds.), *Environmental Sustainability and Climate Change Adaptation Strategies* (pp. 1–31). Hershey, PA: IGI Global. doi:10.4018/978-1-5225-1607-1.ch001

Hasdemir, T. A., & Çetin, S. (2019). Ecological Issues and Public Awareness in Social Media: Greenpeace Mediterranean's Turkey Case. In S. Narula, S. Rai, & A. Sharma (Eds.), *Environmental Awareness and the Role of Social Media* (pp. 1–14). Hershey, PA: IGI Global. doi:10.4018/978-1-5225-5291-8.ch001

Hassanzadeh, M., Saremi, M., Valefi, Z., & Pakseresht, A. H. (2018). Investigation on Improved-Durability Thermal Barrier Coatings: An Overview of Nanostructured, Multilayered, and Self-Healing TBCs. In A. Pakseresht (Ed.), *Production, Properties, and Applications of High Temperature Coatings* (pp. 60–78). Hershey, PA: IGI Global. doi:10.4018/978-1-5225-4194-3.ch003

Henao, J., & Sotelo, O. (2018). Surface Engineering at High Temperature: Thermal Cycling and Corrosion Resistance. In A. Pakseresht (Ed.), *Production, Properties, and Applications of High Temperature Coatings* (pp. 131–159). Hershey, PA: IGI Global. doi:10.4018/978-1-5225-4194-3.ch006

Hernández-Romero, R. M., & Martínez, Y. J. (2018). Electrochemical Techniques as a Tool for Catalysts' Characterization: VIMP and SECM – A Provocative Approach. In S. González-Cortés & F. Imbert (Eds.), *Advanced Solid Catalysts for Renewable Energy Production* (pp. 386–428). Hershey, PA: IGI Global. doi:10.4018/978-1-5225-3903-2.ch012

Idris, S. (2018). Water Security in Pakistan. In S. Idris (Ed.), *Promoting Global Environmental Sustainability and Cooperation* (pp. 58–84). Hershey, PA: IGI Global. doi:10.4018/978-1-5225-3990-2.ch003

Ince, F. (2018). International Businesses and Environmental Issues. In S. Idris (Ed.), *Promoting Global Environmental Sustainability and Cooperation* (pp. 86–111). Hershey, PA: IGI Global. doi:10.4018/978-1-5225-3990-2.ch004

Jayapalan, S. (2018). A Review of Chemical Treatments on Natural Fibers-Based Hybrid Composites for Engineering Applications. In K. Kumar & J. Davim (Eds.), *Composites and Advanced Materials for Industrial Applications* (pp. 16–37). Hershey, PA: IGI Global. doi:10.4018/978-1-5225-5216-1.ch002

Jucha, S. O., & Moskal, G. J. (2018). Thermal Barrier Coatings on the Base of Samarium Zirconates. In A. Pakseresht (Ed.), *Production, Properties, and Applications of High Temperature Coatings* (pp. 79–106). Hershey, PA: IGI Global. doi:10.4018/978-1-5225-4194-3.ch004

Júnior, J. L. (2019). An Introduction to Computational Fluid Dynamics and Its Application in Microfluidics. In H. Santana, J. Lameu da Silva Jr, & O. Taranto (Eds.), *Process Analysis, Design, and Intensification in Microfluidics and Chemical Engineering* (pp. 50–78). Hershey, PA: IGI Global. doi:10.4018/978-1-5225-7138-4.ch002

Júnior, J. L., & Santana, H. S. (2019). Experimental and Numerical Analyses of a Micro-Heat Exchanger for Ethanol Excess Recovery From Biodiesel. In H. Santana, J. Lameu da Silva Jr, & O. Taranto (Eds.), *Process Analysis, Design, and Intensification in Microfluidics and Chemical Engineering* (pp. 167–194). Hershey, PA: IGI Global. doi:10.4018/978-1-5225-7138-4.ch006

Kais, S. M. (2017). Climate Change: Vulnerability and Resilience in Commercial Shrimp Aquaculture in Bangladesh. In W. Ganpat & W. Isaac (Eds.), *Environmental Sustainability and Climate Change Adaptation Strategies* (pp. 162–187). Hershey, PA: IGI Global. doi:10.4018/978-1-5225-1607-1.ch006

Kang, S., Ju, M., & Kim, J. (2018). The Extraction of Metal Contaminants Using Supercritical Carbon Dioxide. *International Journal of Chemoinformatics and Chemical Engineering*, *7*(1), 52–60. doi:10.4018/IJCCE.2018010104

Kapilan, N. (2018). Catalyst for Biodiesel: An Overview of Catalysts Used in Biodiesel Production. In S. González-Cortés & F. Imbert (Eds.), *Advanced Solid Catalysts for Renewable Energy Production* (pp. 156–175). Hershey, PA: IGI Global. doi:10.4018/978-1-5225-3903-2.ch006

Keshavamurthy, R., Naveena, B. E., & Sekhar, N. (2018). Thermal Spray Coatings for Erosion-Corrosion Protection. In A. Pakseresht (Ed.), *Production, Properties, and Applications of High Temperature Coatings* (pp. 246–267). Hershey, PA: IGI Global. doi:10.4018/978-1-5225-4194-3.ch010

Khanna, B. K. (2017). Indian National Strategy for Climate Change Adaptation and Mitigation. In W. Ganpat & W. Isaac (Eds.), *Environmental Sustainability and Climate Change Adaptation Strategies* (pp. 32–63). Hershey, PA: IGI Global. doi:10.4018/978-1-5225-1607-1.ch002

Khanna, B. K. (2017). Vulnerability of the Lakshadweep Coral Islands in India and Strategies for Mitigating Climate Change Impacts. In W. Ganpat & W. Isaac (Eds.), *Environmental Sustainability and Climate Change Adaptation Strategies* (pp. 64–96). Hershey, PA: IGI Global. doi:10.4018/978-1-5225-1607-1.ch003

Khelfi, A. (2019). Toxicocinetic and Mechanisms of Action of Nanoparticles. In R. Nazir (Ed.), *Nanotechnology Applications in Environmental Engineering* (pp. 344–368). Hershey, PA: IGI Global. doi:10.4018/978-1-5225-5745-6.ch014

Kumari, N., & Kumar, K. (2018). Fabrication of Orthotic Calipers With Epoxy-Based Green Composite. In K. Kumar & J. Davim (Eds.), *Composites and Advanced Materials for Industrial Applications* (pp. 157–176). Hershey, PA: IGI Global. doi:10.4018/978-1-5225-5216-1.ch008

Kuppusamy, R. R. (2018). Development of Aerospace Composite Structures Through Vacuum-Enhanced Resin Transfer Moulding Technology (VERTMTy): Vacuum-Enhanced Resin Transfer Moulding. In K. Kumar & J. Davim (Eds.), *Composites and Advanced Materials for Industrial Applications* (pp. 99–111). Hershey, PA: IGI Global. doi:10.4018/978-1-5225-5216-1.ch005

Lallo, C. H., Smalling, S., Facey, A., & Hughes, M. (2017). The Impact of Climate Change on Small Ruminant Performance in Caribbean Communities. In W. Ganpat & W. Isaac (Eds.), *Environmental Sustainability and Climate Change Adaptation Strategies* (pp. 296–321). Hershey, PA: IGI Global. doi:10.4018/978-1-5225-1607-1.ch011

Lamani, M. (2017). Rice Straw Extracted Cellulose Biocompatible Nanofiber. *International Journal of Chemoinformatics and Chemical Engineering*, *6*(2), 1–20. doi:10.4018/IJCCE.2017070101

Lateef, A., & Nazir, R. (2019). Application of Magnetic Nanomaterials for Water Treatment. In R. Nazir (Ed.), *Nanotechnology Applications in Environmental Engineering* (pp. 201–219). Hershey, PA: IGI Global. doi:10.4018/978-1-5225-5745-6.ch009

Li, Y., Wang, J., Wang, S., Li, D., Song, S., Zhang, P., ... Yuan, H. (2019). Immiscible Two-Phase Parallel Microflow and Its Applications in Fabricating Micro- and Nanomaterials. In H. Santana, J. Lameu da Silva Jr, & O. Taranto (Eds.), *Process Analysis, Design, and Intensification in Microfluidics and Chemical Engineering* (pp. 136–166). Hershey, PA: IGI Global. doi:10.4018/978-1-5225-7138-4.ch005

Loi, N. K., Huyen, N. T., Tu, L. H., Tram, V. N., Liem, N. D., Dat, N. L., ... Minh, D. N. (2017). Sustainable Land Use and Watershed Management in Response to Climate Change Impacts: Case Study in Srepok Watershed, Central Highland of Vietnam. In W. Ganpat & W. Isaac (Eds.), *Environmental Sustainability and Climate Change Adaptation Strategies* (pp. 255–295). Hershey, PA: IGI Global. doi:10.4018/978-1-5225-1607-1.ch010

Luo, D., Ma, W., Wu, L., & Nakata, K. (2018). CO2 Reduction by Photocatalysis on TiO2. In S. González-Cortés & F. Imbert (Eds.), *Advanced Solid Catalysts for Renewable Energy Production* (pp. 268–299). Hershey, PA: IGI Global. doi:10.4018/978-1-5225-3903-2.ch010

Madhavi, T. P. (2019). Importance of Social Media to Create Environmental Awareness in Sustainable Higher Education System. In S. Narula, S. Rai, & A. Sharma (Eds.), *Environmental Awareness and the Role of Social Media* (pp. 72–81). Hershey, PA: IGI Global. doi:10.4018/978-1-5225-5291-8.ch004

Mallick, R., & Bajpai, S. P. (2019). Impact of Social Media on Environmental Awareness. In S. Narula, S. Rai, & A. Sharma (Eds.), *Environmental Awareness and the Role of Social Media* (pp. 140–149). Hershey, PA: IGI Global. doi:10.4018/978-1-5225-5291-8.ch007

Manne, A., & Myneni, M. B. (2019). Social Media for Disaster Awareness and Management. In S. Narula, S. Rai, & A. Sharma (Eds.), *Environmental Awareness and the Role of Social Media* (pp. 82–116). Hershey, PA: IGI Global. doi:10.4018/978-1-5225-5291-8.ch005

Marquez, O., & Jairo Márquez, P. (2018). Synthesis of Electrocatalysts for Electrochemistry in Energy. In S. González-Cortés & F. Imbert (Eds.), *Advanced Solid Catalysts for Renewable Energy Production* (pp. 300–385). Hershey, PA: IGI Global. doi:10.4018/978-1-5225-3903-2.ch011

Martinez-Guerra, E., Muppaneni, T., Gude, V. G., & Deng, S. (2018). Non-Conventional Feedstock and Technologies for Biodiesel Production. In S. González-Cortés & F. Imbert (Eds.), *Advanced Solid Catalysts for Renewable Energy Production* (pp. 96–118). Hershey, PA: IGI Global. doi:10.4018/978-1-5225-3903-2.ch004

Mkhize, N., & Ellis, D. (2018). Consumer Cooperation in Sustainability: The Green Gap in an Emerging Market. In S. Idris (Ed.), *Promoting Global Environmental Sustainability and Cooperation* (pp. 112–135). Hershey, PA: IGI Global. doi:10.4018/978-1-5225-3990-2.ch005

Montazeri, A., & Karjibani, M. (2017). Preparation and Characterization of Chitosan - Double Walled Carbon Nanotubes Hydrogels. *International Journal of Chemoinformatics and Chemical Engineering*, *6*(2), 21–30. doi:10.4018/IJCCE.2017070102

Mubarak, S. M. (2019). Nanotechnology for Air Remediation. In R. Nazir (Ed.), *Nanotechnology Applications in Environmental Engineering* (pp. 121–142). Hershey, PA: IGI Global. doi:10.4018/978-1-5225-5745-6.ch006

Mujere, N., & Moyce, W. (2017). Climate Change Impacts on Surface Water Quality. In W. Ganpat & W. Isaac (Eds.), *Environmental Sustainability and Climate Change Adaptation Strategies* (pp. 322–340). Hershey, PA: IGI Global. doi:10.4018/978-1-5225-1607-1.ch012

Mukhopadhyay, A., Barman, T. K., & Sahoo, P. (2018). Electroless Nickel Coatings for High Temperature Applications. In K. Kumar & J. Davim (Eds.), *Composites and Advanced Materials for Industrial Applications* (pp. 297–331). Hershey, PA: IGI Global. doi:10.4018/978-1-5225-5216-1.ch013

Naghiyev, F. B. (2018). Amplitude-Frequency Characteristics of the Oscillations of Methane Gas Bubbles in Oil. *International Journal of Chemoinformatics and Chemical Engineering*, *7*(2), 16–28. doi:10.4018/IJCCE.2018070102

Najim, M., Feddaoui, M., Alla, A. N., & Charef, A. (2019). Heat and Mass Transfer Characteristics of Evaporating Falling Films: Application to Thermal Protection and Desalination. In H. Santana, J. Lameu da Silva Jr, & O. Taranto (Eds.), *Process Analysis, Design, and Intensification in Microfluidics and Chemical Engineering* (pp. 251–276). Hershey, PA: IGI Global. doi:10.4018/978-1-5225-7138-4.ch009

Nasreen, S., & Rafique, U. (2019). Nanotechnology for Water Environmental Application: Functionalized Silica Hybrids as Nano-Sorbents. In R. Nazir (Ed.), *Nanotechnology Applications in Environmental Engineering* (pp. 171–200). Hershey, PA: IGI Global. doi:10.4018/978-1-5225-5745-6.ch008

Nazir, R. (2019). Introduction to Environmental Nanotechnology: E-Nano. In R. Nazir (Ed.), *Nanotechnology Applications in Environmental Engineering* (pp. 1–27). Hershey, PA: IGI Global. doi:10.4018/978-1-5225-5745-6.ch001

Ngubevana, L. (2018). Process Synthesis and the G-H Space for Analysing and Optimising Biomass Gasification. *International Journal of Chemoinformatics and Chemical Engineering*, *7*(1), 13–41. doi:10.4018/IJCCE.2018010102

Noman, D., & Owens, G. (2019). Wastewater Treatment and Role of Green Synthesized Metal Oxide Nanocomposites. In R. Nazir (Ed.), *Nanotechnology Applications in Environmental Engineering* (pp. 268–307). Hershey, PA: IGI Global. doi:10.4018/978-1-5225-5745-6.ch012

Nosheen, S. (2019). Nanomembrane Applications in Environmental Engineering. In R. Nazir (Ed.), *Nanotechnology Applications in Environmental Engineering* (pp. 103–120). Hershey, PA: IGI Global. doi:10.4018/978-1-5225-5745-6.ch005

Odetola, P. I., Popoola, P. A., & Oladijo, P. (2018). Thin Coating Deposition by Magnetron Sputtering. In A. Pakseresht (Ed.), *Production, Properties, and Applications of High Temperature Coatings* (pp. 403–428). Hershey, PA: IGI Global. doi:10.4018/978-1-5225-4194-3.ch015

Pandey, J. (2019). Recent Progresses in Membranes for Proton Exchange Membrane Fuel Cell (PEMFC) for Clean and Environmentally Friendly Applications. In R. Nazir (Ed.), *Nanotechnology Applications in Environmental Engineering* (pp. 308–343). Hershey, PA: IGI Global. doi:10.4018/978-1-5225-5745-6.ch013

Purohit, P., & Vagge, S. T. (2018). Oxidation Behavior of LTA/YSZ Intermixed Layer Coating. In A. Pakseresht (Ed.), *Production, Properties, and Applications of High Temperature Coatings* (pp. 107–130). Hershey, PA: IGI Global. doi:10.4018/978-1-5225-4194-3.ch005

Ramdani, N., & Azibi, M. (2018). Polymer Composite Materials for Microelectronics Packaging Applications: Composites for Microelectronics Packaging. In K. Kumar & J. Davim (Eds.), *Composites and Advanced Materials for Industrial Applications* (pp. 177–211). Hershey, PA: IGI Global. doi:10.4018/978-1-5225-5216-1.ch009

Ramesh, D., Muniraj, I. K., Thangavelu, K., & Karthikeyan, S. (2018). Pretreatment of Lignocellulosic Biomass Feedstocks for Biofuel Production. In S. González-Cortés & F. Imbert (Eds.), *Advanced Solid Catalysts for Renewable Energy Production* (pp. 33–60). Hershey, PA: IGI Global. doi:10.4018/978-1-5225-3903-2.ch002

Ramona, B. (2018). Global Implications of Sustainability and E-Society Infrastructure in Developing Economies. In S. Idris (Ed.), *Promoting Global Environmental Sustainability and Cooperation* (pp. 162–183). Hershey, PA: IGI Global. doi:10.4018/978-1-5225-3990-2.ch007

Rangel, M. D., Pereira, A. L., Marchetti, G., Querino, P. S., & Albornoz, A. (2018). Case Study: Effect of Zirconium on the Textural and Catalytic Properties of Magnetite. In S. González-Cortés & F. Imbert (Eds.), *Advanced Solid Catalysts for Renewable Energy Production* (pp. 216–241). Hershey, PA: IGI Global. doi:10.4018/978-1-5225-3903-2.ch008

Rodríguez-Sulbarán, P. J., Lugo, C. A., Perez, M. A., Gonzalez-Cortes, S. L., D'Angelo, R., Rondon, J., ... Del Castillo, H. L. (2018). Dry Reforming of Methane on LaSrNiAl Perovskite-Type Structures Synthesized by Solution Combustion. In S. González-Cortés & F. Imbert (Eds.), *Advanced Solid Catalysts for Renewable Energy Production* (pp. 242–266). Hershey, PA: IGI Global. doi:10.4018/978-1-5225-3903-2.ch009

Rosa, R. L., & Seabra, A. C. (2019). Design Techniques for Microfluidic Devices Implementation Applicable to Chemical Analysis Systems: Theoretical and Experimental Analysis of Microcomponents. In H. Santana, J. Lameu da Silva Jr, & O. Taranto (Eds.), *Process Analysis, Design, and Intensification in Microfluidics and Chemical Engineering* (pp. 195–222). Hershey, PA: IGI Global. doi:10.4018/978-1-5225-7138-4.ch007

Sabharwal, D. (2019). Study of Consumers' Attitudes Towards Marketing of Environmentally Friendly Products. In S. Narula, S. Rai, & A. Sharma (Eds.), *Environmental Awareness and the Role of Social Media* (pp. 150–168). Hershey, PA: IGI Global. doi:10.4018/978-1-5225-5291-8.ch008

Sadegh, H., Ali, G. A., Nia, H. J., & Mahmoodi, Z. (2019). Nanomaterial Surface Modifications for Enhancement of the Pollutant Adsorption From Wastewater: Adsorption of Nanomaterials. In R. Nazir (Ed.), *Nanotechnology Applications in Environmental Engineering* (pp. 143–170). Hershey, PA: IGI Global. doi:10.4018/978-1-5225-5745-6.ch007

Sadri, E., & Ashrafizadeh, F. (2018). High Temperature Nanocomposite Coatings by Plasma Spraying for Friction and Wear Applications. In A. Pakseresht (Ed.), *Production, Properties, and Applications of High Temperature Coatings* (pp. 216–245). Hershey, PA: IGI Global. doi:10.4018/978-1-5225-4194-3.ch009

Sahoo, S. (2018). Laminated Composite Hypar Shells as Roofing Units: Static and Dynamic Behavior. In K. Kumar & J. Davim (Eds.), *Composites and Advanced Materials for Industrial Applications* (pp. 249–269). Hershey, PA: IGI Global. doi:10.4018/978-1-5225-5216-1.ch011

Sanchez-Claros, M., Ortega-Casanova, J., & Galindo-Rosales, F. J. (2019). 2D Numerical Study of a Micromixer Based on Blowing and Vortex Shedding Mechanisms. In H. Santana, J. Lameu da Silva Jr, & O. Taranto (Eds.), *Process Analysis, Design, and Intensification in Microfluidics and Chemical Engineering* (pp. 79–113). Hershey, PA: IGI Global. doi:10.4018/978-1-5225-7138-4.ch003

Seyedin, S., Maghsoodloo, S., & Mottaghitalab, V. (2017). A Quantitative Study on Simultaneous Effects of Governing Parameters in Electrospinning of Nanofibers using Modified Neural Network using Genetic Algorithm. *International Journal of Chemoinformatics and Chemical Engineering*, *6*(1), 20–38. doi:10.4018/IJCCE.2017010102

Sezgin, H., & Berkalp, O. B. (2018). Textile-Reinforced Composites for the Automotive Industry. In K. Kumar & J. Davim (Eds.), *Composites and Advanced Materials for Industrial Applications* (pp. 129–156). Hershey, PA: IGI Global. doi:10.4018/978-1-5225-5216-1.ch007

Shahien, M. (2018). Reactive Plasma Spray: A Method for Nitride Coatings Deposition in Thermal Spray. In A. Pakseresht (Ed.), *Production, Properties, and Applications of High Temperature Coatings* (pp. 299–332). Hershey, PA: IGI Global. doi:10.4018/978-1-5225-4194-3.ch012

Shahzadi, P. (2019). Properties of Nanomaterials and Environment. In R. Nazir (Ed.), *Nanotechnology Applications in Environmental Engineering* (pp. 28–43). Hershey, PA: IGI Global. doi:10.4018/978-1-5225-5745-6.ch002

Sharma, N., & Kumar, K. (2018). Fabrication of Porous NiTi Alloy Using Organic Binders. In K. Kumar & J. Davim (Eds.), *Composites and Advanced Materials for Industrial Applications* (pp. 38–62). Hershey, PA: IGI Global. doi:10.4018/978-1-5225-5216-1.ch003

Sheikholeslami, M. (2019). *Applications of Nanofluid Transportation and Heat Transfer Simulation* (pp. 1–692). Hershey, PA: IGI Global. doi:10.4018/978-1-5225-7595-5

Singh, R., & Dutta, S. (2018). Visible Light Active Nanocomposites for Photocatalytic Applications. In K. Kumar & J. Davim (Eds.), *Composites and Advanced Materials for Industrial Applications* (pp. 270–296). Hershey, PA: IGI Global. doi:10.4018/978-1-5225-5216-1.ch012

Singh-Ackbarali, D., & Maharaj, R. (2017). Mini Livestock Ranching: Solution to Reducing the Carbon Footprint and Negative Environmental Impacts of Agriculture. In W. Ganpat & W. Isaac (Eds.), *Environmental Sustainability and Climate Change Adaptation Strategies* (pp. 188–212). Hershey, PA: IGI Global. doi:10.4018/978-1-5225-1607-1.ch007

Stone, R. J. (2017). Modelling the Frequency of Tropical Cyclones in the Lower Caribbean Region. In W. Ganpat & W. Isaac (Eds.), *Environmental Sustainability and Climate Change Adaptation Strategies* (pp. 341–349). Hershey, PA: IGI Global. doi:10.4018/978-1-5225-1607-1.ch013

Tahir, L. (2019). Green Approaches to Environmental Sustainability. In R. Nazir (Ed.), *Nanotechnology Applications in Environmental Engineering* (pp. 81–101). Hershey, PA: IGI Global. doi:10.4018/978-1-5225-5745-6.ch004

Tilwankar, V., Rai, S., & Bajpai, S. P. (2019). Role of Social Media in Environment Awareness: Social Media and Environment. In S. Narula, S. Rai, & A. Sharma (Eds.), *Environmental Awareness and the Role of Social Media* (pp. 117–139). Hershey, PA: IGI Global. doi:10.4018/978-1-5225-5291-8.ch006

Valange, S., Chatel, G., Amaniampong, P. N., Behling, R., & Jérôme, F. (2018). Ultrasound-Assisted Synthesis of Nanostructured Oxide Materials: Basic Concepts and Applications to Energy. In S. González-Cortés & F. Imbert (Eds.), *Advanced Solid Catalysts for Renewable Energy Production* (pp. 177–215). Hershey, PA: IGI Global. doi:10.4018/978-1-5225-3903-2.ch007

Verma, R., Suri, N. M., & Kant, S. (2018). Slurry Sprayed Mullite Coatings and Their Corrosion Performances. In A. Pakseresht (Ed.), *Production, Properties, and Applications of High Temperature Coatings* (pp. 187–214). Hershey, PA: IGI Global. doi:10.4018/978-1-5225-4194-3.ch008

Xie, W. (2018). Thermal Spray Methods and Splat Formation. In A. Pakseresht (Ed.), *Production, Properties, and Applications of High Temperature Coatings* (pp. 1–24). Hershey, PA: IGI Global. doi:10.4018/978-1-5225-4194-3.ch001

Yang, K. C., Kang, Y., & Wang, R. (2019). Integrating Social and Mobile Media in Environmental Marketing Communications in China: Opportunities and Challenges. In S. Narula, S. Rai, & A. Sharma (Eds.), *Environmental Awareness and the Role of Social Media* (pp. 43–71). Hershey, PA: IGI Global. doi:10.4018/978-1-5225-5291-8.ch003

Yaseen, T., & Yaseen, A. (2019). Wastewater Treatment in Removal of Heavy Metals: Nanotechnology Applications in Environmental Engineering. In R. Nazir (Ed.), *Nanotechnology Applications in Environmental Engineering* (pp. 220–240). Hershey, PA: IGI Global. doi:10.4018/978-1-5225-5745-6.ch010

Zaher, A. A. (2019). Smart Temperature Sensors: History, Current Practices, and Future Trends. In H. Santana, J. Lameu da Silva Jr, & O. Taranto (Eds.), *Process Analysis, Design, and Intensification in Microfluidics and Chemical Engineering* (pp. 223–250). Hershey, PA: IGI Global. doi:10.4018/978-1-5225-7138-4.ch008

Zhang, Y. (2018). Covalent Radii and New Applications. *International Journal of Chemoinformatics and Chemical Engineering*, *7*(1), 42–51. doi:10.4018/IJCCE.2018010103

Zhizhin, G. V. (2017). Systematics of Anomalies in the Filling of Electron Orbitals of Chemical Elements. *International Journal of Chemoinformatics and Chemical Engineering*, *6*(1), 48–62. doi:10.4018/IJCCE.2017010104

Zhizhin, G. V. (2018). *Chemical Compound Structures and the Higher Dimension of Molecules: Emerging Research and Opportunities*. Hershey, PA: IGI Global. doi:10.4018/978-1-5225-4108-0

Zhizhin, G. V. (2018). The Metric Structure of Nucleic Acids and the Higher Dimension of Their Constituents. *International Journal of Chemoinformatics and Chemical Engineering*, *7*(2), 1–15. doi:10.4018/IJCCE.2018070101

Zhu, S., & Wang, F. (2018). Nanocrystalline, Enamel and Composite Coatings for Superalloys. In A. Pakseresht (Ed.), *Production, Properties, and Applications of High Temperature Coatings* (pp. 160–186). Hershey, PA: IGI Global. doi:10.4018/978-1-5225-4194-3.ch007

Zindani, D., & Kumar, K. (2018). Industrial Applications of Polymer Composite Materials. In K. Kumar & J. Davim (Eds.), *Composites and Advanced Materials for Industrial Applications* (pp. 1–15). Hershey, PA: IGI Global. doi:10.4018/978-1-5225-5216-1.ch001

Zindani, D., Maity, S. R., & Bhowmik, S. (2018). A Decision-Making Approach for Material Selection of Polymeric Composite Bumper Beam. In K. Kumar & J. Davim (Eds.), *Composites and Advanced Materials for Industrial Applications* (pp. 112–128). Hershey, PA: IGI Global. doi:10.4018/978-1-5225-5216-1.ch006

About the Contributors

Mohammed C. Al-Kinany holds Ph.D. in heterogeneous and homogeneous catalysis from university of Manchester for science and technology (UK) since 1984. From 1988 till date he has been working as a Research Specialist in the National Center for Petrochemicals Technology, KACST. He has worked at the department of industrial polymer chemistry-Lancaster university (UK) between 1986-1988, in the field of thermoplastics, polycarbonate and antifouling organometallic polymers. He has also worked at UMIST (UK), between 1982-1986, in a joint research project with Imperial Chemical Industries (UK) in the field of alkylation, isomerization, transalkylation, disproportionation, and hexane cracking. He has also worked on transformer oils, biomass to oil, practical class lecturer. He has over 35 years of experience in the development and the use of homogeneous and heterogeneous catalysts (nano- and traditional) for the field petroleum and petrochemical industry, such as, alkylation, isomerization, transalkylation, oligomerization, oxidative dehydrogenation, cracking, hydrogenation, hydroxylation, clean fuel production, fuel additives, dehydrogenation, and hydrodesulphurization. He has extensive experience in design and operation bench scale pilot plant, feedstock, and product characterization; catalyst preparation and characterization, testing and evaluation. Dr. Al-Kinany is an author of more than (62) research articles, (6) patents, (12) books, translated (2) books; and he designed and constructed data base of toxic and hazardous chemicals used in Saudi Arabia.

Saud Aldrees was born in Riyadh-Saudi Arabia. He received his Bachelor in Chemical Engineering at King Saud University (Riyadh, Saudi Arabia). After eight years of working as assistant-researcher at KACST he moved to the UK where he received his Master in Heterogeneous Catalysis from University of Oxford in 2011, he back to KACST in 2012 as Academic Researcher for 6 years. Now he is reading for DPhil in Inorganic Chemistry at the University of Oxford. Eng. Saud is an author of more than (25) research articles and (3) patents.

* * *

Saeed Al-Shihri received his D.Phil. degree in Chemical Engineering from Imperial College London in the United Kingdom in 2016. He is currently the manager of national centre for Petrochemical. His research area is in applied catalysis and catalytic reaction engineering. Key themes are the upgrading of Natural gas to fuels and chemicals, development of nanostructured materials for different reaction systems, capturing and conversion of carbon dioxide to fuel and chemicals, and catalytic conversion of heavy oil to olefins. He is a technical member in the challenge of Carbon Capture, Storage and Utilization in Mission Innovation. He is the principal investigator of the project of capturing and conversion of CO2 to high value chemicals in collaboration with University of California, Berkeley. He is also the principal investigator of the project of turning CO2 to synthetic fuel in collaboration with university of Oxford. He has represented KACST in conferences, meetings and workshops that related to CCUS. With colleagues in the National Centre for Petrochemical technology, he was awarded the S.I.T.S Gold Medal for the invention of Design a system containing six micro reactors sets working in parallel for catalysts evaluation (2009), and the ITEX Bronze medal for innovation in chemical process design (2009).

Gomaa A. M. Ali works as Assistant Professor at Chemistry Department, Faculty of Science, Al-Azhar University Egypt. He obtained his BSc (2006) and MSc (2012) in Physical Chemistry from Al-Azhar University, Faculty of Science. He was awarded his PhD (2015) in Advanced Nanomaterials for Energy Storage from UMP, Malaysia. He is a reviewer and editor for many journals. His research interests: Materials Science, Nanocomposite, Graphene, Supercapacitors, Watertreatment,

Faisal Alotaibi obtained his Bs and Ms degrees in general chemistry from King Saud university, Riyadh in 1998 and 2006 respectively. He received a PhD degree in heterogeneous catalysis from the University of Manchester, School of chemical engineering and analytical sciences, UK in 2012, and its topic was "comparative study of the time-on-stream stability of bifunctional nanoporous-based catalysts in n-heptane hydroisomerisation". He worked for SPIMCO pharmaceutical company in research and development department as analyst for one year since 1998, and then he worked for King Faisal specialist hospital and research center in radio-nucleoside and cyclotron department as radio-chemist for four years since 1999. Since 2003, he joined Saudi Basics Industries Co. (SABIC) in research and technology center as analytical chemist then lead scientist until 2013. Recently, Dr. Faisal has joined KACST at national center for petrochemical technology as an assistant research professor. He has published a number of research papers, and one patent granted. Research interest includes: Catalytic cracking, isomerisation, and oxidation.

Yousef Ibrahim Alrashed was born on March 7th, 1985 in Riyadh, Saudi Arabia. He received his bachelor degree in Chemical Engineering on February 2011 from King Saud University (KSU) in Riyadh, Saudi Arabia. In August 2011, he joined the National Petrochemical Center at King Abdulaziz city for science and technology. He completed training at the National Water Company for three months. He worked on catalysts synthesis, characterization, and performance testing of several industrial processes, such as hydrocracking and hydrotreating. He earned his master's degree in Chemical Engineering and Technology on December 2017 from Beijing University and Technology, Beijing, China.

Feras Alshehri worked for KACST since 2003. He got Master's degree in Chemistry from University of Glasgow 2013. He also got PhD in Chemistry from University of Glasgow in 2017.

G. Janet Joshiba is a Junior research fellow in the Department of Chemical Engineering at SSN College of Engineering, TamilNadu. She has completed her post graduation in Environmental Science and Technology at SSN College of Engineering, Kalavakkam in 2016 and she has finished her undergraduate in biotechnology at Prathyusha Institute of Technology and Management, Thiruvallur in 2014. She is doing her Ph.D. in the field of water and wastewater treatment systems and her area of interest in research include adsorption, photocatalysis, biological treatment, microbiology, etc. She has published review papers and research papers in well-renowned journals and also has published six book chapters in Springer.

Vikas Gupta currently serves as Chief Manager, Facilities, Planning and Engineering at Hindustan Petroleum Corporation Limited (HPCL), a Fortune 500 & Forbes 2000 ranked company. He has diverse experience of working at the corporation for over two decades in the downstream segment. His work ranges from project conceptualization to implementation, Operations, Research/Analysis and currently working on Green Biofuels. He is an engineer by profession and a gold medalist in Business Administration. He is TUV-certified professionals in 'Functional Safety'; certified as Project Management Expert from 'U-21 Singapore' and certified professional in Global Sustainable Energy from University of Virginia-Darden School, USA. He is a subject matter expert in Automation systems, wireless technology, safety systems, & Environment sustainability & biofuels. He has presented papers on innovative solutions, conceptualized and successfully implemented at HPCL, at various International and National Fora. He is also on board of various high impact Journals in the field of Sustainability & Economics, as Associate Editor and reviewer.

Nibedita Nath was born in Odisha, India, and obtained her B.Sc. in chemistry, M. Sc. in advanced organic chemistry and M.Phil in physical chemistry from Sambalpur University in 2005, 2007, 2011 respectively and obtained her Ph.D. degree in chemistry from Ravenshaw University in the area of polymer-supported catalysts. Her research interests include heterogeneous catalysts, the development of novel metal complexes for ROP of cyclic esters, and biodegradable polymers.

Senthil Kumar Ponnusamy, Associate Professor in the Department of Chemical Engineering, was born on 08th May, 1982 in Madurai, Tamilnadu, India. He has 9 years of teaching experience, 6 year of research experience and 1 year of industrial experience. He is a recognized Anna University Research Supervisor to guide Ph.D./M.S. candidates. Under his guidance, 3 Ph.D scholars have completed and 11 Ph.D. scholars are doing their active research in the area of Waste Water Treatment and Alternative Fuels. He has published more than 113+ research papers in highly reputed International Journals and 70 research papers in National/International Conferences. He has organized/attended many externally/internally funded conferences/workshops/seminars at national and international level.

Index

H

I

L

R

Z

www.ingramcontent.com/pod-product-compliance
Lightning Source LLC
LaVergne TN
LVHW081324300926
849161LV00008B/1873